Irrigation Engineering

Irrigation Engineering

Edited by **Davis Twomey**

R CALLISTO REFERENCE

New York

Published by Callisto Reference,
106 Park Avenue, Suite 200,
New York, NY 10016, USA
www.callistoreference.com

Irrigation Engineering
Edited by Davis Twomey

International Standard Book Number: 978-1-63239-766-9 (Hardback)

The publisher's policy is to use permanent paper from mills that operate a sustainable forestry policy. Furthermore, the publisher ensures that the text paper and cover boards used have met acceptable environmental accreditation standards.

Trademark Notice: Registered trademark of products or corporate names are used only for explanation and identification without intent to infringe.

Printed in the United States of America.

Contents

Preface

This book has been a concerted effort by a group of academicians, researchers and scientists, who have contributed their research works for the realization of the book. This book has materialized in the wake of emerging advancements and innovations in this field. Therefore, the need of the hour was to compile all the required researches and disseminate the knowledge to a broad spectrum of people comprising of students, researchers and specialists of the field.

Irrigation is the artificial employment of water to soil or land. Irrigation engineering mainly deals with drains, canal, barrage, dams and corresponding systems employed for the similar purposes. There are different types of irrigation systems present worldwide such as drip irrigation, sprinkler systems, localized irrigation, surface irrigation, in-ground irrigation, etc. The various studies that are constantly contributing towards advancing technologies and evolution of this field are examined in detail. It aims to shed light on some of the unexplored aspects of irrigation engineering and the recent researches revolving around it. This book elucidates the concepts and innovative models around prospective developments with respect to irrigation engineering. The readers would gain knowledge that would broaden their perspective about this field.

At the end of the preface, I would like to thank the authors for their brilliant chapters and the publisher for guiding us all-through the making of the book till its final stage. Also, I would like to thank my family for providing the support and encouragement throughout my academic career and research projects.

Editor

Effect of Crop Root on Soil Water Retentivity and Movement

Kozue Yuge[1*], Keiki Shigematsu[2], Mitsumasa Anan[3], Shinogi Yoshiyuki[1]

[1]Faculty of Agriculture, Kyushu University, Fukuoka, Japan; [2]Graduate School of Bioresource and Bioenvironment Science, Kyushu University, Fukuoka, Japan; [3]Takasaki Sogo Consultant Co., Ltd., Kurume, Japan.
Email: *yuge@bpes.kyushu-u.ac.jp

ABSTRACT

The objective of this study was to clarify the effect of crop root on soil water retentivity and movement to improve the crop growth environment and irrigation efficiency. To simulate soil water movement considering the crop root effect on the physical properties of soil, a numerical model describing the soil water and heat transfers was introduced. Cultivation experiments were conducted to clarify the effect of the crop root on soil water retentivity and verify the accuracy of the numerical model. The relationship between soil water retentivity and the root content of soil samples was clarified by soil water retention curves. The soil water content displayed a high value with increasing crop root content in the high volumetric water content zone. The experimental results indicated that the saturated water content increased with the crop root content because of the porosity formed by the crop root. The differences of the soil water retentivity became smaller when the value of the matric potential was over pF 1.5. To verify the accuracy of the numerical model, an observation using acrylic slit pot was also conduced. The temporal and spatial changes of the volumetric water content and soil temperature were measured. Soil water and heat transfers, which considered the effect of the crop root on the soil water retentivity clarified by the soil water retention curves, were simulated. Simulated volumetric water content and temperature of soil agreed with observed data. This indicated that the numerical model used to simulate the soil water and heat transfer considering the crop root effect on soil water retentivity was satisfactory. Using this model, spatial and temporal changes of soil water content were simulated. The soil water condition of the root zone was relatively high compared with the initial conditions. This indicated that the volumetric water condition of the root zone increased with the soil water extraction and high soil water conditions was maintained because the soil water retentivity of root zone increased with the root effect.

Keywords: Water Consumption; Soil Water; Heat Transfer; Numerical Model; Irrigation Water Saving

1. Introduction

Irrigation scheduling is one of the most important factors for healthy breeding of crops. Quantification of water consumption is necessary for both adequate crop breeding and improved irrigation efficiency. Mechanism of water consumption and soil water movement is affected by crop roots because the soil structure and physical properties are changed by crop root physiological activities, including growth or water extraction. To quantify the water consumption in crop fields, the crop root effects on soil physical properties should be clarified.

Various researches have been conducted to clarify the biochemical and physical effects of soil on crop root growth. Drew (1975) [1] studied the adequate external concentrations of nitrogen and phosphorus required by root growth. Effects of various chemical materials of soil

on crop root growth have been clarified [2-9]. Iijima et al. (1991) [10] determined the effects of soil compaction on the development of root system components of rice and maize. A combined root growth and water extraction model was introduced by Bengough (1997) [11]. Crop root cellular response to soil physical stress was evaluated by Bengough et al. (2006) [12]. Effects of the soil water content and bulk density on crop root development processes were investigated by Becel et al. (2012) [13].

Although the effects of soil biochemical and physical conditions on crop root growth have been extensively studied, the effects of the crop root on the physical properties and water consumption of soil have not been clarified.

Studies have been conducted to clarify soil water movement and quantify water consumption in the crop fields [14,15]. However, the crop root effect on the physical properties of soil was not considered in these

---

studies, as a method to evaluate soil water movement considering the effect of the crop root on the soil physic properties has not been established.

The objective of this study is to clarify the effects of the crop root on soil water retentivity and soil water movement. A numerical model was introduced to simulate the soil water and heat transfer considering the crop root effect on soil water retentivity. Cultivation experiments were conducted to clarify the relationship between soil water retentivity and crop root content and to verify the accuracy of the numerical model.

2. Methodology

2.1. Governing Equations of Soil Water and Heat Transfer

To estimate the soil water transport considering the crop root effect on soil water retentivity and hydraulic conductivity, a numerical model was introduced. The governing equation describing soil water and heat transfers can be described as follows:

$$\frac{\partial \theta}{\partial t} = \frac{\partial}{\partial x}\left(D_w \frac{\partial \theta}{\partial x}\right) + \frac{\partial}{\partial z}\left(D_w \frac{\partial \theta}{\partial z}\right) \\ + \frac{\partial}{\partial x}\left(D_T \frac{\partial T}{\partial x}\right) + \frac{\partial}{\partial z}\left(D_T \frac{\partial T}{\partial z}\right) + \frac{\partial K}{\partial x} + S \tag{1}$$

$$C_v \frac{\partial T}{\partial t} = \frac{\partial}{\partial x}\left(\lambda \frac{\partial T}{\partial x}\right) + \frac{\partial}{\partial z}\left(\lambda \frac{\partial T}{\partial z}\right) \\ + L\rho_l \left\{\frac{\partial}{\partial x}\left(D_{wv} \frac{\partial \theta}{\partial x}\right) + \frac{\partial}{\partial z}\left(D_{wv} \frac{\partial \theta}{\partial z}\right)\right\} \tag{2}$$

where C_v is the volumetric heat capacity (J·m^{-3}·°C^{-1}), D_θ is the isothermal water diffusivity (m^2·s^{-1}), $D_{\theta v}$ is the isothermal vapor diffusivity (m^2·s^{-1}), D_T is the thermal water diffusivity (m^2·s^{-1}·°C^{-1}), K is the hydraulic conductivity (m·s^{-1}), L is the latent heat of water vaporization (J·kg^{-1}), S is the sink(m^3·m^{-3}·s^{-1}), T is the soil temperature (°C), t is the time(s), λ is the thermal conductivity (W·m^{-1}·°C^{-1}), ρ_l is the water density (kg·m^{-3}), and θ is the volumetric soil water content (m^3·m^{-3}).

2.2. Boundary Conditions

The energy budget on the soil surface at the crop field can be described as follows:

$$R_n = E + H + G \tag{3}$$

where R_n is the net radiation (W·m^{-2}), E is the latent heat flux (W·m^{-2}), H is the sensible heat flux (W·m^{-2}), and G is the ground heat flux (W·m^{-2}).

The net radiation R_n can be estimated using the following equation considering the shortwave and longwave radiation balance.

$$R_n = (1-\alpha)R_s + L_c + L_{sky} - L_{soil} \tag{4}$$

where R_s is the shortwave radiation on the soil surface (W·m^{-2}), L_c is the longwave radiation from the crop body (W·m^{-2}), L_{sky} is the longwave radiation from the sky (W·m^{-2}), and L_{soil} is the longwave radiation from the soil surface (W·m^{-2}).

The sensible heat flux and the latent heat flux on the soil surface can be estimated as follows

$$H = c_p \rho \frac{T_s - T_a}{r} \tag{5}$$

$$LE = \frac{c_p \rho}{\gamma} \frac{e_s - e_a}{r} \tag{6}$$

where T_s is the soil surface temperature (°C), c_p is the specific heat of the air (J·kg^{-1}·°C^{-1}), e_a is the air vapor pressure (hPa), e_s is the vapor pressure on the soil surface (hPa), r_a is the diffusion resistance (s·m^{-1}), α is the albedo, γ is the psychrometer constant (hPa·°C^{-1}), and ρ_a is the air density (kg·m^{-3}).

The diffusion resistance can be calculated using the following equation (Chamberlain, 1968):

$$r_a = \frac{1}{\kappa u_*} \ln\left(\frac{z}{z_0}\right) + a\left(\frac{u_* \xi}{v}\right)^b \left(\frac{v}{D_v}\right)^c \tag{7}$$

where D_v is the molecular diffusion coefficient (m^2·s^{-1}), u_* is the friction velocity (m·s^{-1}), z is the height of the measurement of the wind velocity (m), z_0 is the roughness length (m), ξ is the effective soil surface roughness (m), and v is the kinematic viscosity of air (m^2·s^{-1}). The constants a, b, and c are reported as 0.52, 0.45, and 0.8, respectively, by Chamberlain (1968) [16].

Using energy balance estimated by Equations (3)-(7), boundary conditions on the soil surface can be described as follows:

$$E = L\rho_w \left(-D_w \frac{\partial \theta}{\partial z} - D_T \frac{\partial T}{\partial z} - K\right) \tag{8}$$

$$G = -\lambda \frac{\partial T}{\partial z} - L\rho_w D_{wv} \frac{\partial \theta}{\partial z} \tag{9}$$

2.3. Model Structure

Figure 1 shows the numerical model describing water and heat transfers in the soil. To solve the two-dimensional transfers of water and heat, the finite-differential method was used. As the bottom boundary condition, the soil water potential was set as constant. The matric potential and hydraulic conductivity were set considering the root content for an interior node. The sink was set using the transpiration rate.

3. Cultivation Experiments

A cultivation experiment was conducted to evaluate the

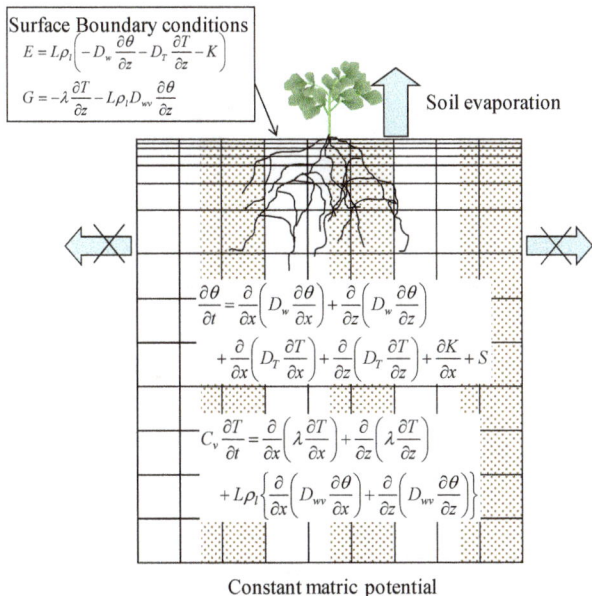

Figure 1. Schematic view of the numerical model describing the water and heat transfers in soil.

Figure 2. Condition of the cultivation experiment to verify model accuracy.

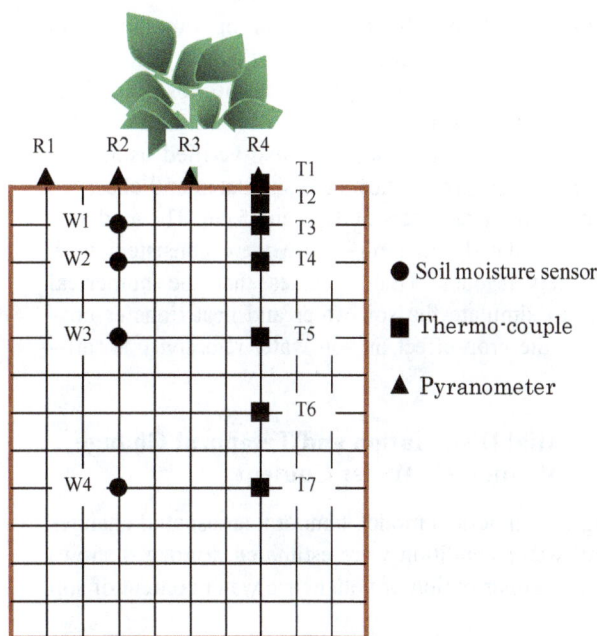

Figure 3. Schematic view of the observation.

effect of the crop root on soil water retentivity. The soil containing the crop root was sampled. Soil moisture characteristic curves were estimated, and the volumetric root contents of soil samples were measured to clarify the relationship between the soil water retentivity and root contents.

To verify the numerical model accuracy, an observation using acrylic slit pot was also conduced. **Figure 2** shows the condition of the experiment. Broccoli was planted in the acrylic slit pot, at a size of 0.5 m × 0.6 m × 0.1 m. The ballasts were paved at the bottom of the acrylic slit pot, and the weathered granite soil was filled at a depth of 0.48 m. The volumetric water content and soil temperature were measured by soil moisture sensors (SM200, Delta-T) and thermo-couples at the depths shown in **Figure 3**. The solar radiations on the soil surface were measured by pyranometers (LI-200, LI-COR) to calculate the net radiation by Equation (4). In addition, the air temperature and humidity were measured to estimate the sensible and latent heat fluxes by Equations (5) and (6). The crop root content in 5 cm × 5 cm soil portion was measured by imaging analysis using the cross-sectional photograph taken from the front side of the acrylic slit pot.

4. Results and Discussion

4.1. Relationship between Soil Water Retentivity and Root Content

Figure 4 shows the relationship between the soil water retention curves and the crop root content in the soil sample. This figure indicates that the soil water retentive-

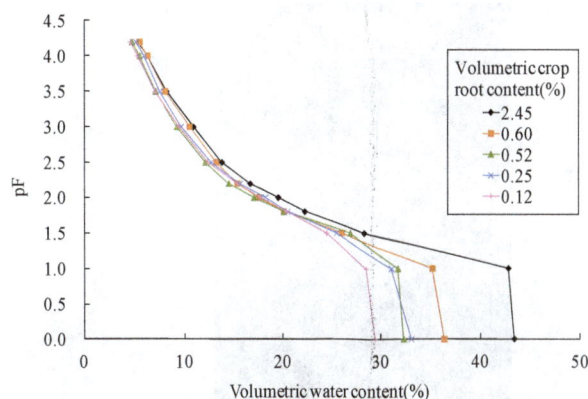

Figure 4. Relationship between the soil water retention curves and crop root content.

ity varied with the root content of the soil sample. In the high volumetric water content zone, the soil water content showed a high value with increasing root content. The experimental results indicate that the saturated water content increased with the crop root content because of the porosity generated by the crop root. The differences of the soil water retentivity became smaller when the matric potential was over pF 1.5.

4.2. Model Accuracy

Using the numerical model described in **Figure 1**, the soil water movement was estimated. **Figure 5** shows the distribution of root content measured by the cultivation experiment. Using these data, the matric potential for interior nodes was given using the soil water retention curves considering the root content in the simulation procedure.

Figure 6 shows the comparison of simulated and observed volumetric water contents at depths of 5 cm and 20 cm. Simulated values agree with the observed data. As the soil moisture movement was minimal in this period, the model accuracy was also verified using soil temperature. **Figure 7** shows the simulated and observed soil temperature at depths of 0 cm and 5 cm. The tendency of the simulated soil temperature were consistent with the observed data. This indicates that the numerical model to simulate the soil water and heat transfer considering the crop effect on soil water retentivity is satisfactory.

4.3. Spatial Distribution and Temporal Change of Volumetric Water Content

Using the numerical model, temporal and spatial changes of soil water condition were estimated. **Figure 8** shows the spatial distribution of volumetric water content of soil

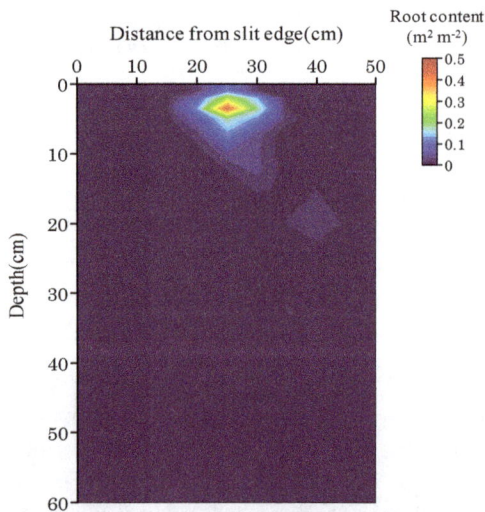

Figure 5. Spatial distribution of the crop root content.

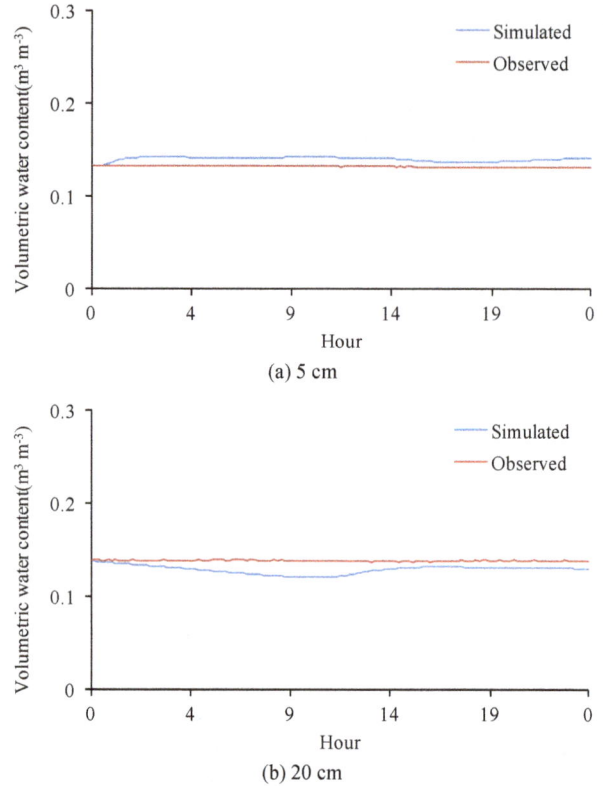

Figure 6. Comparison of simulated and observed volumetric water contents.

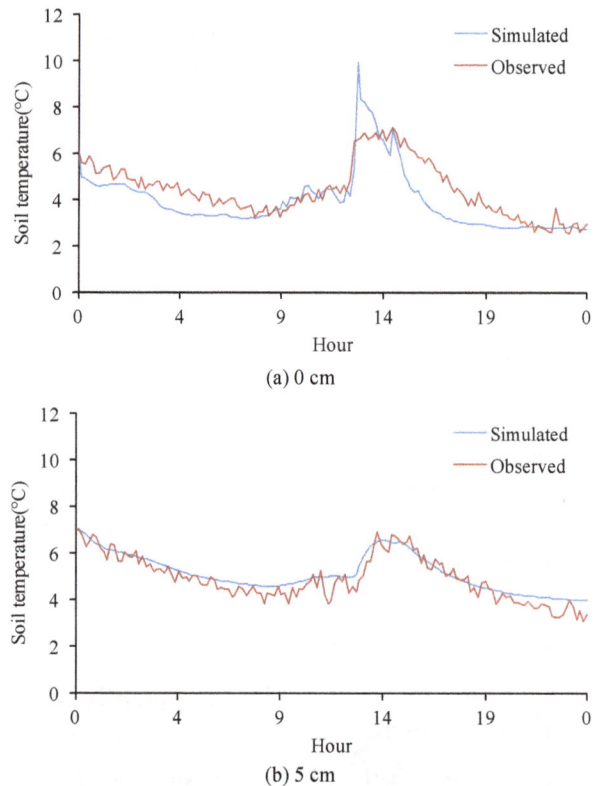

Figure 7. Comparison of simulated and observed soil temperature.

(a) 0:00

(b) 10:00

(c) 14:00

(d) 18:00

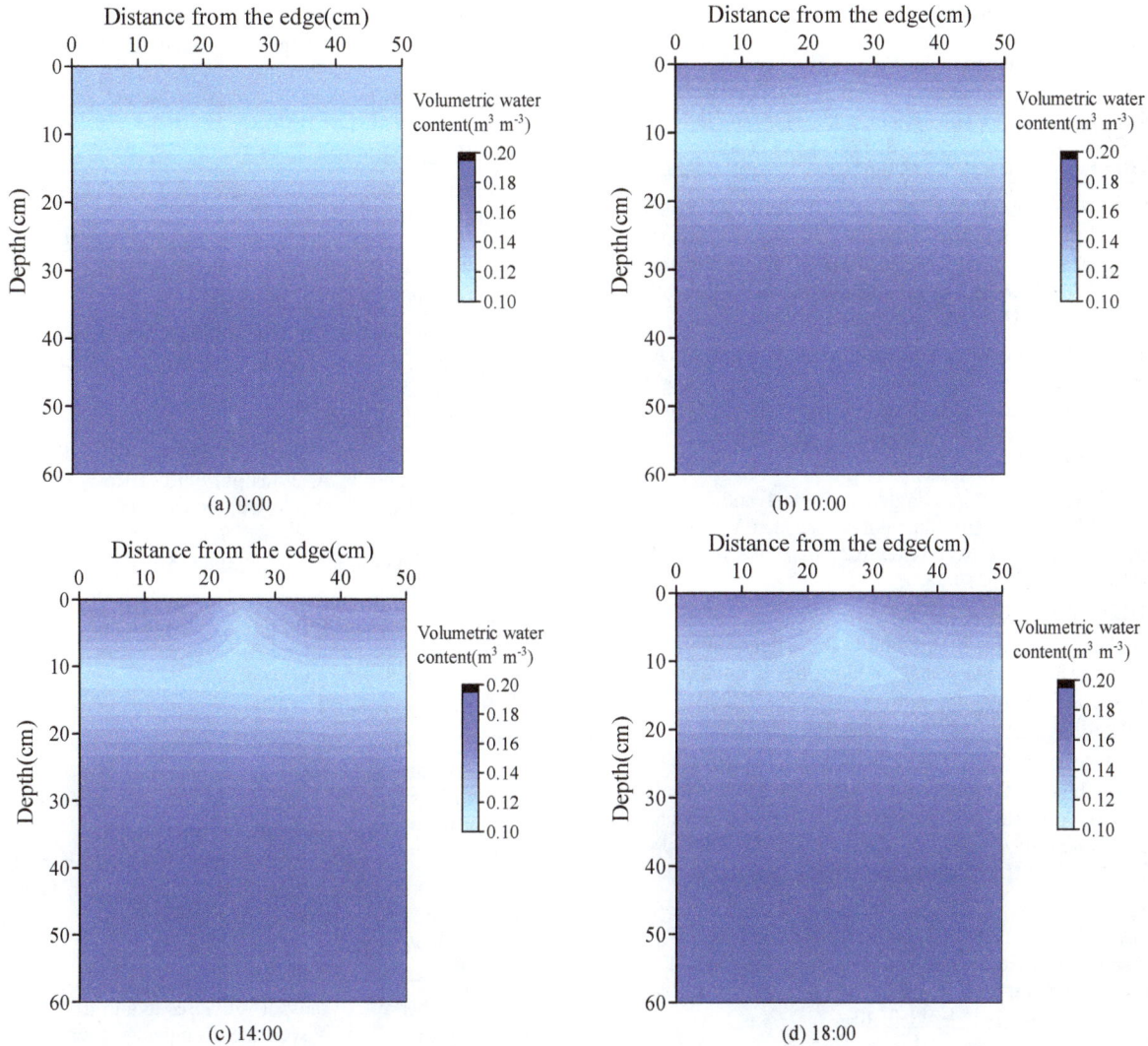

Figure 8. Comparison of simulated and observed soil temperature(continued).

every four hours. The distribution of the volumetric water content around root zone changed from 10 a.m. because of the soil water extraction by the crop root. The soil water condition of root zone was relatively high compared with the initial condition shown in **Figure 8(a)**. This result indicates that the volumetric water condition of root zone increased with the soil water extraction, and high soil water condition was maintained because the soil water retentivity of root zone increased with the root content.

5. Conclusion

To simulate soil water movement considering the crop root effect on soil water retentivity, a numerical model describing soil water and heat transfers was introduced. Cultivation experiments were conducted to clarify the effect of crop roots on soil water retentivity and to verify the accuracy of the numerical model. The relationship between soil water retentivity and root content was clarified by soil water retention curves. The soil water content was directly related to root content in the high volumetric water content zone. The differences in soil water retentivity decreased when the matric potential was over pF 1.5. Simulated volumetric water content and temperature of soil agreed with observed data. Using this model, spatial and temporal changes of soil water content were simulated. The soil water condition of the root zone was relatively high compared with initial conditions. This result indicates that the volumetric water condition of the root zone increased with soil water extraction, and high soil water condition was maintained because the soil water retentivity of root zone increased with the root content. The method introduced here is effective to estimate more accurately water consumption in crop fields. Considering the crop root effect on soil water conditions will allow scientists and farmers to correctly irrigate for crop growth and save irrigation water.

REFERENCES

[1] M. C. Drew, "Comparison of Effects of a Localized Supply of Phosphate, Nitrate, Ammonium and Potassium on Growth of Seminal Root System, and Shoot, in Barley," *New Phytologist*, Vol. 75, No. 3, 1975, pp. 479-490. doi:10.1111/j.1469-8137.1975.tb01409.x

[2] I. J. Bingham and E. A. Stevenson, "Control of Root-Growth-Effects of Carbohydrates on the Extension, Branching and Rate of Respiration of Different Fractiations of Wheat Roots," *Physiologia Plantarum*, Vol. 88, No. 1, 1993, pp. 149-158.

[3] A. E. S. Macklon, L. A. Mackiedawson, A. Sim, C. A. Shand and A. Lilly, "Soil-P Resources, Plant-Growth and Rooting Characteristics in Nutrient Poor Upland Grasslands," *Plant and Soil*, Vol. 163, No. 2, 1994, pp. 257-266. doi: 10.1007/BF00007975

[4] V. M. Dunbabin, A. J. Diggle, Z. Rengel and R. van Hugten, "Modelling the Interactions between Water and Nutrient Uptake and Root Growth," *Plant and Soil*, Vol. 239, No. 1, 2002, pp. 19-38. doi:10.1023/A:1014939512104

[5] M. J. Hutchings and E. A. John, "The Effects of Environmental Heterogeneity on Root Growth and Root/Shoot Partitioning," *Annals of Botany*, Vol. 94, No. 1, 2004, pp. 1-8. doi:10.1093/aob/mch111

[6] Q. Y. Tian, F. J. Chen, J. X. Liu, F. S. Zhang and G. H. Mi, "Inhibition of Maize Root Growth by High Nitrate Supply Is Correlated with Reduced IAA Levels in Roots," *Journal of Plant Physiology*, Vol. 165, No. 9, 2008, pp. 942-951. doi:10.1016/j.jplph.2007.02.011

[7] F. G. Fernandez, S. M. Brouder, J. J. Volenec, C. A. Beyrouty and R. Hoyum, "Soybean Shoot and Root Response to Localized Water and Potassium in a Split-Pot Study," *Plant and Soil*, Vol. 344, No. 1-2, 2011, pp. 197-212. doi:10.1007/s11104-011-0740-z

[8] C. Stritsis, B. Steingrobe, and N. Claassen, "Shoot Cadmium Concentration of Soil-Grown Plants as Related to Their Root Properties," *Journal of Plant Nutrition and Soil Science*, Vol. 175, No. 3, 2012, pp. 456-465. doi:10.1002/jpln.201100336

[9] Y. Wang, P. Marschner and F. S. Zhang, "Phosphorus Pools and Other Soil Properties in the Rhizosphere of Wheat and Legumes Growing in Three Soils in Monoculture or as a Mixture of Wheat and Legume," *Plant and Soil*, Vol. 354, No. 1-2, 2012, pp. 283-298. doi:10.1007/s11104-011-1065-7

[10] M. Iijima, Y. Kono, A. Yamauchi and J. R. Paedales, "Effects of Soil Compaction on the Development of Rice and Maize Root Systems," *Environmental and Experimental Botany*, Vol. 31, No. 3, 1991, pp. 333-342. doi:10.1016/0098-8472(91)90058-V

[11] A. G. Bengough, "Modelling Rooting Depth and Soil Strength in a Drying Soil Profile," *Journal of Theoretical Biology*, Vol. 186, No. 3, 1997, pp. 327-338. doi:10.1006/jtbi.1996.0367

[12] A. G. Bengough, M. F. Bransby, J. Hans, S. J. McKenna, T. J. Roberts and T. A. Valentine, "Root Responses to Soil Physical Conditions; Growth Dynamics from Field to Cell," *Journal of Experimental Botany*, Vol. 57, No. 2, 2006, pp. 437-447. doi:10.1093/jxb/erj003

[13] C. Becel, G. Vercambre and L. Pages, "Soil Penetration Resistance, a Suitable Soil Property to Account for Variations in Root Elongation and Branching," *Plant and Soil*, Vo. 353, No. 1-2, 2012, pp. 169-180. doi:10.1007/s11104-011-1020-7

[14] K. Yuge, M. Ito, Y. Nakano, M. Kuroda and T. Haraguchi, "Soil Moisture and Temperature Changes Affected by Isolated Plant Shadow," *Journal of Agricultural Meteorology*, Vol. 60, No. 5, 2005, pp. 717-720.

[15] K. Yuge, T. Haraguchi, Y. Nakano, M. Kuroda and M. Anan, "Quantification of Soil Surface Evaporation under Micro-Scale Advection in Drip-Irrigated Fields," *Paddy Water Environment*, Vol. 3, No. 1, 2005, pp. 5-12. doi:10.1007/s10333-004-0058-z

[16] A. C. Chamberlain, "Transport of Gases to and from Surface with Bluff and Wave-Like Roughness Elements," *Quarterly Journal of the Royal Meteorological Society*, Vol. 94, No. 401, 1968, pp. 318-332. doi:10.1002/qj.49709440108

2

Response of Peach, Plum and Almond to Water Restrictions Applied during Slowdown Periods of Fruit Growth

Rachid Razouk[1,2*], Jamal Ibijbijen[2], Abdellah Kajji[1], Mohammed Karrou[3]

[1]Department of Agronomy, National Institute of Agronomic Research, Meknès, Morocco; [2]Department of Biology, Faculty of Sciences, University of Moulay Ismail, Meknès, Morocco; [3]Integrated Water and Land Management Program, International Center for Agricultural Research in the Dry Areas, Allepo, Syria.
Email: [*]razouk01@yahoo.fr

ABSTRACT

Water restrictions management for fruit rosaceous during slowdown periods of fruit growth can increase water use efficiency and improve fruit quality without reducing significantly their yield. In this context, two water restriction levels were tested during four consecutive seasons (2007-2011) in peach, plum and almond trees during slowdown periods of fruit growth corresponding to Stage II for peach and plum and to Stages II and III for almond. Water was applied by drip irrigation to produce different water-application treatments of 50% ETc (T_{50}) and 75% ETc (T_{75}) of non-stressed trees irrigated at 100% ETc (T_{100}). The response of trees is presented only for the last season of the experiment (2010-2011) where the effect of the applied water stress is more pronounced. Results show that the effect of water restrictions varied depending on the species. Yield and fruit size were reduced significantly for peach only under treatment T_{50}. Fruit quality was improved for this species with an increase of brix refractometric index and a decrease of acidity. These parameters were evolved in the same manner for plum but the observed differences were not significant. For almond, kernel quality remained unaffected by water restriction at T_{75}. However, the epidermal wrinkles of kernels were more embossed, in response to treatment T_{50}, which affected their appearance. Except leaf area, the evolution of shoots growth, chlorophyll content and leaf temperature showed that the physiology of all species was affected by water stress created by the application of the two irrigation treatments but without profound influence, particularly in plum. In conclusion, irrigation-water may be economized during slowdown periods of fruit growth without major negative effect up to 25% ETc for peach and almond and up to 50% ETc for plum.

Keywords: Prunus Persica; Prunus Domestica; Prunus Dulcis; Water Stress; Fruit Yield; Fruit Quality; Vegetative Growth

1. Introduction

In Morocco, fruit rosaceous area is more than 208,000 hectares from which 84% are rosaceous with pits where almond, plum and peach trees represent 85%, 5% and 3% respectively. Production of these three species in the last three years ranged between 227,000 and 257,000 t/year [1]. The great variability in production is due to the occurrence of stressful climatic factors such as frost and drought [2,3]. These species, in particular almond, have known a real development during the last years thanks to efforts of the government (promotion of hydro-agricultural investments, distribution of plants, subsidies) and private initiatives [4].

Reduction of water resources and constant increase of

water requirements in agriculture, due to the competition with the other sectors, such as industry and drinking water [5], have lead to the concern of water savings. Therefore, it is necessary to develop techniques for improving plant-water use efficiency, especially for more water requiring species, like the majority of rosaceous trees [6,7], especially in regions where drought events are frequent, such as the case of Morocco [8]. This can be achieved through the effective management of irrigation, which consumes in Morocco 80 at 90% of available water resources [9].

Regulated deficit irrigation (RDI) is commonly used in fruit trees to reduce the amount of irrigation water applied without—or with only very small—reductions in yield [10]. RDI imposes a period of water stress that is controlled in terms of its intensity and the period of

---

[*]Corresponding author.

application [11]. This period corresponds generally to slow phases of fruit growth where a tree is relatively most tolerant to water deficit [12]. In almond tree, this phase occurs during Stages II and III of fruit development. However, in peach and plum trees, it is situated during stage II only [13].

RDI, if imposed judiciously, minimizes water use, decreases vegetative growth and pruning cost, and may improve fruit quality [14]. Studies of RDI in rosaceous trees remains very limited under Moroccan conditions. The adoption of the findings obtained in similar experiments conducted in other countries [15-17] is not justified because the results are not conclusive, probably, because of different experimental conditions and used genotype. The studies on RDI are specific to a particular ecosystem. It must consider the productions levels and their stability, physiological behavior of trees and fruits quality. For these reasons the aim of this paper was to test various levels of water stress applied during slow-down periods of fruit growth of mature peach, plum and almond trees. The evaluation of trees response was based on measurement of yield, fruits quality, biometrics characteristics of fruits and vegetative growth.

2. Materials and Methods

2.1. Experimental Design

The experiment was carried out during four consecutive seasons (2007-2011) in the Taoujdate experiment station of the National Institute of Agronomic Research (INRA) located 40 km North of Meknes city in Morocco at 33°56'E, 5°13'N; 499 m. Meteorological data of the site are presented in **Figure 1** where it is shown that rainfall deficit is more marked between May and September with a peak in July and August. The soil is sandy clay with an average of 3% $CaCO_3$, rich in organic matter, with an average of 2.51% in the top soil surface layer (0 - 30 cm). The soil pH is slightly alkaline (7.7), the soil is not saline (average EC around 0.07 ms·cm^{-1} in the top 60 cm).

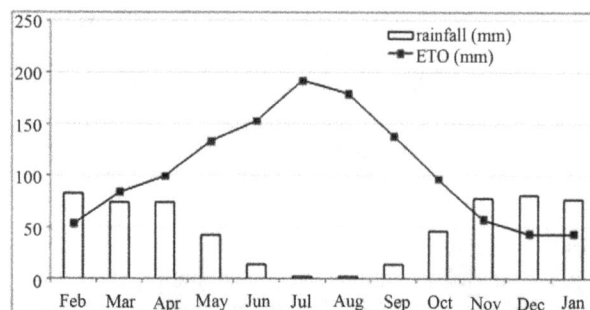

Figure 1. Monthly rainfall in 2010-2011 and Hargreaves evapotranspiration calculated using data for 11 last years from meteorological station in field.

For each species, peach (*Prunus Persica*, cv, JH-Hall), plum (*Prunus Domestica*, cv, Stanley) and almond (*Prunus Dulcis*, cv, Tuono), planted in 2004 in parallel lines spaced by 5 × 3 m, 15 trees were used: The trees were trimmed as goblet canopy shape. During the experiment, all the trees of each species were pruned, fertilized and managed similarly, except for irrigation where different water levels were applied.

Crop evapotranspiration (ETc) was estimated as the product of reference evapotranspiration (ETo) obtained with the Hargreaves model [18] and the crop coefficients recommended by FAO adjusted to planting density and foliage dimensions using a reduction coefficient (Kr) recommended for almond tree: $Kr = \pi D^2 N/20000$ where "D" is the average of foliage diameters and "N" is planting density [19]. ETo was determined using climate data of the last eleven years, collected from the INRA meteorological station located in the experimental field.

The irrigation treatments were applied during four consecutive seasons (2007-2011) at slowdown period of fruit growth for each species by supplying different fractions of crop water requirements (ETc). This period was determined for each species under full-irrigation (100% ET_0) by weekly *in situ* measurement of fruits diameter, on six fruiting branches, from fruit set to fruit maturity during three seasons (2007-2010). In parallel, shoots elongation was measured per linear meter on the same fruiting branches to provide explanations of results, especially for vegetative growth. In the fourth season (2010-2011), two irrigation treatments, 50% ETc (T_{50}) and 75% ETc (T_{75}) were compared to 100% ETc (T_{100}), imposed during slowdown period of fruit-growth for each species. Irrigation water was applied daily using drip system with two emitters per plant. Water quantity was regulated by watering duration.

For each species, the experimental design was a randomized complete block, with three replications. Each of the three block consisted of five trees. The three central trees of each block were selected for application of water treatments, while the surrounding trees were considered as "guard tree" borders.

2.2. Measurements

Generally, response of adult trees to regulated deficit irrigation is not detected in the first year, but it appears after a few years, because of reserves mobilized in wood and large volume explored by roots [20,21]. For this reason, different parameters describing these effects were measured during the fourth season of stress, in 2010-2011.

2.2.1. Vegetative Growth and Physiological Measurements

Effect of the applied water stress to vegetative growth

was evaluated by measurement of annual shoot elongation and leaf area in early November. Average of annual shoot elongation was estimated on all shoots worn by four fruiting two-year-old branches per replication (twelve branches per treatment) chosen at the same height in four sides of trees. Shoot length was reported in linear meter of fruiting branches for eliminate variability that may due to their vigor. Leaf area was measured on all leaves cut without petioles from ten shoots with almost the same length selected randomly per replication. After cutting, leaves were placed in plastic bags and were transported immediately to the laboratory. The area of each leaf was measured using a leaf area meter (adc, bioscientific Ltd) calibrated to 0.01 cm^2.

Leaf temperature and chlorophyll content index were measured weekly, in the morning at 11 h, from the beginning of water treatments to harvest on six marked leaves per replication for each species, using an infrared thermometer and SPAD chlorophyll-meter. The leaves were selected on shoots exposed to the north and having approximately equal lengths.

2.2.2. Biometric Parameters of Fruit and Yield
At fruit maturity of each species, samples of fruits, approximately 3 kg each, were collected from randomly ten selected fruiting branches per replication to evaluate the following parameters: fruit and pit weight, fruit and pit dimensions (length and width). This method of sampling fruits takes into consideration the variability of fruit size in a tree. After sampling fruits, each tree was manually harvested and weighed in the field. For eliminating the differences in yield due to variability of tree vigor, yield values were determined per cm^2 of trunk section area estimated by measuring trunk-circumference.

2.2.3. Fruit Quality Measurements
Effect of the water treatments on fruit quality was evaluated for peach and plum by measurements of sugar content, acidity, pH and water content. While for almond kernel, it was assessed by measurement of number and relief of epidermal wrinkles. All the parameters were measured on ten mature fruits per replication.

On pulp of peach and plum, sugar content was determined in drops of fruit juice by measurements of refractometric index (°Brix) using a refractometer. Acidity was determined by titration of free acids and measurements of pH following methods reported by Lichou [22]. Indeed, 5 g of pulp is mixed in 50 ml of distilled water and titrated by sodium hydroxide 0.1 N under continuous agitation until the pH value reached 8.1. Thus, acidity (Ac) is given by the relationship: Ac = V × C × 100/m where "V" is the volume of sodium hydroxide used in ml, "C" is the concentration of sodium hydroxide in $mol·l^{-1}$ and "m" is the mass of fresh pulp dosed in g. Whereas pulp

pH was determined directly by soaking the pH-meter electrode in crushed pulp. Water content was measured by drying fruits pulp at 80°C during 48 h. On almond kernels, epidermal wrinkles were counted visually and their relief has been evaluated by awarding points from 0 to 5.

2.2.4. Statistical Analysis
Data were analyzed by analysis of variance (ANOVA) using the SPSS software (version 17.0). Mean comparisons were performed using Dunett's test to compare deficit irrigation to full irrigation and student's test to compare between T_{50} and T_{75}.

3. Results and Discussion

3.1. Slow Growth Periods

In peach and plum trees, fruit-growth slowdown corresponds to pit hardening stage (stage II). However in almond tree, it included pit hardening and final stage of fruit growth (stages II and III) [23]. Based on monitoring of fruit diameter, this period is located at different periods in the three species (**Figure 2**). In almond tree, fruit-growth slowdown was observed during four months, from April 30 to harvest in September 04. In peach tree, fruit-growth slowdown was recorded during two weeks only, between May 25 and June 10. While in plum tree, fruit-growth slowdown was observed during five weeks, from May 25 to July 01. During the slowdown periods, outer dimensions of the fruit change little, increasing significantly by different magnitudes depending to species. Indeed during these periods, fruit diameter was increased under full irrigation T_{100} by 26% for peach, 12% for plum and 8% for almond, equivalent respectively to a daily growth of $1.73\%·d^{-1}$, $0.34\%·d^{-1}$ and $0.06\%·d^{-1}$.

Shoot growth began when fruit diameter reached 17% for almond, 27% for peach and 13% for plum. This fruit production came from floral receptacle and previous reserves of trees. During slowdown period of fruit growth, shoot growth is however rapid. During this period, shoot grows by around 133%, 46% and 88% under full irrigation T_{100} respectively in peach, plum and almond. Slowdown period of shoot growth began after two weeks of pit hardening stage in the three species. At this date, fruit diameter reached 92% for peach, 85% for plum and 99% for almond.

Based on duration of slowdown period of fruit growth, it appears clear that application of RDI during these periods seems more economically important for almond and plum. However for peach, two weeks of saving water seems insufficient period to generate a considerable economic impact. The impact would be important using later varieties, for which kernel hardens during a longer period [24]. Fruit growth rate during slowdown periods

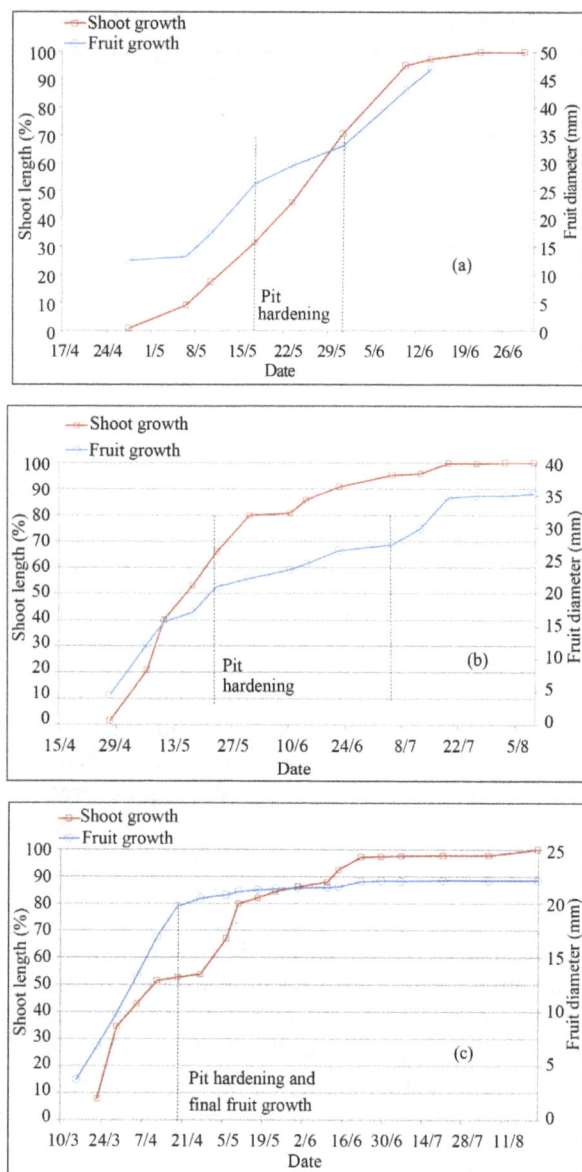

Figure 2. Shoot and fruit growth for peach (a), plum (b) and almond (c) under full irrigation (T$_{100}$).

may indicate the percentage of fruit weight on which act water restrictions, showing that peach remains the most sensitive fruit, followed by plum and almond, whose around quarter of fruit size is determined during this period. The coincidence of slow fruit growth with rapid shoot growth makes that the RDI strategy in experimentation may be used to control excessive vegetative growth of trees. This finding has been concluded by other authors [15].

3.2. Shoot Length and Leaf Area

Shoot growth of all tested species was very affected by RDI treatments (**Table 1**). The differences between values of final shoot length under RDI treatments and those

obtained under treatment T$_{100}$ showed that the applied water stress had a more pronounced effect on almond, followed by plum and then by peach. Based on Student's test for each species, the two RDI treatments reduced shoot length by the same magnitude, by an average of 63%, 45% and 42% respectively in almond, plum and peach. However, Dunett's test revealed that the effect is significantly more pronounced under treatment T$_{50}$ in peach and plum trees for which shoot length was reduced in comparison to treatment T$_{75}$ respectively by 19% and 7%. For leaf area, although its values decreased under the applied water stress, the effect was not significant.

Therefore, it is clear that moderate water stress applied during the slowdown period of fruit growth restricts shoots growth. This finding is important because it means that water stress applied under these conditions will limit effects of competition exerted by shoot growth against fruit growth. The effect observed does not corroborate with those obtained by other authors, it appears that RDI effect on shoot elongation varies with climatic conditions and used cultivar. In Spain, it is found that shoot length in peach cv. Sudanell under water treatment similar to T$_{50}$ does not exceed 25% [15]. In Italy, there was no significant effect of RDI applied during stage II of plum development, cv. Fortune, on shoot growth [25]. However, limitation of shoot growth does not significantly affect leaf area. RDI effect on leaf growth is compensated by reduction of shoot length implicitly reducing their number.

Shoot growth reduction is an adaptive mechanism which reduces tree transpiration [26]. It results from several reversible mechanisms such as decrease of cell division speed [27], rigidity of cellular wall limiting cell growth [28] and decrease of cell turgor [29]. However, there is a consensus that reduction of vegetative growth is not a passive consequence of water deficiency in cells, but is rather controlled by trees [30]. It emerges that shoot growth reduction does not necessarily imply a water stress in cells.

Table 1. Shoot length and leaf area under different irrigation treatments.

		Peach	Plum	Almond
Shoot length (cm·Lm^{-1})	T$_{100}$	295.0 ± 2.8 a	200.3 ± 6.0 a	298.3 ± 88.6 a
	T$_{75}$	189.0 ± 48.4 b	117.3 ± 19.3* b	114.3 ± 28.5* b
	T$_{50}$	153.3 ± 0.3* b	102.0 ± 12.7** b	103.2 ± 21.9* b
Leaf area (cm^2)	T$_{100}$	50.6 ± 4.8	19.4 ± 1.2	11.8 ± 0.7
	T$_{75}$	49.8 ± 2.9	19.4 ± 1.3	10.5 ± 2.3
	T$_{50}$	46.7 ± 4.0	19.3 ± 0.9	9.7 ± 0.6

Lm: linear meter; **: Significant difference at 99% using Dunett's test in comparison to treatment "T$_{100}$"; *: Significant difference at 95% using Dunett's test in comparison to treatment "T$_{100}$".

3.3. Chlorophyll Content

Chlorophyll content is an important indicator to assess the effect of water stress on the physiological behavior of trees because of its role in solar energy absorption which is necessary for photosynthesis [31]. Chlorophyll concentration index measured using the chlorophyll meter indicated that there were significant differences in variation of chlorophyll content, induced by water restrictions. After the application of water restrictions, chlorophyll content decreased very much for all species and began to increase to reach the initial values at the end of water restrictions for peach and before this date for plum and almond (**Figure 3**). The decrease of chlorophyll content was more pronounced under treatment T_{50} whose effect was extended to stage III of fruit growth.

Certainly, reduction of chlorophyll content under water stress conditions is related to decrease in assimilation and translocation of nitrogen [32]. Indeed, water deficit induced a nitrogen deficit which comes mainly from reductions in nitrogen flow at the roots, and secondarily from capacity reductions of root absorption and reduction of transport between leaves and roots due to transpiration feebleness [30].

3.4. Leaf Temperature and Stress Degree Day

Leaf temperature features prominently among biophysical parameters commonly used to evaluate plant water status [33]. Water restrictions increased significantly leaf temperature for all species during and even after their application. For peach, the average of increase was 3.2°C under treatment T_{50} and 2.1°C under treatment T_{75}, while for plum and almond, the applied water restrictions increased leaf temperature with the same magnitude of 2.2°C (**Table 2**). Therefore, it is clear that the water deficit is associated with thermal stress because of reductions in transpiration.

The linear relationship between leaf temperature and air temperature was also influenced by water restrictions. Under full irrigation T_{100}, the relationship between the two temperatures is not significant, but under the water restrictions the correlation coefficient changes to significant values. This same finding was obtained by Helyes *et al.* in haricot [34]. In fact, under full irrigation, the leaf temperature was not very influenced by air temperature because of importance of steam water in leaves surface. However, under water restriction, transpiration and steam water in leaves surface weakened and make accordingly leaves more exposed to air temperature changes.

Stress degree day values (SDD), corresponding to the difference between leaf temperature and air temperature, showed that in some days under water stress, particularly where air temperature was relatively high, steam water of transpiration was unable to cool leaf surface making leaf temperature exceeded air temperature. Accumulated SDD values at harvest were high and significant for all species under the two tested water restrictions. Particularly for peach, where fruit yield was affected by treatment T_{50}, it was estimated that a 1°C higher SDD value might cause 105.4 kg·ha^{-1} yield losses. However, for plum and almond, SDD values variation did not have a significant effect on fruit yield.

Figure 4 shows the evolution of accumulated SDD values from the beginning of water restrictions to harvest. The curves obtained indicate that the effect of water stress started during the first week of its application; but, it increased rapidly after one week for peach and after three weeks for plum and almond even during the period following the end of water restrictions.

3.5. Fruit Weight, Yield and Water Use Efficiency

The effect of water restrictions on final fruit weight and dimensions was evaluated using Dunett's test (**Table 3**).

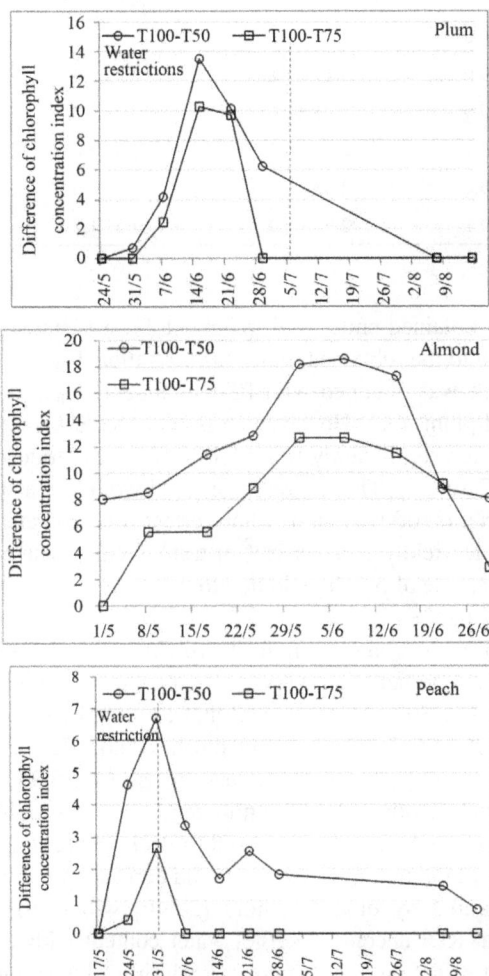

Figure 3. Effect of water restrictions on chlorophyll concentration index for peach, plum and almond.

Table 2. Effect of irrigation treatments on SDD values and on relationship between leaf and air temperatures for peach, plum and almond trees.

		Number of days	Air temp. (°C)	Leaf temp. (°C)	SDD (°C)	Cumulative SDD values (°C)	Regression function	r^2
Peach	T_{100}	29	34.5	29.2 c	5.3 a	-	$y = 0.030x + 28.1$	0.310
	T_{75}	29	34.5	31.3 b	3.1 b	49.0	$y = 0.254x + 22.6$	0.651
	T_{50}	29	34.5	32.4 a	2.0 c	73.3	$y = 0.743x + 6.8$	0.810**
Plum	T_{100}	101	34.5	26.3 c	8.2 a	-	$y = 1.353x - 20.3$	0.518
	T_{75}	101	34.5	28.0 a	6.5 b	223.6	$y = 1.496x - 23.6$	0.623
	T_{50}	101	34.5	29.0 a	5.5 b	109.3	$y = 1.409x - 19.6$	0.667*
Almond	T_{100}	123	34.5	26.8 b	7.6 a	-	$y = 0.984x - 7.1$	0.432
	T_{75}	123	34.5	28.4 a	6.0 b	252.1	$y = 1.558x - 25.2$	0.554
	T_{50}	123	34.5	29.6 a	4.8 b	135.2	$y = 1.238x - 13.0$	0.631*

Table 3. Peach, plum and almond fruit characteristics under different irrigation treatments.

		Fruit weight (g)	Pit weight (g)	Pit/fruit (g/g)	Length (cm)	Width (cm)
Peach	T_{100}	118.7 ± 3.7	6.54 ± 0.4	0.06 ± 0.0	5.80 ± 0.5	6.31 ± 1.0
	T_{75}	106.4 ± 4.9	6.07 ± 0.2	0.06 ± 0.0	5.68 ± 1.1	5.87 ± 1.2*
	T_{50}	90.2 ± 4.5**	4.60 ± 0.2**	0.06 ± 0.0	5.32 ± 0.8**	5.52 ± 0.9**
Plum	T_{100}	37.7 ± 1.1	2.09 ± 0.1	0.06 ± 0.0	5.20 ± 0.3	3.58 ± 0.7
	T_{75}	35.4 ± 1.3	2.00 ± 0.6	0.06 ± 0.0	5.10 ± 0.8	3.44 ± 0.5
	T_{50}	35.0 ± 0.8	1.78 ± 0.1	0.06 ± 0.0	5.06 ± 0.4	3.46 ± 0.5
Almond	T_{100}	2.65 ± 0.0	1.12 ± 0.0	0.42 ± 0.1	2.86 ± 0.5	1.54 ± 0.2
	T_{75}	2.63 ± 0.0	1.04 ± 0.0	0.39 ± 0.0	2.82 ± 0.5	1.52 ± 0.1
	T_{50}	2.57 ± 0.0	1.01 ± 0.3	0.39 ± 0.1	2.93 ± 0.2	1.56 ± 0.2

Width measured with the suture in the middle.

The difference was significant only for peach under treatment T_{50} which reduced fruit and pit weights respectively by 24% and 29.6%. Reduction of peach weight exceeded the rate of fruit growth observed during pit hardening stage (**Figure 2**), proving that the effect of this water stress level was also extended to the final stage of fruit growth. This reduction resulted certainly from simultaneous regression of pulp and pit growth confirmed by the non-significant differences in the ratio of "pit weight/fruit weight". Sensitivity of peach to water restriction T_{50} may be linked to fruit growth rate during stage II where water stress was applied, which is relatively high for peach compared to plum and almond (**Figure 2**). In plum and almond, there was a slight tendency for diminished fruit weight under RDI treatments but the differences are not significant. This result was related to fruit growth during period of water stress (**Figure 2**) which is less important for these two species compared to peach.

The variations in yield levels were linked in large part to fruit weight variation because water treatments were started after fruit set and that there were no differences in the physiological downfall of fruits. In all tested species,

fruit yield obtained under RDI treatment T_{75} was statistically equal to that obtained under full irrigation T_{100}. The same result was obtained with RDI treatment T_{50} in almond and plum trees. However, in peach, this last treatment affected significantly fruit yield which was reduced by 41% (**Table 4**). This decrease of yield would be also due to variation of trees vigor whose effect was separated according to yield values per cm^2 of trunk section. Thereby, the decrease of peach yield due to water stress only is estimated at 25%.

The observed changes in fruit weight and yield of peach are in contradiction with results obtained by others authors [15,35,36] which reported that RDI applied during stage II of fruit development did have no effect on fruit weight and yield even at a very stressful level 35% ETc. In others studies, there is an increase in fruit weight and yield when trees were subjected to RDI during stage II [37,38]. For almond, the results found corroborate with these obtained by others authors [39,40] which concluded that RDI decreased kernel water content without significant effect on fruit weight at maturity. This result is in contradiction with these obtained by others researchers which reported that almond productivity de-

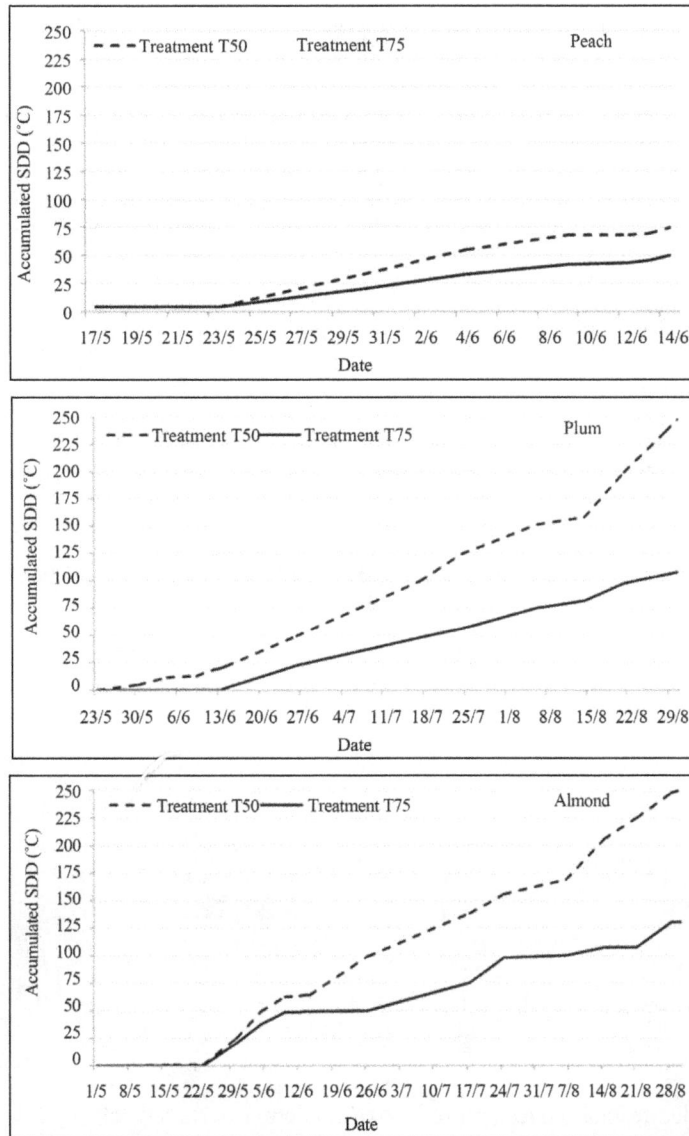

Figure 4. Accumulated SDD values from pit hardening to harvest under water restrictions for peach, plum and almond (SDD = Tl − Ta, where the positives values were accumulated).

crease when trees received 30% less water than full irrigation [41,42]. Likewise, for plum, the same result was obtained by Battilani [25] and Intrigliolo and Castel [43]. However, in other experiments, it was concluded that fruit weight and yield increase when trees subjected to RDI during stage II [44]. The contradictory results may be due to differences in soil texture, soil depth and water capacity of the soil.

Generally, water use efficiency (WUE) was improved significantly by application of RDI treatments. In peach tree, under treatment T_{75}, which had no effect on fruit yield, ensured an improvement of WUE by 36% compared to treatment T_{100} based on values per cm^2 of trunk section. The same improvement of WUE was ensured by treatment T_{50} under which fruit yield decreased. In plum

tree, the tow treatments T_{75} and T_{50} have improved significantly WUE by the same amplitude by an average of 41%. Contrary, in almond tree, the two treatments affected differently WUE values. Indeed, it was improved under treatment T_{75} by 15% and significantly better under treatment T_{50} by 30%.

3.6. Fruit Quality

Table 5 shows that water content for peach and plum was not affected by the imposed water restrictions. This same finding was obtained by few authors [35,44]. RDI decreases fruit water content during stage where it is applied, but after its elimination, water content regains values obtained under full irrigation [41]. No changes in fruit water content indicate that variations in fruit

Table 4. Fruit yield and water use efficiency under different irrigation treatments on peach, almond and plum trees.

		Fruit yield (kg·tree^{-1})	Yield efficiency (g·cm^{-2}·TS)	WUE (m^3·kg^{-1}·tree^{-1})	WUE (L·g^{-1}·cm^{-2}·TS)
	T_{100}	28.3 ± 2.3 a	142.9 ± 7.8 a	0.11 ± 0.01 a	16.9 ± 0.6 a
Peach	T_{75}	26.7 ± 1.2 a	133.9 ± 6.0 a	0.07 ± 0.03* b	14.1 ± 0.6 a
	T_{50}	16.7 ± 2.0* b	107.2 ± 4.9* b	0.07 ± 0.01* b	13.8 ± 0.9* b
	T_{100}	33.9 ± 4.6	223.9 ± 32.8	0.17 ± 0.01a	16.4 ± 2.1a
Plum	T_{75}	31.7 ± 1.6	234.0 ± 15.4	0.10 ± 0.04* b	15.2 ± 2.5* b
	T_{50}	30.1 ± 1.2	219.6 ± 36.3	0.10 ± 0.05* b	14.8 ± 0.9* b
	T_{100}	10.6 ± 0.9	60.8 ± 2.6	0.33 ± 0.03 a	58.2 ± 31.0 a
Almond	T_{75}	10.0 ± 1.0	57.7 ± 7.4	0.28 ± 0.02* b	50.7 ± 26.9* b
	T_{50}	9.8 ± 1.7	54.3 ± 9.5	0.23 ± 0.03* c	46.9 ± 34.8* c

cm^{-2}·TS: cm^2 of trunk section; WUE: water use efficiency.

Table 5. Fruit quality parameters for peach, plum and almond under different irrigation treatments.

		T_{100}	T_{75}	T_{50}
	°Brix	12.23 ± 0.1 b	13.50 ± 0.1** a	13.66 ± 0.1** a
Peach	Acidity (meq.100gFM^{-1})	22.01 ± 0.1 a	20.01 ± 0.1** b	16.40 ± 0.1** c
	pH	7.2 ± 0.0 a	7.1 ± 0.0 a	6.9 ± 0.0* b
	Humidity (%)	81.5 ± 0.7	81.2 ± 1.2	81.2 ± 1.2
	°Brix	22.76 ± 0.3	24.66 ± 1.3	25.23 ± 0.7
Plum	Acidity (meq.100gFM^{-1})	5.30 ± 0.0	4.83 ± 0.2	4.63 ± 0.3
	pH	7.1 ± 0.1	6.6 ± 0.1	6.6 ± 0.1
	Humidity (%)	74.9 ± 0.8	74.9 ± 0.8	74.9 ± 0.8
Almond	Wrinkles/kernel	10.0 ± 0.5	11.0 ± 0.2	11.1 ± 0.5
	Wrinkles relief (points/5)	2.2 ± 0.1 b	2.2 ± 0.1 b	3.0 ± 0.0** a

chemical parameters were not due to their concentration ratio with water content, but rather to changes in fruit metabolic activity.

The acidity and sugar content remained unchanged with the variation of irrigation treatments in plum, but significantly affected in peach. The relationship between increase of sugar content in peach and those of water stress level is not linear. In fact, the two stress treatments T_{50} and T_{75} increased sugar content by the same rate, by an average of 1.35 °Brix. However, the acidity decreased almost linearly with water stress increase, with significant correlation coefficient r of 0.78. The decrease rate was 9.1% under treatment T_{75} and 25.4% under treatment T_{100}. Changes observed in peach quality corroborate with results found by others authors [15,45,46] who reported that RDI applied during Stage II induces an improvement of fruit quality by increasing sugar concentration accompanied by a decrease of organic acids concentration.

Almond kernel quality, evaluated based on observation of epidermal wrinkles, changed with variation of water stress level. Wrinkles did not change in number, but their relief was increased, damaging almond nut quality by the deterioration of their appearance. However, significant variation of wrinkles relief was obtained only under the most stressful treatment T_{50} increasing the relief by 36%. The accentuation of wrinkles relief under RDI treatment is related to slightly dehydration of kernel, especially around harvest.

4. Conclusion

Based on evolution of observed parameters under water restrictions, it is concluded that RDI treatment T_{75} may be adopted for irrigation of peach tree, allowing a net improvement of WUE and fruit quality without significant reductions in productivity and fruit weight. This water regime may be also adopted for almond tree giving the same production level compared to full irrigation T_{100} and avoiding epidermal wrinkles on almond kernel. However, for plum, the RDI treatment T_{50} may be adopted without fear to affect yield and fruit quality. Furthermore, RDI regimes permit to control excessive vegetative growth of trees. The negative effects noted on physiological parameters may be minimized by scheduling restrictions using the curves of SDD values.

5. Acknowledgements

Publication cost was supported by the MCA project (USA).

The authors would like to acknowledge the assistance of the technicians: Lahlou Mohammed, Bouichou Lhoussain, Khalfi Chems Doha and Alghoum Mohamed.

REFERENCES

[1] Anonymous, "Situation de l'agriculture marocaine," Ministry of Agriculture and Marine Fisheries, Morocco, 2011.

[2] A. Agoumi, "Vulnérabilité des pays du Maghreb face aux changements climatiques," International Institute for Sustainable Developement, Manitoba, 2003.

[3] R. Balaghi, M. Jliben, H. Kamil and H. Benaouda, "Etude cadre de l'impact environnemental et social du changement climatique," Institut National de la Recherche Agronomique, Morocco, 2011.

[4] Anonymous, "Les aides financières de l'état pour l'encouragement des investissements agricoles," Ministry of Agriculture and Marine Fisheries, Morocco, 2011.

[5] G. Thivet and M. Blinda, "Améliorer l'efficience d'utilisation de l'eau pour faire face aux crises et pénuries d'eau en méditerranée," Sophia Antipolis, Alpes-Maritimes, France, France, 2007.

[6] P. Vaysse, P. Soing and P. Peyremorte, "L'irrigation des arbres fruitiers," Centre Technique Interprofessionnel des Fruits et Légumes, Paris, 1990.

[7] S. Henry, "Irrigation for the Farm, Garden and Orchard," O. Judd Company, New York, 2010.

[8] L. Stour and A. Agoumi, "Climatic Drought in Morocco during the Last Decades," Hydroécololgie Appliquée, Vol. 16, 2008, pp. 215-232.

[9] M. Azouggagh, "Matériel d'irrigation: choix, utilisation et entretien," Bulletin Mensuel d'Information et de Liaison du Programme National de Transfert de Technologie en Agriculture, No. 81, Rabat, Morocco, 2001.

[10] P. E. Kriedemann and I. Goodwin, "Regulated Deficit Irrigation and Partial Rootzone Drying," Land and Water Australia, Canberra, 2003.

[11] I. Goodwin and A. M. Boland, "Scheduling Deficit Irrigation of Fruit Tree for Optimizing Water Use Efficiency," In: Anonymous, Ed., Deficit Irrigation Practices—FAO Water Reports, Food and Agriculture Organization, No. 22, 2002, pp. 67-78.

[12] M. Kathleen and W. Thomas, "Tree Fruit Irrigation: A Comprehensive Manual of Deciduous Tree Fruit Irrigation Needs," Good Fruit Growers, Wenatchee, 1994.

[13] J. Bretaudeau and Y. Fauré, "Atlas d'arboriculture fruitière," Techniques et Documentation, Paris, Vol. 3, 1991.

[14] E. J. Wickson, "Irrigation in Fruit Growing," Bastian Books, Toronto, 2008.

[15] J. Girona, M. Mata, A. Arbonès, S. Alegre, J. Rufat and J. Marsal, "Peach Tree Response to Single and Combined Regulated Deficit Irrigation under Shallow Soils," Journal of the American Society for Horticultural Science, Vol. 128, No. 3, 2003, pp. 432-440.

[16] M. Gelly, I. Recasens, J. Girona, M. Mata, A. Arbones, J. Rufat and J. Marsal, "Effects of Stage II and Postharvest Deficit Irrigation on Peach Quality during Maturation and after Cold Storage," Journal of the Science of Food and Agriculture, Vol. 84, No. 6, 2004, pp. 561-568. doi:10.1002/jsfa.1686

[17] A. Naor, "Irrigation Scheduling of Peach—Deficit Irrigation at Different Phenological Stages and Water Stress Assessment," Acta Horticulturae, No. 713, 2006, pp. 339-350.

[18] G. H. Hargreaves, "Defining and Using Reference Evapotranspiration," ASCE Journal of Irrigation Drainage Engineering, Vol. 120, No. 6, 1994, pp. 1132-1139. doi:10.1061/(ASCE)0733-9437(1994)120:6(1132)

[19] E. Fereres, W. O. Pruitt, J. A. Beutel, D. W. Henderson, E. Holzapfel, H. Shulbach and K. Uriu, "ET and Drip Irrigation Scheduling," In: E. Fereres and F. K. Aljibury, Eds., Drip Irrigation Management, Cooperative Extension University of California, 1981, pp. 8-13.

[20] E. Fereres and M. A. Soriano, "Deficit Irrigation for Reducing Agricultural Water Use," Journal of Experimental Botany, Vol. 58, No. 2, 2007, pp. 147-159. doi:10.1093/jxb/erl165

[21] P. Monney and E. Bravin, "Irrigation des arbres fruitiers," Station de Recherche Agroscope Changins-Wadenswil ACW, Nyon, 2010.

[22] J. Lichou, "Abricot: les variétés, mode d'emploi," Centrex, Paris, 1998.

[23] K. M. Maib, P. K. Andrews, G. A. Lang and K. Mullinix, "Tree Fruit Physiology: Growth and Development: A Comprehensive Manual for Regulating Deciduous Tree Fruit Growth and Development," Good Fruit Growers, Wenatchee, 1996.

[24] C. Hilaire, P. Giauque, V. Mathieu, P. Soing, A. Osaer, D. Scandella, J. Lichou, F. Maillard and C. Hutin, "Le pêcher," Centre Technique Interprofessionnel des Fruits et Légumes, Paris, 2003.

[25] A. Battilani, "Regulated Deficit of Irrigation Effects on Growth and Yield of Plum Tree," Acta Horticulturae, No. 664, 2004, pp. 55-62.

[26] A. Mahhou, T. M. Dejong, T. Cao and K. S. Shackel, "Water Stress and Crop Load Effects on Vegetative and Fruit Growth of 'Elegant Lady' Peach [Prunus persica (L.) Batch] Trees," Fruits, Vol. 60, No. 1, 2005, pp. 55-68. doi:10.1051/fruits:2005013

[27] C. Granier, D. Inzé and F. Tardieu, "Spatial Distribution Cell Division Rate Can Be Deduced from That of P34 Kinase Activity in Maize Leaves Grown in Contrasting Conditions of Temperature and Water Status," Plant Physiology, Vol. 124, No. 3, 2000, pp. 1393-1402. doi:10.1104/pp.124.3.1393

[28] D. J. Coscrove, "Growth of the Cell Wall," Molecular

Cell Biology, Vol. 6, No. 11, 2005, pp. 850-861.

[29] O. Bouchabke, F. Tardieu and T. Simounno, "Leaf Growth and Turgor in Growing Cells of Maize (*Zea mays* L.) Respond to Evaporative Demand in Well-Watered But Not in Water Saturated Soil," *Plant, Cell and Environment*, Vol. 29, No. 6, 2006, pp. 1138-1148. doi:10.1111/j.1365-3040.2005.01494.x

[30] F. Tardieu, P. Cruiziat, J. L. Durand, E. Triboi and M. Zivy, "Perception de la sécheresse par la plante, conséquences sur la productivité et sur la qualité des produits récoltés," In: sécheresse et agriculture, Ed., Unité ESCo. INRA, Paris, 2006, pp. 49-67.

[31] R. Kauser, H. R. Athar and M. Ashraf, "Chlorophyll Fluorescence: A Potential Indicator for Rapid Assessment of Water Tolerance in Canola," *Pakistan Journal of Botany*, Vol. 38, No. 5, 2006, pp. 1501-1509.

[32] B. Bojović and A. Marković, "Correlation between Nitrogen and Chlorophyll Content in Wheat (*Triticum aestivum* L.)," *Kragujevac Journal of Science*, No. 31, 2009, pp. 69-74.

[33] D. Luquet, A. Vidal, J. Dauzat, A. Begue, A. Olioso and P. Clouvel, "Using Directional Tir Measurements and 3d Simulations to Assess the Limitations and Opportunities of Water Stress Indices," *Remote Sensing of Environment*, Vol. 90, No. 1, 2004, pp. 53-62. doi:10.1016/j.rse.2003.09.008

[34] L. Helyes, Z. Pek and B. Mcmicheal, "Relationship between Stress Degree Day Index and Biomass Production and the Effect and Timing of Irrigation in Snap Been (*Phaseolus vulgaris*. Var. Nanus) Stands: Result of a Long-Term Experiments," *Acta Botanica Hungarica*, Vol. 48, No. 3-4, 2006, pp. 311-321. doi:10.1556/ABot.48.2006.3-4.6

[35] T. Sotiropoulos, D. Kalfountzos, I. Aleksiou, S. Kotsopoulos and N. Koutinas, "Response of a Clingstone Peach Cultivar to Regulated Deficit Irrigation," *Scientia Agricola*, Vol. 67, No. 2, 2010, pp. 164-169. doi:10.1590/S0103-90162010000200006

[36] A. M. Boland, P. H. Jerie, P. D. Mitchell and I. Goodwin, "Long-Term Effects of Restricted Root Volume and Regulated Deficit Irrigation on Peach: II. Productivity and Water Use," *Journal of the American Society for Horticultural Science*, Vol. 125, No. 1, 2000, pp. 143-148.

[37] D. J. Chalmers, P. D. Mitchell and L. Van Heek, "Control of Peach Tree Growth and Productivity by Regulated Water Supply, Tree Density and Summer Pruning," *Journal of the American Society for Horticultural Science*, No. 106, 1981, pp. 307-312.

[38] P. D. Mitchell and D. J. Chalmers, "The Effect of Reduced Water Supply on Peach Tree Growth and Yields," *Journal of the American Society for Horticultural Science*, No. 107, 1982, pp. 853-856.

[39] M. Valverde, R. Madrid and A. L. Garcia, "Effect of the Irrigation Regime, Type of Fertilization, and Culture Year on the Physical Proprieties of Almond (cv. Guara)," *Journal of Food Engineering*, Vol. 76, No. 4, 2006, pp. 584-593. doi:10.1016/j.jfoodeng.2005.06.009

[40] D. A. Goldhamer, E. Fereres and M. Salinas, "Can Almond Trees Directly Dictate Their Irrigation Needs," *California Agriculture*, Vol. 57, No. 4, 2003, pp. 138-144. doi:10.3733/ca.v057n04p138

[41] J. Girona, M. Mata and J. Marsal, "Regulated Deficit Irrigation during the Kernel-Filling Period and Optimal Irrigation Rates in Almond," *Agricultural Water Management*, Vol. 75, No. 2, 2005, pp. 152-167. doi:10.1016/j.agwat.2004.12.008

[42] I. F. Garcia-Tegero, V. H. Duràn-Zuazo, L. M. Vélez, A. Hernàndez, A. Salguero and J. L. Muruel-Fernàndez, "Improving Almond Productivity under Deficit Irrigation in Semiarid Zones," *The Open Agriculture Journal*, No. 5, 2011, pp. 56-62. doi:10.2174/1874331501105010056

[43] D. S. Intrigliolo and J. R. Castel, "Performance of Various Water Stress Indicators for Prediction of Fruit Size Response to Deficit Irrigation in Plum," *Agricultural Water Management*, Vol. 83, No. 1, 2006, pp. 173-180. doi:10.1016/j.agwat.2005.12.005

[44] B. D. Lampinen, K. A. Shackel, S. M. Southwick, B. Olson and J. T. Yeager, "Sensitivity of Yield and Fruit Quality of French Prune to Water Deprivation at Different Fruit Growth Stage," *Journal of the American Society for Horticultural Science*, Vol. 120, No. 2, 1995, pp. 139-140.

[45] D. J. Chalmers, P. D. Mitchell and P. H. Jerie, "The Relation between Irrigation, Growth and Productivity of Peach Tree," *Acta Horticulturae*, No. 173, 1985, pp. 283-288.

[46] M. Gelly, I. Recasens, M. Mata, A. Arbones, J. Rufat, J. Girona and J. Marsal, "Effect of Water Deficit during Stage II of Peach Fruit Development and Postharvest on Fruit Quality and Ethylene Production," *Journal of Horticultural Science and Biotechnology*, Vol. 78, No. 3, 2003, pp. 324-330.

3

Irrigation Scheduling Using Remote Sensing Data Assimilation Approach

Baburao Kamble[1], Ayse Irmak[1], Kenneth Hubbard[1], Prasanna Gowda[2]
[1]University of Nebraska-Lincoln, Lincoln, USA
[2]USDA-ARS, Bushland, USA
Email: bkamble3@unl.edu

ABSTRACT

Remote sensing and crop growth models have enhanced our ability to understand soil water balance in irrigated agriculture. However, limited efforts have been made to adopt data assimilation methodologies in these linked models that use stochastic parameter estimation with genetic algorithm (GA) to improve irrigation scheduling. In this study, an innovative irrigation scheduling technique, based on soil moisture and crop water productivity, was evaluated with data from Sirsa Irrigation Circle of Haryana State, India. This was done by integrating SEBAL (Surface Energy Balance Algorithm for Land)-based evapotranspiration (ET) rates with the SWAP (Soil-Water-Atmosphere-Plant), a process-based crop growth model, using a GA. Remotely sensed ET and ground measurements from an experiment field were combined to estimate SWAP model parameters such as sowing and harvesting dates, irrigation scheduling, and groundwater levels to estimate soil moisture. Modeling results showed that estimated sowing, harvesting, and irrigation application dates were within ±10 days of observations and produced good estimates of ET and soil moisture fluxes. The SWAP-GA model driven by the remotely sensed ET moderately improved surface soil moisture estimates suggesting that it has the potential to serve as an operational tool for irrigation scheduling purposes.

Keywords: Artificial Neural Network; Genetic Algorithms; SEBAL; Remote Sensing; Groundwater; Crop Growth Modeling

1. Introduction

Water scarcity is causing more pressure on utilization of fresh water resources in irrigated agriculture [1]. Therefore, a paradigm shift is necessary from a supply driven into a more demand driven water management. It is recognized that appropriate irrigation scheduling should lead to improvements in water management performance, especially at a farm level [2]. Evapotranspiration (ET) is one of the components of water use efficiency and resulting crop productivity. Periodic information of ET based on remote sensing would be very useful to reduce uncertainty in the crop model parameters and subsequent accurate estimation of the water balance. Several algorithms have been developed to utilize remote sensing data for quantifying ET [3-9]. Researchers have also reviewed different ET algorithms [10] and used remotely sensed data in conjunction with crop or hydrological models via data assimilation for improving soil moisture estimation [11-14]. Researchers also used the Ensemble Kalman Filter (EnKF) with daily microwave observa-

tions over an eight-day period, as well as through a cropping season for estimating soil moisture fluxes [11]. Studies were showed the concept of optimal downscaling for a case where soil moisture estimates were required at scales smaller than that of the microwave observations [14]. An extensive review was conducted on soil water simulation model that uses remotely sensed data to predict moisture in soil profiles [12,15-17]. Ines and Honda developed an assimilation methodology [12] for the Soil, Water, Atmosphere, and Plant (SWAP) simulation model [18] with remote sensing data using Genetic Algorithm (GA) [19]. Similar work was done at spatial scale [20] with an objective to fuse remote sensing data.

Given the above background, the emphasis of this study was to develop a comprehensive data assimilation approach for scheduling on-demand irrigation using SWAP model predictions and SEBAL (Surface Energy Balance Algorithm for Land) based ET [5]. This methodology was implemented using a GA to estimate values for SWAP's sensitive input parameters and by updating

SWAP-ET predictions with SEBAL-ET. This was achieved by (a) estimating ET from MODIS (Moderate Resolution Imaging Spectroradiometer) data using SEBAL, (b) developing an ET data assimilation scheme with a GA to optimize SWAP input parameters for scheduling irrigation, (c) validating optimized parameters with the observations made at the experimental site, (d) evaluating the potential use of optimized parameters for irrigation scheduling with two separate runs of SWAP model with and without optimizer, and (e) simulating and comparing yield and water use efficiency under different irrigation scenarios for an irrigated cotton field in the Sirsa Irrigation Circle (SIC) located in Haryana, India (**Figure 1**). The irrigation circle is an administrative irrigation unit managed by the Haryana Irrigation Department [21,22].

2. Materials and Methods

2.1. Study Area

The proposed approach was tested using a dataset on irrigated cotton field in the SIC (**Figure 1**). Data used in this study was collected as part of another study conducted by the Wageningen Agricultural University, The Netherlands during 2002 for calibrating the SWAP model. The hourly meteorological measurements (**Figure 2**) included air temperature, wind speed, solar radiation, and precipitation from a weather station installed at the Indian Council of Agricultural Research-Cotton Research Institute (ICAR-CRS) (latitude 29°35' North; longitude 75°08' East) located within the SIC (**Figure 1**). The SIC is located in the extreme western part of Haryana between latitudes 29.1° and 30.0° North and longitudes 74.2° and 75.3° East. The climate of the SIC area is characterized by semi-arid and short, mild, variably wet monsoons. The climate of this SIC is characterized by its dryness and extremes temperatures and scanty rainfall. Based on long-term records, January is coldest month with daily minimum temperature 5°C and May/June is hottest month with temperature rises up-to 45°C. The average annual precipitation in Sirsa Irrigation Circle varies from 100 to 400 mm, which is only 10% - 25% of the potential evapotranspiration of common crop rotations. The precipitation mainly occurs during monsoon months of July to September [21]. Ground surface elevations vary from 192 to 207 m above mean sea level. Rice-wheat cropping system is the major cropping system in the SIC. Use of comparative short-duration (100 - 120 days after transplanting) of rice and wheat (135 to 150 days) wheat varieties has offered a unique opportunity for extension of area under a two -crops-a-year. [17, 21-23].

The total area of the SIC is 44,200 km^2 with about 82% of the area under cultivation. At present, only 40% of the total cultivated area is under surface (canal) water irrigation. Water management in the SIC, like any other arid or semi-arid regions in the developing world, is very complex in nature. Key characteristics of the SIC are: a) scarce and erratic precipitation with no perennial rivers in and around the area, b) high evaporative demand, c) marginal to poor quality groundwater in most parts, d) rising groundwater levels with occasional flooding, and e) low water-holding capacity of soils.

Other factors affecting water use efficiency and crop production include fluctuations in canal water supply,

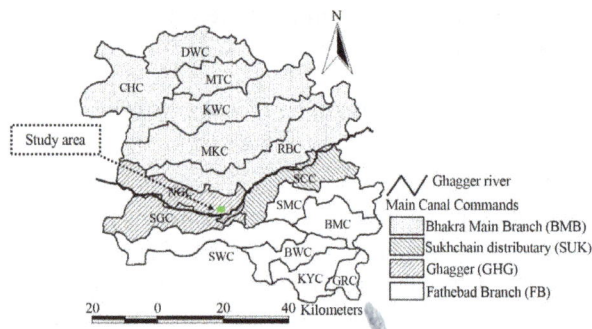

Figure 1. The location of the study area (green rectangle) within Sirsa Irrigation Circle (CIS), Haryana, India.

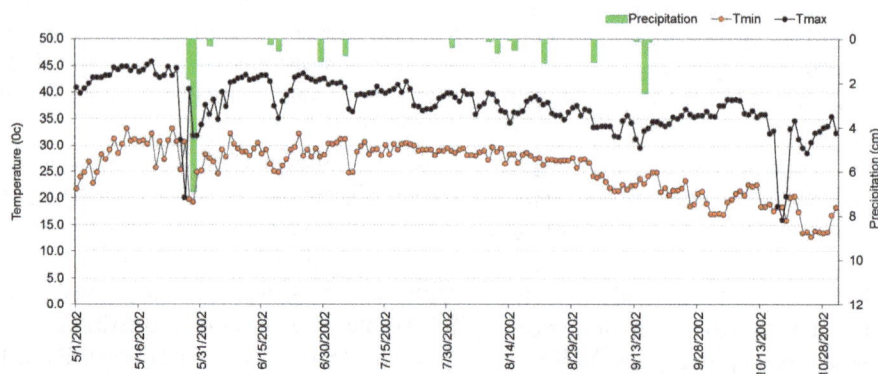

Figure 2. Measured daily values of minimum and maximum temperature, humidity and precipitation, in Sirsa district during the agricultural year 2002.

low irrigation application efficiency due to light textured soils, and conveyance losses from the irrigation system [21,23]. A summary of the physical properties of soil at the time of sowing of cotton is presented in the **Table 1**. The soil at the experimental site was a typical sandy loam with 8% - 17% clay, 10% - 16% silt and 60% - 80% sand with 0.5% organic matter, electrical conductivity rate of 0.15 dS/m, and pH of 8.5 with bulk density 1.65 gm/cc with low water-holding characteristics. **Table 1** shows soil physical properties measured in the experimental field. Soil water content measurements were

made in the top two soil layers (0 - 15 cm and 15 - 30 cm) using a gravimetric sampling method for seven times during 2002.

There were about 15 - 20 soil samples takenfor gravimetric analysis on each measurement day and soil moisture was estimated based on volumetric basis (% moisture on dry weight basis × bulk density). The study field was irrigated from a tube-well with a discharge of 64.76 m^3/hr.

2.2. Satellite Data Processing

The MODIS Level 1B (L1B) images (radiometrically corrected) of the Indo-gangatic area covering the SIC were downloaded from the Earth Observing System Data Gateway of NASA (National Aeronautics and Space Administration). Although MODIS images were available for every 1 - 2 days, there were only 13 cloud-free MODIS images from eight day composites available to estimate seasonal ET for the Kharif (summer) growing season (5/2002-10/2002). **Table 2** shows the details of MODIS products used in the study. A subset image for the study area was extracted for better visualization and computationally efficient analysis of satellite data. The MOD11 L2 data comprised of two thermal bands with a 1 km resolution and was used to estimate surface temperature and emissivity. Extraction of the binary files was performed for two visible (bands 1 and 2), five short-wave infrared (bands 3-7) and two thermal (31 and

Table 1. Physical properties of soil of field at the time of sowing of cotton.

Parameter	Unit	Depth (cm)			
		0 - 15	15 - 30	30 - 60	60 - 90
Textural class	Name	Sandy loam	Loamy sand	Loamy sand	Sandy loam
Clay	(%)	11.15	9.82	8.94	10.31
Silt	(%)	11.77	10.74	10.07	13.04
Sand	(%)	77.08	79.44	80.99	76.29
Soil moisture at saturation	(%)	31.3	31.2	31.5	35.7
Saturated hydraulic conductivity	cm/hr	4.932	4.012	5.037	5.011
Bulk density	g/cc	1.65	1.69	1.65	1.63

Table 2. MODIS data products used in the analysis.

Data set	Data Type	Fill Value	Valid range	Scale Factor
MOD09: MODIS Terra Surface Reflectance (500 m)				
Surface Reflectance Band 1 (620 - 670 nm)	16-bit signed integer	−28672	−100 - 18,000	0.0001
Surface Reflectance Band 2 (841 - 876 nm)	16-bit signed integer	−28672	−100 - 18,000	0.0001
Surface Reflectance Band 3 (459 - 479 nm)	16-bit signed integer	−28672	−100 - 18,000	0.0001
Surface Reflectance Band 4 (545 - 565 nm)	16-bit signed integer	−28672	−100 - 18,000	0.0001
Surface Reflectance Band 5 (1230 - 1250 nm)	16-bit signed integer	−28672	−100 - 18,000	0.0001
Surface Reflectance Band 6 (1628 - 1652 nm)	16-bit signed integer	−28672	−100 - 18,000	0.0001
Surface Reflectance Band 7 (2105 - 2155 nm)	16-bit signed integer	−28672	−100 - 18,000	0.0001
Solar Zenith Angle	16-bit signed integer	0	0 - 18000	0.01
Granule Time	16-bit signed integer	0	0 - 2355	1
MOD11: MODIS Land Surface Temperature and Emissivity (1000 m)				
Land Surface Temperature	16-bit signed integer	0	7500 − 65,535	0.02
Band 31 emissivity	8-bit unsigned integer	0	1 - 255	0.002
Band 32 emissivity	8-bit unsigned integer	0	1 - 255	0.002
Local solar time of Land-surface Temperature observation	8-bit unsigned integer	0	0 - 240	0.1

32) bands. The original MODIS data was provided in HDF (Hierarchical Data Format). A HEG conversion tool (http://gcmd.nasa.gov/ records/HEG.html) was used to convert the HDF files into Geo TIFF images. Individual images of each band were created for each day by converting their corresponding HDF files. Our experiment field size was about 4386 m^2, which was less than one pixel (1 × 1 km) in MODIS thermal image. MODIS LST (Land Surface Temperature) product was downscaled to 250 m using a cubic convolution technique to be consistent with the spatial resolution of MODIS visible and near infrared (MOD09) data. For geo-rectification, we have changed the projection from sinusoidal to UTM with a WGS84 datum. This ended up with gridded data for which both the geographic coordinate system and the projected coordinate system are defined in terms of the WGS84 ellipsoid.

2.3. Evapotranspiration Mapping with SEBAL

SEBAL is a remote sensing based algorithm that computes a complete surface energy balance along with resistances for momentum, heat and water vapor transport for each pixel [5]. Land surface parameters such as surface albedo, vegetation index, emissivity, and surface temperature were derived from MODIS data using the SEBAL. The key input data for SEBAL consists of spectral radiance in the visible, near-infrared and thermal infrared part of the electromagnetic spectrum. In addition to MODIS data, the SEBAL requires routine weather data parameters (wind speed, humidity, solar radiation and air temperature). Under the absence of advection, the energy balance in SEBAL is calculated at an instant time t for each satellite overpass by the following equation:

$$R_n = \lambda E + H + G \qquad (1)$$

where R_n is the net radiation (W/m^2), G is the soil heat flux (W/m^2), H is the sensible heat flux (W/m^2), and λE is the latent heat flux which is the energy necessary to vaporize water (W/m^2). The λE under given atmospheric conditions can be calculated as a residual of the energy balance components in Equation (1). The instantaneous evaporative fraction (Λ, dimensionless) is an expression to obtain the actual ET when the atmospheric moisture conditions are in equilibrium with the soil moisture conditions. The Λ is used to calculate the daily value based on the assumption that the evaporative fraction is constant during daytime hours under non-advective conditions [5]:

$$\Lambda = \frac{\lambda E}{\lambda E + H} = \frac{\lambda E}{R_n - G} \qquad (2)$$

where daily actual evapotranspiration (ET$_{24}$) is calculated from the Λ, and the R_n integrated over the 24-h period (R_{n24}). According to assumptions made in the SEBAL model, net available energy ($R_n - G$) reduces to R_n at daily timescales. ET_{24} is computed as:

$$ET_{24} = \frac{86400 X 10^3}{\lambda \rho_w} R_{n24} \qquad (3)$$

where R_{n24} is the 24-h averaged net radiation (W/m^2), λ is the latent heat of vaporization (J/kg), ρ_w is the density of water (kg/m^3) and ET_{24} is daily actual ET (mm/day).

2.4. Soil-Water-Atmosphere–Plant (SWAP) Model

An intermediate version of the SWAP model (SWAP-GA) [12] was used in this study. The SWAP is a physically based one-dimensional model that simulates vertical transport of water flow, solute transport, heat flow and crop growth at the field scale level [16]. It requires inputs including management practices and environmental conditions to compute a daily soil water balance and crop growth. The major processes taken into account are phenological development, assimilation, respiration and ET. The SWAP model uses Richard's equation [24] to simulate vertical soil water movement in variably saturated soils as follows:

$$\frac{\delta\theta}{\delta t} = \frac{\partial}{\partial z}\left[K(\psi)\left(\frac{\partial \psi}{\partial z} + 1\right)\right] \qquad (4)$$

where K is the hydraulic conductivity (cm·d^{-1}), ψ is the pressure head (cm), z is the elevation above a vertical datum (cm), θ is the water content (cm^3·cm^{-3}), and t is time (d). The soil hydraulic functions in the model are defined by the Mualem-Van Genuchten (MVG) equations [25] which describe the capacity of the soil to store, release and transmit water under different environmental and boundary conditions. Darcy's law is used to determine potential soil evaporation in wet soil conditions. Root water extraction at various depths in the root zone is calculated from potential transpiration, root length density and possible reductions due to wet, dry, or saline conditions. The SWAP also integrates the basic WO FOST (World Foods Tudies) crop growth model and was frequently used to study the effect of the climate change on crop production [12,17,20-22]. Water requirements of a crop depend mainly on crop growth stage and environmental conditions. Root water uptake estimated by model does not depend on the rooting density but only on the actual rooting depth and available soil water. Different crops have different water-use requirements under the same weather conditions. SWAP model simulation gives the balance of water inputs from precipitation and from addition of water to root zone by root growth and water losses computed by crop transpiration, soil evaporation, and percolation to deep soil layers, which gives a complete picture of the water availability and water consumption in particular cropping system [18].

2.5. Optimization Scheme

The SWAP-GA model relies heavily on assimilation of land surface data, which has shown significant potential to improve the realistic representation of the land surface condition. The objective of data assimilation is to obtain the best estimate of the state of the system by combining observations with the forecast model's first guess. Genetic algorithms (GA) technique is a function of optimization derived from the principles of evolutionary theory. It is designed to search, discover, and emphasize optimum solutions by applying selection and crossover techniques, inspired by nature, to supply solutions [19,26]. GA operates on pieces of information as nature does on genes in the course of evolution. It has good global search characteristics. Three operators are designed to modify individuals: selection, mutation and crossover [27]. The evolution usually starts from a population of randomly generated individuals and happens in generations. In each generation, the fitness of every individual in the population is evaluated; multiple individuals are stochastically selected from the current population based on their fitness, and recombined and possibly randomly mutated to form a new population. The new population is then used in the next iteration of the algorithm. The strength of GA with respect to other local search algorithms (lookup table method, ant colony etc,) is to derive more strategies which can be adopted together to find individuals to add to the mating pool, both in the initial population phase and in the dynamic generation phase. Thus, a more variable search space can be explored at each step. Based on the above biological evolution idea, a so-called "SWAP-GA" has been developed by researchers [12] to estimate input parameters of SWAP from remote sensing data.

Based on the above biological evolution idea, a so-called "SWAP-GA" [12] to estimate input parameters of SWAP from remote-sensing data. The model was adopted and recoded according to the objectives of this research. Cotton is grown in Kharif (April-October) season in the Haryana State of India. Time of sowing spread over a period of April to first fortnight of June. The op-

timized parameters were planting date, crop growth period, starting date of irrigation scheduling, and the groundwater depth at the start and end of the simulation (**Table 3**). The proposed parameters were fed to SWAP by GA according to the objective function. The GA searches for an optimum crop parameter set, while SWAP tests the proposed parameters simultaneously by using them in forward simulations. We compared the results from GA for different populations and different generations. Best results were obtained by applying the algorithm that was configured for 100 populations and 100 generations with up to five optimized crop growth parameters (emergence day, time extent of crop, start of irrigation scheduling, groundwater at start of season, groundwater at end of season). We also optimized two parameters that represent depths to ground water at start of season, groundwater at end of season. There was no reliable field information available to check the validity of these parameters.

Optimizing groundwater at start of season allowed us to initialize water table at the beginning of the simulation. In general, the introduction of a priori information improves the convergence and accuracy of the derived parameters, even in cases where the a priori information is slightly erroneous.

Consider C as the cost function having (x, y, d) parameters. The x and y define coordinates of a pixel location, with x being the longitude [0-180/E-W], y being the latitude [0-90/N-S] and d is the satellite overpass date $[i,...j]$.

$$C_{xyd} = \sqrt{\frac{\left(ET_{SEBAL} - ET_{SWAP}^{2}\right)}{n}} \tag{5}$$

where ET_{SEBAL} is estimated ET via the SEBAL model using remotely sensed data (cm), as the "observed" data for the experimental field in the SIC. ET_{SWAP} is estimated actual ET from SWAP-GA and based on optimized model parameters, n is the time domain as number of satellite images (sum of i to j = 13) and C_{xyd} is the objective function (root mean square error: RMSE) for the pixel at x, y location (cm) and $i - j$ are satellite image dates. When a minimum-difference defined threshold was reached, SWAP parameters were stored for reconstruct-

Table 3. Definition, unit, minimum, and maximum values of optimized parameters in SWAP-GA.

Optimized parameters	Definition	Unit	Minimum value	Maximum value
DEC	Emergence day	Ordinal day	140	160
TC	Time extent of crop	Ordinal day	100	200
STS	Start of irrigation scheduling	Ordinal day	140	160
GW$_{jan}$	Groundwater at start of season	cm	140	160
GW$_{dec}$	Groundwater at end of season	cm	140	160

tion of *ET* for any required day in the cropping season. We tested the procedure assuming that some degree of error in remote sensing observations (ET_{SEBAL}). The fitness of an individual having *x*, *y* pixel location characteristics is the inverse of the cost function times the constraints aimed at minimizing the RMSE between SWAP ET and target SEBAL ET. Each water balance parameter after optimization is estimated for kharif growing period to calculate water use efficiency based on Yield/Transpiration, Yield/Evapotranspiration, and Yield/Irrigation. This will evaluate the phenomena with respect to the yield of new irrigation scheme.

We used regression analysis, and root mean square error to evaluate the simulation results. Regression analysis gives information on the relationship between the observed ET variable and the predicted ET variable to the extent that information is contained in the data. To evaluate the performance of the soil moisture simulation, coefficient of determination was used as a relative index of model performance, and root mean square error (RMSE) was used to compare the observed soil moisture and predicted soil moisture. This gave an indication of both bias and variance from the 1:1 line. The RMSE provides a good measure of how closely two independent data sets match.

3. Results and Discussion

3.1. Estimation of Evapotranspiration with SEBAL Model

During 2002 Kharif season, the actual ET via SEBAL model has been quantified for a cotton field in the SIC. The selection of study area is from homogenous cropping practice region which is suitable for applying low spatial resolution remote sensing [16,17,20]. Therefore, the signal in the specific pixel of study area represents the actual electromagnetic characteristics of the cotton. **Figure 3** shows the temporal distribution of normalized differ-

ence vegetation index (NDVI) and remote sensing estimated ET over the experimental field for Kharif 2002 growing season. Multidate satellite data provide the information of the different stages of the crop. **Figure 3** shows the NDVI varied from 0.1 around 1 May to 0.3 at flowering stage (early June) where photosynthetic capacity of a cotton leaf depends on its age. Leaf area index (LAI) and NDVI curves demonstrated gradual increase or decrease in their values with changes in precipitation in early and mid-season. The irrigation demand of cotton increases with increase in NDVI or with increasing photosynthetic rate and vice-versa. NDVI gives important information on the amount of area exposed to the atmosphere for photosynthesis. Soil water availability has direct relation with stomatal behavior and is ultimately related to the photosynthetic process of that crop [14]. Changing progression of ET over cotton crop followed the trend in NDVI during the growing season except for satellite overpasses in mid-September, 2002 (**Figure 3**) which shows the low NDVI and high ET values. This might be because of occurrence of precipitation, climatological conditions or changes in land use within the MODIS pixel covering the study field. The ET for the study field was low early in the season and varied from 0.01 to 0.2 cm/day for months when the soil was bare and open. During crop development and mid-season stages, ET varied from 0.3 cm/day to 0.46 cm/day. This is due to available soil water via irrigation and precipitation events occurred during that period. After mid-season, ET varied between 0.2 mm/day and 0.4 mm/day, ET declined to 0.3 mm/day at the time of harvesting. However, ET continued to increase in the experimental field even after the NDVI reached its maxima. This may be due to the saturation of NDVI after reaching a certain leaf area index (LAI). **Figure 3** shows NDVI is declining from 0.5 in mid-September to 0.1 in mid-October and during this period, LAI increased from about 4 to 6. The differences in NDVI may partly be due to change in landuse or dif-

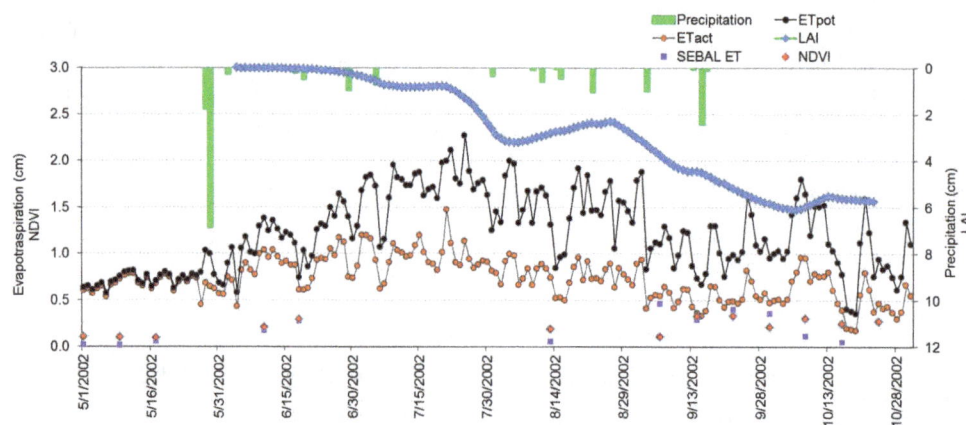

Figure 3. Temporal distribution of LAI, NDVI and evapotranspiration (ET) during the Cotton growing season in the study area.

ferences in timing and amount of irrigation in the surrounding fields that fall within the study pixel.

3.2. Remotely Sensed Evapotranspiration Data Assimilation in Hydrological Model

Figure 4 compares temporal distribution of SEBAL based ET and SWAP-GA based ET, two curves shows similar trends of under and over estimations of actual ET. The SWAP-GA marginally overestimated than the SEBAL based ET in early season when the soil surface was dry and underestimated late in the season when the soil surface was wet and covered by the crop, which influences efficiency of water use, high water productivity and efficient farming activities. The larger ET differences between SWAP-GA and SEBAL were found during May 2002 and June 2002 with a mean absolute difference of 4 mm/day. However, mean absolute difference was increased to 5 mm/day when simulations were made without data assimilation. This difference has huge impact on estimating irrigation demand and scheduling. During early in the growing season, the bias far exceeded the actual values. The main reasons for this bias are over estimation of SEBAL-ET and model considers no transpiration till plant emergence.

The SWAP-GA system tries to minimize the difference between SWAP model and SEBAL-ET and the difference between the SWAP-GA and SEBAL ET minimized to 0.25 mm. On August 13, 2002 (**Figure 4**), the difference between SEBAL-ET and SWAP-GA-ET was about 8 mm and it was because the SWAP-GA model usually overestimates ET right after irrigation application or a precipitation event (any citation). The data assimilation results in **Figure 4** are promising but further refinement is necessary to improve the propagation of the correction to the domain outside the assimilation points caused by mixed pixels and to get better bias estimates. The bias due to the comparison of pixel observations

with the model needs to be explicitly taken into account to prevent unnecessary forcing of the model towards biased observations. In our case, there is a bias due to the comparison of SEBAL pixel observations with the SWAP-GA model.

3.3. Optimization of Crop Growth Parameter Using SWAP-GA Model

SWAT-GA model parameters were optimized by minimizing the RMSE between SWAP-GA-ET and the target SEBAL-ET values and resulting parameter values were used as input for simulating irrigation scheduling. Generally, remote sensing based ET values contains errors due to errors associated atmospheric correction of the reflectance data and due to errors associated with ET algorithms.

Furthermore, coarser resolution, multispectral images such as MODIS have mixed pixel problems which makes it more complicated if the selected pixel exhibits some heterogeneity on the high spatial resolution satellite image. **Table 4** shows the values of optimized parameters as well as data from the experimental field. Optimized parameter values with SWAP-GA were closely matched with field measurements. The simulated cropping period from planting to harvest was 169 days against the actual period of 179 days. The depth of groundwater (water table) about −141 cm to −151 cm is not uncommon in irrigated cotton cropping areas in the SIC [21,22], specially considering an inundated condition at the start of the period of study.

3.4. Soil Moisture Based Irrigation Scheduling Scheme

Figure 5 shows time series observed and simulated soil water contents (cm³/cm³) at 0 - 15 cm and 15 - 30 cm soil depths and at 30 - 60 cm and 60 - 90 cm soil depths in **Figure 6**. About 50 - 60 percent of the total water uptake

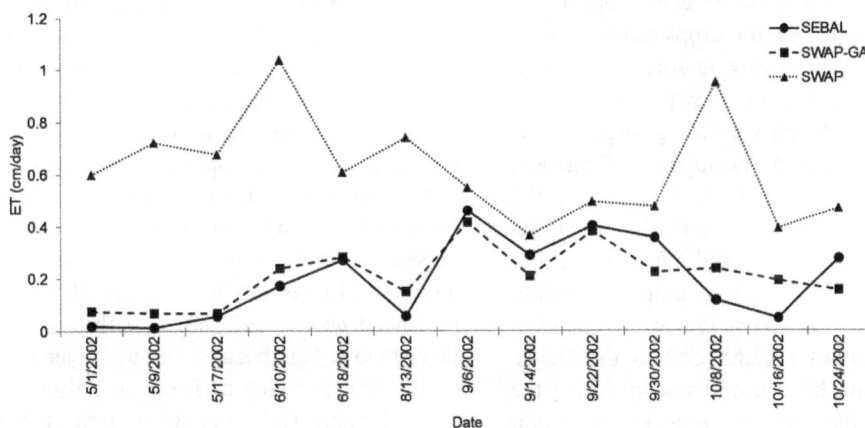

Figure 4. Actual evapotranspiration (cm/day) for the 2002 cotton growing seasons. Observed ET is based on SEBAL algorithms (SEBAL ET) on satellite overpass dates. ET predictions are with original SWAP and SWAP-GA models.

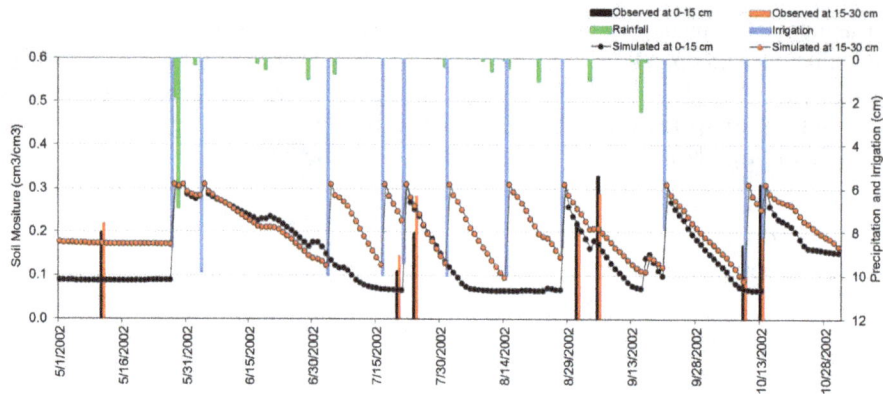

Figure 5. Simulated and observed soil water content (cm³/cm³) at 0 - 15 cm and 15 - 30 cm soil depths by SWAP-GA with optimized parameters, rainfall and on-demand irrigation amounts are also shown.

Table 4. The simulated and observed optimized parameters based on 10 generations & 10 populations.

	Emergence	End of Crop	Start of irrigation scheduling
Simulated Parameter	04-June[*]	23-October	02-June
Observed Parameter	20-May (Actual sowing date)	15-November	18-May

[*]Consider germination period 14 days *i.e.* Emergence = sowing date + germination period.

by the crop occurred within the top 90 cm depth, where more than 90 percent of the total root mass found. **Figures 5** and **6** shows the soil water depletion till 60 percent as the season progresses. SWAP-GA predicted a total of eleven irrigation applications with irrigation demands varying from 7.8 cm to 9.9 cm per application during wheat growth and development period. The major portion of the irrigation demand usually occurs in May, June, August and September months to avoid water stress during critical crop development stages *i.e.* flowering and fruiting. For cotton, the irrigation demand vary from 7.8 cm to 9.9 cm with respect to the irrigation timing, the growth stage of the crop, climate and length of the total growing period. Early development stages show more difference in the actual and potential ET, while mid and late season shows very less difference because of the increased irrigation demand. Cotton crops received regular precipitation during the growing season, and most of it occurred during mid-season. Further, presence of excess water in the root zone early in the growing is expected to restrict root and crop development. **Figures 5** shows that, two consecutive irrigations (8.7 cm and 9.8 cm) during the cotton emergence in early June addition to two precipitation events(1.8 cm and 6.9 cm).The precipitation contribution (17.67 cm) to crop ET mainly during kharif (cotton) which is very low as compared to seasonal irrigation supplies (102.68 cm) to the fields. Irrigation frequencies are high in mid-season during the flowering stage when the leaf area is at its maximum level. Time series of moisture data indicated that the soil water content at top and bottom layers were quite similar

from germination until the date of first precipitation. The top soil layers have slightly lower water contents than lower layers. The model predictions closely matched observations indicating model's ability in simulating soil water content. Root Mean Square Error (RMSE) for simulated and observed soil water content (cm³/cm³) at 0 - 15 cm is 0.08, 15 - 30 cm is 0.01, 30 - 60 cm is 0.001 and 60 - 90 cm is 0.01. The RMSE of simulated and observed soil moisture for four depths provides the minimum possible error. Overall, our results showed that the rainfall contribution to crop ET was very minimal as compared to irrigation supplies to the fields. Although the crest of the soil moisture curve and rainfall matched at some locations, soil moisture tended to rise even though there was no rainfall event.

Figures 5 and **6** shows the large ratio of evaporation to precipitation in July/August and has insignificant impact on the soil moisture due to the relatively small precipitation events and less irrigation combined with low temperatures and the soil moisture can be maintained to a constant level. The top soil layers have slightly lower water contents than lower layers. It is because the top layer forms the sphere of life, which receives moisture in pulses of precipitation and irrigation. From **Figures 5** and **6**, it also reveals that the top 30 cm of the soil experienced greater soil moisture fluctuations than in soil layer below. It is because the top layer forms the sphere of life that receives moisture in pulses of precipitation and irrigation while a major portion of that water is extracted through evaporation and transpiration by plants. The simulated and observed soil moisture levels showed

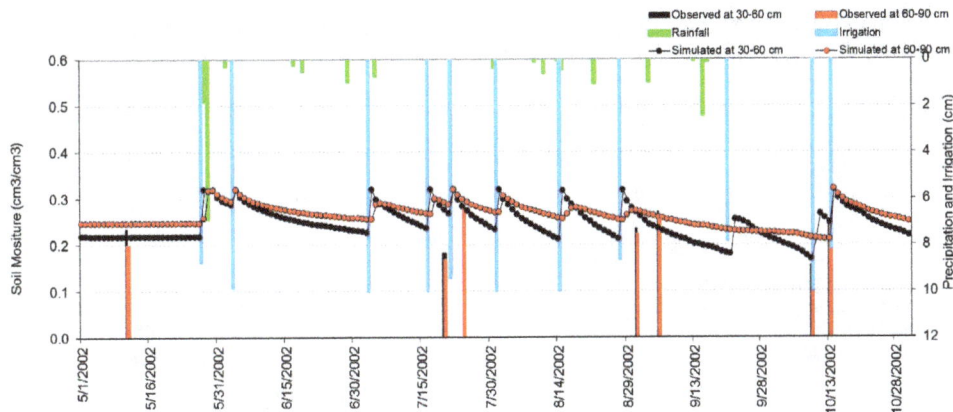

Figure 6. Simulated and observed soil water content (cm³/cm³) at 30 - 60 cm and 60 - 90 cm soil depths by SWAP-GA with optimized parameters, rainfall and on- demand irrigation amounts are also shown.

increasing trend from June to September, and then decreased onwards, which coincided with occurrence of irrigation and precipitation. Lowest level of moisture in the soil profile was simulated for August when plants are transpiring at a maximum rate (**Figure 3**). However, this phenomenon did not occur in the observed data. In SIC, it is very common to have very dry conditions late in the growing season and it can reduce crop yields if soil water content is is not available. As the water table depth increases, the soil layer tends to hold more water with no fluctuations throughout the season. The Sirsa Irrigation Circle has two water tables, the first one at a depth of 5 m from the surface and the second at a depth of 15 m. The effect of capillary rise is expected because of the soil type and deep percolation is the main phenomenon for the water flow into various soil layers. Trends in soil moisture predictions follow that in precipitation although the crest of soil moisture curve and precipitation match at some locations. Cotton plant consumes more water and it has high sensitivity to moisture increase. As the crop matures, soil moisture depletion allowances can be greater. The Sirsa soil have storage reserves of 25 to 100 cm of water which mainly depends on rooting depth of crops grown in this area which makes use of more soil moisture to minimize risk of leaching.

3.5. Evaluation of Optimized Parameters for Yield Estimation under on-Demand Irrigation Scheduling Scheme

Table 5 presents a comparison between SWAP-GA predicted and observed crop yield. A general progressive yield was observed with respect to the simulation criteria (with or without On-Demand irrigation). Current cotton yields in the Haryana state is approximately 3500 kg/ha under irrigated conditions [21] while SWAP-GA simulated with optimized parameters for on-demand irrigation case showed cotton yield 3686 kg/ha and SWAP simulations with observed irrigation and yield dataset at

Table 5. Estimated and observed water balance parameters (cm) for two irrigation cases.

Parameters of interest	Case 1[1]	Case 2[2]
Transpiration (cm)	44.2	79.5
Evapotranspiration (cm)	52.2	89.2
Irrigation (cm)	34.1	102.7
Crop dry mass (kg/ha)	17,170	18,701
Harvesting Index	21.47	21.47
Total production of cotton (Seed+Lint)	3687	4015.1
Ginning percentage	37.1	37.11
Lint weight (kg/ha)	1368	1490

[1]SWAP-GA simulations with optimized parameters for on-demand irrigation. [2]SWAP simulations with observed irrigation and yield dataset at farmer's field.

farmer's field showed 4015 kg/ha.

The SWAP-GA simulations show that crop water needs supplied by on-demand irrigation increased lint yield response to the progressively greater irrigation capacity treatments. **Table 6** shows estimated and observed water balance parameters (cm) for two irrigation cases. The models indicate that the water use efficiency of cotton increased from 0.15 to 0.4 kg/m³ based on irrigation while 0.17 to 0.26 kg/m³ by ET (**Table 6**) which indicating considerable variation and scope exists for improvements in WUE based on calibrated parameters. In Haryana, successful crop production is not possible without supplemental irrigation because of erratic precipitation events. Irrigation application by the calibrated model and on-demand irrigation, the water use efficiency obtained from the on-demand is increased considerably without water deficit. Factors responsible for the low WUE-values include both the relatively high fractions of soil evaporation in the ET term and of water percolation from the irrigation water applied.

Table 6. Water use efficiency (WUE, kg·m^{-3}) for two irrigation case studies.

Water Use Efficiency[3]	Case 1[1]	Case 2[2]
WUE$_T$ (kg·m^{-3})	0.31	0.19
WUE$_{ET}$ (kg·m^{-3})	0.26	0.17
WUE$_{IR}$ (kg·m^{-3})	0.40	0.15

[1]SWAP-GA simulations with optimized parameters for on-demand irrigation. [2]SWAP simulations with observed irrigation and yield dataset at farmer's field. [3]WUE$_T$, WUE$_{ET}$, and WUE$_{IR}$ are water use efficiencies based on transpiration, evapotranspiration, and irrigation, respectively (kg·m^{-3}).

4. Summary

A conceptual modeling methodology was tested in this research to schedule irrigation based on the on-demand strategy using SWAP crop growth model with a genetic algorithm as an optimizer. We used remote sensing ET data to characterize our model via a stochastic data assimilation approach, and then the optimized crop growth parameters were used as inputs to agro-hydrological model. The strength of an integrated data assimilation approach was shown explicitly in the scenario analysis. It was shown that there is a strong relationship between irrigation scheduling, ET, soil water availability, and groundwater table. The effects of ET on the water balance have been demonstrated. It does show that there is a lot of scope for reducing errors in estimated ET to improve water balance estimates. Parameter estimations are successful and the ability of the SWAP-GA to produce ET and soil moisture values accurately in relations with precipitation and irrigation were promising, although the general performance of the model can be described as reasonable. In summary, this study has explored the potential of genetic algorithm to estimate the crop parameters for improving characterization of water management options for predicting soil water content to schedule irrigation.

In this study, the numerical case of 100 generations and 100 populations showed that the GA was able to characterize the terms included in the fitness function very well and parameters could be predictable reasonably well. It has also demonstrated the potential of the data assimilation approach as used in this study is a powerful tool in crop and water parameter estimation for irrigation scheduling. Our results also indicate that the soil moisture profile estimates obtained from this particular synthetic experiment are as good as realistic data. In practice, the feasibility of retrieving subsurface moisture profiles from surface measurements depends on the accuracy and the physical realism of the land surface model and the associated error statistics. Since the subsurface states cannot be remotely sensed at the pixel scale, they can only be estimated by using the hydrologic model to propagate information downward from the surface.

REFERENCES

[1] D. W. Seckler, U. Amarasinghe, D. Molden, R. De Silva, and R. Barker, "World Water Demand and Supply, 1990 to 2025: Scenarios and Issues," International Water Management Institute (IWMI), Colombo, Research Report No. 19.

[2] B. Kamble and A. Irmak "Combining Remote Sensing Measurements and Model Estimates through Data Assimilation," *IEEE International*, Vol. 3, 2008, p. 1036.

[3] R. G. Allen, M. Tasumi and R. Trezza. "Satellite-Based Energy Balance for Mapping Evapotranspiration with Internalized Calibration (METRIC)-Model," *Journal of Irrigation and Drainage Engineering*, Vol. 133, No. 4, 2007, pp. 380-394. doi:10.1061/(ASCE)0733-9437(2007)133:4(380)

[4] M. Anderson, J. Norman, G. Diak, W. Kustas and J. R. Mecikalski. "A Two-Source Time Integrated Model for Estimating Surface Fluxes Using Thermal Infrared Remote Sensing," *Remote sensing of Environment*, Vol. 60, 1997, pp. 195-216. doi:10.1016/S0034-4257(96)00215-5

[5] W. G. M. Bastiaanssen, M. Menenti, R. A. Feddes and A. A. M. Holtslag. "A Remote Sensing Surface Energy Balance Algorithm for Land (SEBAL): 1. Formulation," *Journal of Hydrology*, Vol. 212-213, 1998, pp. 198-213. doi:10.1016/S0022-1694(98)00253-4

[6] W. P. Kustas and J. M. Norman, "Use of Remote Sensing for Evapotranspiration Monitoring over Land Surfaces," *Hydrological Sciences Journal*, Vol. 41, No. 4, 1996, pp. 495-516. doi:10.1080/02626669609491522

[7] S. P. Loheide and S. M. Gorelick, "A High-Resolution Evapotranspiration Mapping Algorithm (ETMA) with Hydroecological Applications at Riparian Restoration Sites," *Remote Sensing of Environment*, Vol. 98, 1998, pp. 182-200.

[8] Z. Su, "The Surface Energy Balance System (SEBS) for Estimation of Turbulent Heat Fluxes," *Hydrology and Earth System Sciences*, Vol. 6, No. 1, 2002, pp. 85-89. doi:10.5194/hess-6-85-2002

[9] B. Seguin, E. Assad, J. P. Freaud, J. P. Imbernon, Y. Kerr, and J. P. Lagouarde, "Use of Meteorological Satellite for Precipitation and Evaporation Monitoring," *International Journal of Remote Sensing*, Vol. 10, 1989, pp. 1001-1017.

[10] P. H. Gowda, J. L. Chavez, P. D. Colaizzi, S. R. Evett, T. A. Howell and J. A. Tolk, "ET Mapping for Agricultural Water Management: Present Status and Challenges," *Irrigation Science*, Vol. 26, No. 3, 2008, pp. 223-237. doi:10.1007/s00271-007-0088-6

[11] D. Entekhabi, H. Nakamura and E. G. Njoku, "Solving the Inverse-Problem for Soil Moisture and Temperature Profiles by Sequential Assimilation of Multifrequency Remotely Sensed Observations," *IEEE Transactions on Geoscience and Remote Sensing*, Vol. 32, No. 2, 1994, pp. 438-448. doi:10.1109/36.295058

[12] A. V. M. Ines, "Improved Crop Production Integration GIS and Genetic Algorithms," Doctoral Thesis, Asian In-

stitute of Technology, Bangkok, 2002.

[13] B. Kamble, "Evapotranspiration Data Assimilation with Genetic Algorithms and SWAP Model for On-Demand Irrigation", AIT Thesis, 2007.

[14] R. H. Reichle, D. Entekhabi and D. B. McLaughlin, "Downscaling of Radiobrightness Measurements for Soil Moisture Estimation: A Four-Dimensional Variational Data Assimilation Approach," *Water Resources Research*, Vol. 37, 2001, pp. 2353-2364. doi:10.1029/2001WR000475

[15] M. R. Smith and R. W. Newton, "The Prediction of Root Zone Soil Moisture with a Water Balance—Microwave Emission Model," Master's Thesis, Texas A&M University, Texas, 1983.

[16] A. Irmak and B. Kamble, "Evapotranspiration Data Assimilation with Genetic Algorithms and SWAP Model for On-Demand Irrigation," *Irrigation Science*, Vol. 28, No. 1, 2009, pp. 101-112. doi:10.1007/s00271-009-0193-9

[17] B. Kamble and A. Irmak, "Remotely Sensed Evapotranspiration Data Assimilation for Crop Growth Modeling," In: L. Labedzki, Ed., *Evapotranspiration*, InTech, 2011. doi:10.5772/13990

[18] J. C. Van Dam, J. Huygen, J. G. Wesseling, R. A. Feddes, P. Kabat, P. E. V. Van Waslum, P. Groenendjik and C. A. Van Diepen, "Theory of SWAP Version 2.0: Simulation of Water Flow and Plant Growth in the Soil-Water-Atmosphere-Plant Environment," Wageningen Agricultural University and DLO Win and Staring Centre, Wageningen, Technical Document 45, 1997.

[19] D. E. Goldberg, "Genetic Algorithms in Search, Optimization, and Machine Learning," *Machine Learning*, Vol. 3, No. 2-3, 1988, pp. 95-99. doi:10.1023/A:1022602019183

[20] Y. Chemin, K. Honda, "Spatiotemporal Fusion of Rice Actual Evapotranspiration with Genetic Algorithms and an Agrohydrological Model," *IEEE Transactions on Geosciences and Remote Sensing*, Vol. 44, No. 11, 2006, pp. 3462-3469. doi:10.1109/TGRS.2006.879111

[21] R. S. Malik, R. Kumar, D. S. Dabas, A. S. Dhindwal, S. Singh, U. Singh, D. Singh, J. Mal, R. Singh and J. J. E. Bessembinder, "Measurement program and description database," In: J. C. Van Dam and R. S. Malik, Eds., *Water Productivity of Irrigated Crops in Sirsa District, India. Integration of Remote Sensing, Crop and Soil Models and Geographical Information Systems*, WATPRO Final Report, 2003, pp. 29-39.

[22] R. Singh, J. C. van Dam and R. A. Feddes, "Water Productivity Analysis of Irrigated Crops in Sirsa District, India," *Agricultural Water Management*, Vol. 82, No. 3, 2006, pp. 253-278. doi:10.1016/j.agwat.2005.07.027

[23] K. B. Singh, P. R. Gajri and V. K. Arora, "Modelling the Effect of Soil and Water Management Practices on the Water Balance and Performance of Rice," *Agricultural Water Management*, Vol. 49, No. 2, 2001, pp. 77-95. doi:10.1016/S0378-3774(00)00144-X

[24] L. A. Richards, "Capillary Conduction of Liquids through Porous Mediums," *Journal of Applied Physics*, Vol. 1, No. 5, 1931, pp. 318-333. doi:10.1063/1.1745010

[25] M. Th Van Genuchten, "A Closed-Form Equation for Predicting the Hydraulic Conductivity of Unsaturated Soils," *Soil Science Society of America Journal*, Vol. 44, No. 5, 1980, pp. 892-898. doi:10.2136/sssaj1980.03615995004400050002x

[26] J. H. Holland, "Adaptation in Natural and Artificial Systems: An Introductory Analysis with Applications to Biology, Control, and Artificial Intelligence," The University of Michigan Press, Ann Arbor, 1975.

[27] S. Schulze-Kremer, "Molecular Bioinformatics-Algorithms and Applications," Walter de Gruyter, Berlin, 1996, pp. 13-108.

4

Root Characters of Maize as Influenced by Drip Fertigation Levels

Anitta Fanish Sundara Raj, Purushothaman Muthukrishnan, Pachamuthu Ayyadurai

Department of Agronomy, Tamil Nadu Agricultural University, Tamil Nadu, India.
Email: fanish_agri@yahoo.co.in

ABSTRACT

The efficient use of water by modern irrigation systems is becoming increasingly important in arid and semi-arid regions with limited water resources. Field experiments were conducted during 2008-2010 to study the effect of drip fertigation with water soluble fertilizer on root growth of maize under maize based intercropping system. The experiment was laid out in strip plot design with three replications. The treatment consists of nine fertigation levels in main plots and four inter crops in sub plots. Root spread and root dry mass were increased under drip fertigation practices while rooting depth was more under surface irrigation. Drip fertigation with water soluble fertilizer improved the root system by inducing new secondary roots which are succulent and actively involved in physiological responses. Drip fertigation has pronounced effect on the root architecture especially in the production of highly fibrous root system.

Keywords: Drip Fertigation; Maize Based Intercropping System; Root Spread; Root Dry Weight

1. Introduction

The sustainability of any production system requires optimum utilization of resources be it water, fertilizer or soil. Because of its highly localized application and the flexibility in scheduling water and chemical applications, drip irrigation has gained widespread popularity as an efficient and economically viable method for fertigation. Drip irrigation with fertigation offers a vast potential for optimum utilization of water and fertilizers [1,2]. Application of fertilizers through an efficient irrigation system, known as fertigation, results in more accurate and timely crop nutrition, thus, leading to increased yield besides considerable savings in fertilizers [3]. Fertilizers applied under traditional methods are generally not utilized efficiently by the crop. In fertigation, nutrients are applied through emitters directly into the zone of maximum root activity and consequently fertilizer-use efficiency can be improved over conventional method of fertilizer application. Generally crop response to fertilizer application through drip irrigation has been excellent and frequent nutrient applications have improved the fertilizer-use efficiency [4]. The crop response to fertilizer applications is expected to vary markedly with the type of fertilizer used.

Intercropping, the growing of two or more crop species simultaneously in the same field during a growing season is important for the development of sustainable food production systems, particularly in cropping systems with limited external inputs. This may be due to some of the potential benefits for intercropping system such as high productivity and profitability, improvement of soil fertility, efficient use of resources, risk minimization, efficient use of labour, reducing damage caused by pests, diseases and weeds, erosion control and food security [5,6]. Further, when the intercrop provides a good soil cover, soil temperature will stay relatively low. This prevents burning of the organic matter in the soil and loss of nutrients. It also provides a microclimate that can be favourable for associated crops. Selection of appropriate and compatible component crops for intercropping helps in reducing inter- and intra-plant competition for resources, thereby enhancing the productivity with optimal use of available and applied resources. Due to ever increasing human population especially in India leading diminishing land sizes, intercropping, with its advantages of risk minimization, reduction of soil erosion, increased food security should be practiced.

India ranks second (next only to China) in the production of vegetables contributing to 12 per cent of world production. But the production (87 mt) and consumption (145 g/head/day) are rather low and inadequate to meet the unfulfilled demand (230 g/head/day) of vegetables of rising population in the country with practically no scope for horizontal expansion of the area under cultivation. The crop intensification in the form of inter and multiple cropping provides greater opportunity for enhancing the production per unit area and time. Apart from encouraging large scale cultivation of vegetables in general, technology also needs to be generated to include vegetables in the cropping systems particularly in an intercropping system with regular crops. This will not only go a long way in augmenting vegetable production in the country but also helps in efficient utilization of natural and scarce resources namely land and water. Maize is one such crop which provides opportunity for inclusion of intercrop, the wide row space provided to the crop and also because of the plasticity of the crop to row spacing. In recent years, vegetables are grown as intercrops for higher income by the farmers of India. Vegetables such as beet root, onion, cabbage, cucumber, radish, snap bean, and broccoli are intercropped into double rows of field maize planted on raised soil beds. In intercropping systems with different canopy heights the crop in the under story needs to be shade tolerant for the plant to be productive [7].

Maize has three possible uses viz., food, feed for livestock and raw material for industry. Maize grown in subsistence agriculture continues to be a basic food crop. In recent years, in developing countries like India maize is becoming a commercial crop due to its high demand as an animal feed ingredient. Among the different factors influencing the productivity of maize, water and nutrients occupy prime position. Maize crop responds very well to water and nutrient application. Maize is one of the amenable crops for drip irrigation system, which is an efficient method of irrigation [8]. Many scientists reported enhanced growth and development of maize under drip irrigation [9,10]. Maize is a nutrient loving crop, which pumps out more quantity of nutrients from soil.

Inter-row space in maize during the initial slow growth period provides ample scope to cultivate the compatible crop in between two rows of maize and increase the productivity per unit area and time. Performance of radish and beetroot [11-13], coriander [14] and onion [15] as with maize are well documented. These systems would provide early, periodic and high economic return besides ensuring stability. Productivity of the system can be enhanced by judicious selection of vegetable crops differing in duration, canopy architecture and growth rhythm so as to adjust the demands of the above and underground resource like light, moisture and nutrient during different growth stages.

In spite of having many economic and other advantages over the method of flood irrigation, the coverage of area under micro-irrigation is not appreciable in India. Among the various reasons for the slow progress of adoption of this new technology, its capital-intensive nature seems to be one of the main deterrent factors. The drip system installed for maize crop can be used for intercrops too simultaneously and helps to reduce the payback period and increase the income. A vegetable

crop fetches higher market values than agricultural field crops in shorter period. Inclusion of vegetable crops in wide spaced field crops under drip irrigation helps to gets more remuneration within shorter period.

Research works on drip irrigation under intercropping situation is very limited. Input information on optimal schedules for micro-irrigation and fertigation to maize and planting geometry for micro-irrigation will have to be generated from the current levels thus enabling the option of micro-irrigation under intercropping situation. Therefore, the present investigation was conducted to study the effect of fertigation involving the source and rate of fertilizers, methods of fertilizer application like through soil and drip irrigation on maize under different intercropping situation

2. Materials and Methods

2.1. Seasons and Weather Data

Field experiments were conducted at the Department of Agronomy, Agricultural College and Research Institute, Coimbatore, India from 2008 (winter) until 2009 (winter). The region is characterized as semi-arid tropical (SAT) climate, located at 11°8'N latitude and 77°8'E longitude. The mean annual rainfall (83 years) at Coimbatore is 674 mm distributed over about 50 rainy days with a 30% annual coefficient of variation. The rainfall is monsoon type, with a south-west monsoon from June to September and a north-east monsoon from October to December. The mean annual maximum and minimum temperature ranged from 29.2°C to 35.2°C and from 17.9°C to 23.8°C, respectively. The mean relative humidity ranged from 50 to 77 percent. The mean pan evaporation per day ranged from 3.1 to 6.7 mm.

The field was uniformly levelled and formed into raised flat beds for maize. The lateral spacing between two raised flat beds was 1.5 m with furrow in-between of 30 cm width and 15 cm depth. In the surface irrigated plots, ridges and furrows were formed at 60 cm apart. Maize along with intercrops viz., radish, beetroot, onion and vegetable coriander were raised during winter (August, 2008-November, 2008) under raised bed lay out of drip system. A confirmation study with the same set of treatments was conducted in the same seasons during 2009. The pre-sowing composite soil samples collected from the experimental soil was analyzed for physical-chemical characteristics. The experimental soil was sandy clay with 1.36 - 1.42 (2008) and 1.34 - 1.41 (2009) $g \cdot cc^{-1}$ of bulk density, field capacity of 25.2% - 26.3% (2008) and 25.1% - 26.1% (2009), and permanent wilting point of 12.5% - 13.7% (2008) and 12.4% - 13.6% (2009). The nutrient status was low (216 - 220 $kg \cdot ha^{-1}$), medium (16 - 17 $kg \cdot ha^{-1}$) and high (425 - 428 $kg \cdot ha^{-1}$) for available nitrogen (N), phosphorus (P) and potassium (K), respectively.

Maize crop along with intercrops were raised during first year in winter (August 2008-November 2008) received 437 mm of rainfall in 26 rainy days (**Table 1**). Weekly mean pan evaporation ranged from 3.6 to 9.6 mm per day. Maximum mean weekly temperature was 31.5°C and minimum was 21.8°C. Relative humidity ranged between 67% and 98% and 41% and 71% at 07:22 and 14:22 h, respectively.

Maize crop along with intercrops were raised during second year in winter (July 2009 to October 2009) received 188 mm of rainfall in 22 rainy days. Weekly mean pan evaporation ranged from 3.9 to 6.6 mm per day. Maximum mean weekly temperature was 28.71°C - 36.2°C and minimum was 17.1°C - 22.6°C. Relative humidity ranged between 76% to 94% and 38% to 71% at 07:22 and 14:22 h, respectively.

2.2. Lay Out and Experimentation

The experiment was laid out in strip plot design with three replications. The treatment consists of nine fertigation levels in main plots and four inter crops in sub plots. The treatments were allotted to the plots by randomization. Recommended dose of fertilizer were 150:75:75 kg NPK·ha^{-1}. Treatments are as follow

Main Plot: Fertigation Level (T)	
T_1:	Surface irrigation (furrow) with soil application of 100% RDF
T_2:	Drip irrigation with soil application of 100% RDF
T_3:	Drip fertigation with 75% of RDF (CF)
T_4:	Drip fertigation with 100% of RDF (CF)
T_5:	Drip fertigation with 125% of RDF (CF)
T_6:	Drip fertigation with 150% of RDF (CF)
T_7:	Drip fertigation with 50% of RDF (50% of P and K through WSF; remaining through CF)
T_8:	Drip fertigation with 75% of RDF (50% of P and K through WSF; remaining through CF)
T_9:	Drip fertigation with 100% of RDF (50% of P and K through WSF; remaining through CF)
Sub plots: Inter crops (S)	
S_1:	Vegetable coriander
S_2:	Radish
S_3:	Beet root
S_4:	Onion

RDF: Recommended Dose of Fertilizer; WSF: Water Soluble Fertilizer; CF: Conventional Fertilizer.

The experimental field was ploughed with tractor drawn disc plough followed by two ploughings with cultivator and the clods were broken with rotavator. The field was uniformly levelled and raised flat beds were formed in the dimension of 120 cm width, 30 cm furrow

and 15 cm height. The lateral spacing between two raised flat beds was 1.5 m. In the surface irrigated plots, ridges and furrows were formed at 60 cm apart. Buffer channels were formed to control the lateral seepage of water from one plot to another.

In surface irrigation treatment, maize was sown at a spacing of 60 × 20 cm. Paired row planting system was adopted under drip irrigation with a spacing of 75 × 20 cm. In between the two rows of maize two rows of inter crops were sown by adopting a spacing of 30 cm between rows. The planting patterns under surface and drip irrigation are given in **Figure 1**.

Good quality seeds of maize, radish, beetroot coriander and onion bulbs were used for the experimental study. All seeds were placed by hand dibbling with specified spacing. Sowing irrigation was uniformly given to all treatments. The drip irrigation scheduling was done based on wetted area concept and irrigation system was operated once in three days.

$$\text{Water requirement or ETc(lpd)}$$

$$= CPE \times Kp \times Kc \times Wp \times S$$

Table 1. Seasonal rainfall distribution.

Crop stages (DAS)	2008		2009	
	Rainfall (mm)	Rainy days (No.)	Rainfall (mm)	Rainy days (No.)
0 - 25	74	2	23	3
26 - 45	18	3	28	4
46 - 70	222	14	127	12
71 - 90	123	7	10	3
Total	437	26	188	22

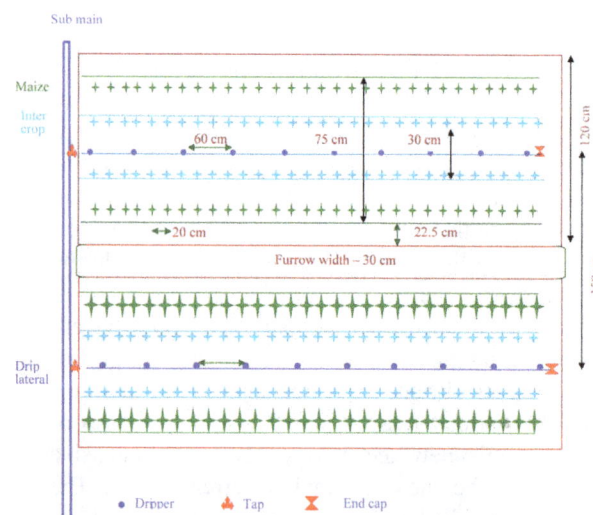

Figure 1. Layout of individual plot for maize based intercropping system.

where, ETc, crop evapotranspiration; CPE, cumulative pan evaporation (mm); Kp, pan factor (0.8); Kc, crop co-efficient; Wp, wetting area percentage (80%); S, crop spacing (0.60 m × 0.20 m for maize).

Irrigation water was pumped through electric motor and conveyed to the main line of 63 mm OD (outer diameter) PVC (poly vinyl chloride) pipes after filtering through sand filter. In the main line, venturi was installed for fertigation. From the main, sub mains of 40 mm OD PVC pipes were drawn and from the sub-main, laterals of 12 mm LLDPE (low linear density polyethylene) pipes were installed at an interval of 1.5 m. Each lateral was provided with individual tap control for imposing respective irrigation schedules. Along the laterals, inline drippers with a discharge capacity of 4 lph were spaced at 0.6 m. Sub-mains and laterals were closed at the end with end cap. After installation, trial run was conducted to assess mean dripper discharge and uniformity coefficient. This was taken into account while fixing the irrigation water application time. During the irrigation period an average of 90% - 95% uniformity was observed. The tabulated Kc values were adjusted according to the local climatic conditions (**Table 2**). Furrow irrigation was applied based on IW/CPE (IW = irrigation water; CPE = cumulative pan evaporation) ratio of 0.80. The depth of irrigation was fixed as 5 cm per irrigation.

The quantity of irrigation water supplied through drip was 173 (2008) and 198 (2009) and the effective rainfall received during the cropping period was 158 mm (2008) and 130 mm (2009). The total water used under the drip irrigation treatments was 331 mm and 328 mm during 2008 and 2009, respectively. Under furrow irrigation method, irrigation was given immediately after sowing followed by life irrigation at 5 cm depth thereafter irrigation was given as per the IW/CPE ratio of 0.8. Quantity of water applied was 300 and 350 mm during winter 2008 and winter 2009, respectively. An effective rainfall of 192 and 161 mm was received during crop period and totally 492 and 511 mm of water was consumed by surface irrigated crop.

Healthy crop stand was ensured by adopting recommended package of practices and need based plant protection measures. Recommended dose of 150:75:75 kg of NPK·ha^{-1} were applied to maize.

Table 2. Kc values.

Crop stage	Duration (days)	Kc value
Initial	20	0.40
Crop development	30	0.80
Mid season	40	1.15
Late season	10	0.70
Total	100	

The recommended doses of inorganic fertilizers were applied directly to soil for the treatments T_1 and T_2. Fertilizer sources used for supplying NPK were urea (46% N), di ammonium phosphate (18% N and 46% P_2O_5) and muriate of potash (60% K_2O) respectively. The entire quantity of phosphorus was applied as basal in treatments T_1 to T_6 in the form of di ammonium phosphate one day before sowing. In the treatments T_1 and T_2 involving soil application of fertilizers, recommended dose of nitrogen and potassium were applied in the form of urea in three splits (basal, 25 DAS and 45 DAS) and muriate of potash in two splits.

2.3. Fertigation

For treatments T_3 to T_9, fertilizers were given through drip fertigation. In treatments T_3 to T_6 normal fertilizer was used as sources for supplying NPK through drip irrigation. Normal fertilizers *viz.*, urea and muriate of potash were used to supply N and K respectively. For the treatments T_7 to T_9 50 per cent P and K were supplied through water soluble fertilizer and remaining through normal fertilizer. Mono ammonium phosphate (12:61:0) and multi-K (13:0:46) were used as water soluble fertilizer for supplying P and K respectively. The fertilizer solution was prepared by dissolving the required quantity of fertilizer with water in 1:5 ratio and injected into the irrigation system through ventury assembly. Considering the nutrient uptake pattern at phenological growth phases of maize, the fertigation schedule was worked out and presented in **Table 3**.

2.4. Observation

Soil moisture content was estimated by gravimetric method. Soil samples were taken at a spacing of 0, 10, 20 and 30 cm between lateral and 10, 20 and 30 cm between dripper at a depth of 0 - 10, 10 - 20, 20 - 30, 30 - 40 and 40 - 50 cm for studying soil moisture distribution pattern in each drip irrigation regime. This observation was made in rain free periods. Data were collected before start of the irrigation from two plants in each treatment

Table 3. Fertigation schedule for maize.

Crop stages (DAS)[*]	Quantity (%)			Fertigation frequency
	N	P	K	
Vegetative stage (6 - 30 DAS)	25	25	25	5
Reproductive stage (31 - 60 DAS)	50	50	50	5
Maturity stage (61 - 75 DAS)	25	25	25	3
Total	100	100	100	

[*]DAS: Days after Sowing.

and the average values were expressed in % soil moisture by oven dry weight. Since the same trend was observed in both the years of the study, mean data for soil moisture content at different locations before irrigation was arrived and depicted as graphical representation. Graphical software package SURFER was used to show the three dimensional view of soil moisture distribution vertically and horizontally from the emitter.

The root studies were made by measuring the tap root (rooting depth), root volume and root dry mass (g) at crop maturity stage in three tagged plants and mean value was arrived. Root characters were assessed by modified trench method as suggested by Bohm [16]. Trenches to a convenient depth and sufficient length on both sides of the sampling row were opened by digging with fork without snatching the root lets and separated away from the sampling rows. The plants were carefully excavated until the tip of each plant root was just visible. The soil adhering to the plant was carefully removed by immersing in water tub and the soil was disintegrated and then observations were made. Rooting depth was measured from the collar region to tip of the deepest root and expressed in cm. Root samples were air dried initially followed by oven drying at $65°C \pm 5°C$ till a constant weight is attained and root mass was expressed as g per plant.

2.5. Statistical Analysis

The data pertained to root characters were subjected to statistical analysis by Analysis of Variance (ANOVA) using AGRES (Data Entry Module for AgRes Statistical software version 3.01, 1994 Pascal Intl. Software Solutions). Differences between means were evaluated for significance using least significant differences (LSD) at 5% probability level as suggested by [17].

3. Results and Discussion

3.1. Rooting Depth

Roots are the main component in absorption of water and minerals, which are essential in plant physiological processes.

Maize root went deeper in soil profile under surface irrigation for want of water and registered more rooting depth. The rooting depth of maize at different fertigation levels and intercrops varied widely as presented in **Table 4**. Data collected at harvet stage showed that surface irrigation recorded significantly higher values for rooting depth (45.04 and 47.34 cm during 2008 and 2009, respectively). Rooting depth was increased as water stress increased.

Among the different fertigation treatments, drip fertigation at 100 per cent RDF with 50 per cent P and K as WSF produced lengthier roots (57.16 and 53.43 during 2008 and 2009, respectively) than other treatments.

Regarding intercropping system, maize + vegetable coriander recorded a higher rooting depth (54.54 and 51.19 cm in 2008 and 2009 respectively) of maize followed by maize + onion.

Interaction effect between fertigation level and intercrops was found to be non significant for rooting depth of maize.

3.2. Root Spread

Root spread of maize was significantly affected by the drip fertigation levels and intercrops (**Table 5**). Lateral root spread was more under drip fertigation of 100 per cent RDF with 50 per cent P and K as WSF (T_9) (44.59 and 42.29 cm respectively, during 2008 and 2009) followed by fertigation at 150 per cent RDF (T_6). In contrast to rooting depth, very short and thin lateral roots were found in surface irrigated plants (T_1). Similar trend was noticed during both the years of experimentation.

Root system was manifested favorably under drip fertigation with water soluble fertilizer with good rooting depth, spread and dry weight when compared to surface irrigation. Water stress under surface irrigation increased the rooting depth to obtain water from the deep soil. If roots were unable to find the water in later stages also, stress made it to loose its succulence and became insensitive to moisture deficit then leads to degrade. Under surface irrigation treatments roots have gone to deeper layers for want of water as evidenced by higher values for rooting depth than others. However, if water is unavailable in deeper soil in later stages, degradation of root will occur. In this experiment also degradation of roots for both lateral and vertical direction at wetting front was observed under surface irrigation treatments and reflected in root spread. Kang [18] and Pandey [19] expressed the similar views on maize root development under water stress conditions.

Among the intercrops, vegetable coriander as intercrop system produced more maize root spread (42.69 cm) followed by onion as intercrop system (41.06 cm during 2008). The lesser root spread of maize was observed under beet root as intercrop (39.03 cm) system. The same trend was observed in second year also.

3.3. Root Dry Weight

The root dry mass recorded at harvest stage of maize was significantly influenced by the drip fertigation and intercrops during both years (**Table 6**). Higher values (17.8 and 17.9 g, respectively during 2007 and 2008) for root dry mass were recorded under drip fertigation of 100 per cent RDF with 50 per cent P and K as WSF (T_9) (43.1 and 42.3 cm respectively, during 2008 and 2009) followed by fertigation at 150 per cent RDF (T_6). Water

Table 4. Effect of drip fertigation levels and inter crops on rooting depth of maize.

| Treatments | Root depth (cm) | | | | | | | | | |
| | 2008 | | | | | 2009 | | | | |
	S_1-Coriander	S_2-Radish	S_3-Beetroot	S_4-Onion	Mean	S_1-Coriander	S_2-Radish	S_3-Beetroot	S_4-Onion	Mean
T_1-SI + SA of 100% RDF	45.36	44.72	44.90	45.18	**45.04**	48.34	47.22	47.64	46.16	**47.34**
T_2-DI + SA of 100% RDF	52.21	51.49	51.69	52.01	**51.85**	49.48	48.2	48.46	46.98	**48.28**
T_3-DF + 75% RDF (NF)	53.82	53.08	53.29	53.61	**53.45**	50.26	49.04	49.39	47.83	**49.13**
T_4-DF + 100% RDF (NF)	56.01	55.73	55.95	56.19	**55.34**	50.88	49.63	49.93	48.28	**49.68**
T_5-DF + 125% RDF (NF)	56.11	55.33	55.55	55.89	**55.72**	51.14	49.88	50.21	48.53	**49.94**
T_6-DF + 150% RDF (NF)	56.84	56.04	56.27	56.61	**56.44**	53.48	52.43	52.70	51.19	**52.45**
T_7-DF + 50% RDF (50% P & K- WSF)	55.23	54.47	54.69	55.01	**54.85**	50.60	49.04	49.75	48.41	**49.45**
T_8-DF + 75% RDF (50% P & K- WSF)	57.01	56.41	56.64	56.98	**56.07**	52.14	51.05	51.42	49.87	**51.12**
T_9-DF + 100% RDF (50% P & K-WSF)	57.56	56.76	56.99	57.33	**57.16**	54.42	53.35	53.83	52.12	**53.43**
Mean	**54.54**	**53.78**	**54.00**	**54.32**		**51.19**	**49.98**	**50.37**	**48.82**	
T-Fertigation levels	0.352					0.215				
S-Intercrops	0.210					0.201				
Interaction	NS					NS				

Table 5. Effect of drip fertigation levels and inter crops on root spread of maize.

| Treatments | Root spread (cm) | | | | | | | | | |
| | 2008 | | | | | 2009 | | | | |
	S_1-Coriander	S_2-Radish	S_3-Beetroot	S_4-Onion	Mean	S_1-Coriander	S_2-Radish	S_3-Beetroot	S_4-Onion	Mean
T_1-SI + SA of 100% RDF	38.56	35.99	35.26	37.09	**36.72**	38.21	37.65	37.12	37.86	**37.71**
T_2-DI + SA of 100% RDF	39.55	36.91	36.16	38.04	**37.66**	38.80	38.16	37.55	38.29	**38.20**
T_3-DF + 75% RDF	40.54	37.84	37.07	39.00	**38.61**	39.24	38.63	38.03	38.81	**38.68**
T_4-DF + 100% RDF	42.81	39.96	39.14	41.18	**40.77**	40.23	39.60	38.93	39.75	**39.63**
T_5-DF + 125% RDF	43.80	40.88	40.04	42.13	**41.71**	40.71	40.08	39.41	40.25	**40.11**
T_6-DF + 150% RDF	45.80	42.75	41.87	44.05	**43.62**	42.24	41.72	41.10	41.85	**41.73**
T_7-DF + 50% RDF (50% P & K-WSF)	41.56	38.79	37.99	39.97	**39.58**	39.72	38.94	38.63	39.30	**39.15**
T_8-DF + 75% RDF (50% P & K- WSF)	44.78	41.79	40.94	43.07	**42.64**	41.12	40.57	39.98	40.76	**40.61**
T_9-DF + 100% RDF (50% P & K-WSF)	46.82	43.70	42.81	45.04	**44.59**	42.78	42.25	41.63	42.49	**42.29**
Mean	**42.69**	**39.84**	**39.03**	**41.06**		**40.34**	**39.73**	**39.15**	**39.93**	
T-Fertigation levels	1.851					0.958				
S-Intercrops	1.454					0.657				
Interaction	NS					NS				

stress under surface irrigation (T_1) significantly reduced the root dry mass recorded significantly least values (15.4 and 15.7 g, respectively during 2008 and 2009) when compared to other treatments.

Table 6. Effect of drip fertigation levels and inter crops on root dry weight of maize.

Treatments	Root dry weight (g plant^{-1})									
	2008					2009				
	S_1-Coriander	S_2-Radish	S_3-Beetroot	S_4-Onion	Mean	S_1-Coriander	S_2-Radish	S_3-Beetroot	S_4-Onion	Mean
T_1-SI + SA of 100% RDF	14.01	13.07	12.81	13.47	**13.34**	13.98	12.86	13.28	11.80	**12.98**
T_2-DI + SA of 100% RDF	14.33	13.38	13.10	13.79	**13.65**	14.75	13.47	13.73	12.25	**13.55**
T_3-DF + 75% RDF	14.64	13.66	13.38	14.08	**13.94**	15.02	13.80	14.15	12.59	**13.89**
T_4-DF + 100% RDF	16.15	15.07	14.76	15.53	**15.38**	16.58	15.33	15.63	13.98	**15.38**
T_5-DF + 125% RDF	16.67	15.56	15.24	16.04	**15.88**	17.16	15.90	16.23	14.55	**15.96**
T_6-DF + 150% RDF	17.83	16.64	16.30	17.15	**16.98**	18.16	17.11	17.38	15.87	**17.13**
T_7-DF + 50% RDF (50% P & K-WSF)	14.95	13.96	13.67	14.38	**14.24**	15.81	14.25	14.96	13.62	**14.66**
T_8-DF + 75% RDF (50% P & K-WSF)	16.98	15.85	15.52	16.33	**16.17**	17.56	16.47	16.84	15.29	**16.54**
T_9-DF + 100% RDF (50% P & K-WSF)	18.64	17.40	17.04	17.93	**17.75**	18.67	17.60	18.08	16.37	**17.68**
Mean	**16.02**	**15.05**	**14.65**	**15.51**		**16.41**	**15.20**	**15.59**	**14.04**	
T-Fertigation levels	**0.263**					**0.318**				
S-Intercrops	**0.227**					**0.298**				
Interaction	**NS**					**NS**				

Surface irrigated plants (T_1) produced longer but thinner roots compared to drip irrigated plant. Preferential partitioning of photosynthates to leaf area expansion at the expense of root growth may have been responsible for the lower root dry mass at water stress compared with fully irrigated roots [20]. The same effect of water stress on root dry mass was observed in this study also.

Drip fertigated maize had a higher root biomass than surface irrigated crop. Drip fertigation levels exerted significant influence on root biomass. Drip fertigation with 100 per cent RDF in which 50 per cent P and K through WSF produced greater root biomass of 17.75 and 17.68 g plant^{-1} during 2008 and 2009, respectively. Lowest root biomass of 13.34 (2008) and 12.98 (2009) g plant^{-1} was observed under surface irrigation with soil application of fertilizer. The root biomass was significantly influenced by the intercrops at harvest stage of maize. Vegetable coriander as intercrop recorded more root biomass of maize (16.02 and 16.41 g plant^{-1} during 2008 and 2009, respectively)

3.4. Influence of Drip Fertigation Levels and Intercrops on Root Characters

Plant roots play a vital role in soil water and solute dynamics by modifying the water and solute uptake patterns in the rooting zone [21,22]. At early stages of crop growth, the better availability of nitrogen could have helped to build up sufficient framework. Under trickle fertigation with surface wetting, the production of lengthier roots clearly indicated that the plant tried hard to extract available water from deeper layers to meet its water requirements. This was in conformity with the findings of Kataria and Michael [23]. Growth of the root, in general, is stimulated by phosphorus and applied P which would have encouraged the early root growth. Similar results were also reported by [24,25].

The root characters like root length, root spread and root dry weight were significantly affected by fertigation levels in all the intercropping systems. The root system is the link between the plant and soil. It is responsible for the absorption of water and nutrients, anchorage, synthesis of some plant harmones and storage [26]. In order to achieve proper growth, the root zone of a plant must be well supplied with both water and oxygen. Among the fertigation treatments, 150 per cent RDF and 100 per cent RDF with P and K as WSF resulted in higher root parameters. Adequate quantity of nutrients coupled with adequate moisture might have resulted in higher root proliferation. Application of readily available form of fertilizer particularly in frequent intervals (once in three days) by reducing the quantity of nutrients at one application, the crops could able to utilize maximum quantity of nutrients reducing the leaching and volatilization loss and increasing the nutrient use efficiency which might have resulted in higher root growth. This showed positive response of root characters to higher fertilizer dose producing higher root biomass under favourable moisture and nutrient status as observed by Parthasarathi [27] in radish. Fertigation enhanced overall root activity; improve the mobility of nutritive elements and their uptake, as well as reducing the contamination of surface and ground water [28].

Under surface application of fertilizer with drip irrigation, nearly 75 per cent of total fertilizers were applied as basal, due to which the crop could not able to absorb all the nutrients at a single time and the unutilized nutrients were either fixed in the soil or leached away from the root zone. So the fertilizer use efficiency was low and the uptake also low which might have resulted in lower root growth. Concentration of root biomass was maximum (0 - 30 cm) in the upper soil profile under drip irrigation system compared to surface irrigation.

Under surface fertilized plots, the entire quantity of P (75 kg·ha^{-1}) was applied as basal on the day of sowing, where the P gets fixed up in the soil profile and when the roots emerge and grows, it was able to utilize the available P alone which are very low due to fixation in the soil. Whereas under drip fertigation (T$_7$ to T$_9$ treatments), the P nutrient in the form of mono ammonium phosphate which is easily soluble in water was fertigated from 7th day after sowing and fertigated once in three days up to 75 days. So the crop was able to absorb more P throughout the crop growth period with very minimum loss of nutrients. This might be the reason for higher root growth under drip fertigated treatments when compared to surface applied plots.

Concentration of roots were maximum in the upper soil profile under drip irrigation system, while in surface irrigation, the tap root length was more but the root biomass was less. This was due to the fact that the moisture was fast depleting beyond the root zone and the plants had to extract water from deeper layers.

There was significant response on root character (rooting depth, root volume and root biomass) of maize crop due to intercropping system. Root growth was higher under maize + vegetable coriander intercropping system. Radish, beet root and onion were root crops. The growths of below ground part of these crops affect the growth of root of maize crop. This might be the reason for higher root growth of maize under maize + vegetable coriander.

4. Conclusion

Fertigation levels and intercrops had significant impact on root system of maize. Root spread and root dry mass were increased under drip fertigation practices while rooting depth was more under surface irrigation. Drip fertigation with water soluble fertilizer improved the root system by inducing new secondary roots which are succulent and actively involved in physiological responses. Drip fertigation has pronounced effect on the root archi-

tecture especially in the production of highly fibrous root system. Fertigation promotes the production of intensely branched roots that facilitates nutrient acquisition and foraging capacity. On the other hand, conventional method of irrigation and fertilizer application has exhibited limited root spread in the rhizosphere. Regarding intercropping system, maize + vegetable coriander recorded higher root characters.

REFERENCES

[1] J. N. Raina, B. C. Thakur and M. L. Verma, "Effect of Drip Irrigation and Polyethylene Mulch on Yield, Quality and Water-Use Efficiency of Tomato," *The Indian Journal of Agricultural Sciences*, Vol. 69, No. 6, 1999, pp. 430-433

[2] J. N. Raina, "Drip Irrigation and Fertigation: Prospects and Retrospect's in Temperate Fruit Production," In: K. K. Jindal and D. R. Gautam Eds., *Enhancement of Temperate Fruit Production in Changing Climate*, UHF, Solan, 2002, pp. 296-301.

[3] C. Bussi, J. G. Huguet and H. Defrance, "Fertilization Scheduling in Peach Orchard under Trickle Irrigation," *Journal of Horticulture Science*, Vol. 66, No. 4, 1991, pp. 487- 493.

[4] R. S. Malik, K. Kumar and A. R. Bhandari, "Effect of Urea Application through Drip Irrigation System on Nitrate Distribution in Loamy Sand Soils and Pea Yield," *Journal of the Indian Society of Soil Science*, Vol. 42, No. 1, 1994, pp. 6-10.

[5] P. Banik, A. Midya, B. K. Sarkar and S. S. Ghose, "Wheat and Chickpea Intercropping Systems in an Additive Series Experiment: Advantages and Weed Smothering," *European Journal of Agronomy*, Vol. 24, No. 4, 2006, pp. 325-332. doi:10.1016/j.eja.2005.10.010

[6] P. K. Ghosh, M. Mohanty, K. K. Bandyopadhyay, D. K. Painuli and A. K. Misra, "Growth, Competition, Yields Advantage and Economics in Soybean/Pigeonpea Intercropping System in Semi-Arid Tropics of India II. Effect of Nutrient Management," *Field Crops Research*, Vol. 96, No. 1, 2006, pp. 90-97. doi:10.1016/j.fcr.2005.05.010

[7] J. M. Sogbedji, H. M. Van, J. R. R. Melkonian and R. R. Schindelbeck, "Evaluation of the P. N. M. Model for Simulating Drain Flow Nitrate-N Concentration under Manure Fertilized Maize," *Plant and Soil*, Vol. 282, 2006, pp. 343-360.

[8] C. Zhu, S. Naqvi, S. Gomez-Galera, A. M. Pelacho, T. Capell and P. Christou, "Transgenic Strategies for the Nutritional Enhancement of Plants," *Trends in Plant Science*, Vol. 12, No. 12, 2007, pp. 548-555. doi:10.1016/j.tplants.2007.09.007

[9] F. R. Lamm, T. P. Trooin, H. L. Manges and H. D. Sunderman, "Nitrogen Fertilization for Subsurface Drip-Fertigated Corn," *Transactions of the American Society of Agricultural Engineers*, Vol. 44, No. 3, 2001, pp. 533-542.

[10] K. Mmolawa and D. Or, "Root Zone Solute Dynamics under Drip Irrigation: A Review," *Plant and Soil*, Vol. 222, No. 1-2, 2000, pp. 163-190.

doi:10.1023/A:1004756832038

[11] M. T. Reddy, S. Ismail and Y. N. Reddy, "Performance of Radish (*Raphanus sativus* L.) under Graded Levels of Nitrogen in Ber-Based Intercropping," *Journal of Research ANGRAU*, Vol. 27, No. 3, 2001, pp. 24-28.

[12] S. N. Singh and A. Kumar, "Production Potential and Economics of Winter D1aize Based Intercropping Systems," *Annals of Agricultural Research*, Vol. 23, No. 4, 2002, pp. 532-534.

[13] R. A. Kumar, K. Chillar and R. C. Gautam, "Nutrient Requirement of Winter Maize *(Zea mays)*-Based Intercropping Systems," *Indian Joumal of Agricultural Sciences*, Vol. 76, No. 5, 2006, pp. 315-318.

[14] N. Chellaiah, U. Solaiappan and S. Senthilvel, "Effect of Sowing Time and Intercropping on the Yield of Coriander under Rainfed Condition," *Madras Agricultural Journal*, Vol. 88, No. 10-12, 2002, pp. 684-689.

[15] M. Mahadevaswamy, "Studies on Intercropping of Aggregated Onion (*Allium cepa var. Aggregatum*) in Wide Spaced Sugarcane," Ph.D. Thesis, TNAU, Coimbatore, 2001.

[16] W. Bohm, H. Maduakor and H. M. Taylor, "Comparison of Five Methods for Characterizing Soybean Rooting Density and Development," *Agronomy Journal*, Vol. 69, No. 3, 1977, pp. 415-419.
doi:10.2134/agronj1977.00021962006900030021x

[17] K. A. Gomez and A. A. Gomez, "Statistical Procedures for Agricultural Research," 2nd Edition, John Wiley and Sons, New York, 1984, p. 680.

[18] S. Kang, W. Shi and J. Zhang, "An Improved Water-Use Efficiency for Maize Grown under Regulated Deficit Irrogation," *Field Crops Research*, Vol. 67, No. 3, 2000, pp. 207-214. doi:10.1016/S0378-4290(00)00095-2

[19] V. K. Pandey, J. P. Bharat and H. D. Bagde, "Water Requirement of Bitter Gourd under Pressurised Irrigation System," *Journal of Indian Water Resource Society*, Vol.

28, No. 2, 2009, pp. 15-19.

[20] M. Pace, J. J. Cole, S. R. Carpenter and J. F. Kitchell, "Trophic Cascades Revealed in Diverse Ecosystems," *Trends in Ecology & Evolution*, Vol. 14, No. 12, 1999, pp. 483-488. doi:10.1016/S0169-5347(99)01723-1

[21] K. Mmolawa and D. Or, "Root Zone Solute Dynamics under Drip Irrigation: A Review," *Plant and Soil*, Vol. 222, No. 1-2, 2000, pp. 163-190.
doi:10.1023/A:1004756832038

[22] T. Selvakumar, "Performance Evaluation of Drip Fertigation on Growth, Yield and Water Use in Hybrid Chilli (*Capsicum annuum* L.)," Ph.D. Thesis, TNAU, Coimbatore, 2006.

[23] D. P. Kataria and A. M. Michael, "Comparative Study of Drip and Furrow Methods," *Proceedings of XI International Congress on the Use of Plastic in Agriculture*, New Delhi, 26 February-2 March 1990, pp. 19-27.

[24] S. U. Ahmed and H. K. Saha, "Effect of Different Levels of N, P and K on the Growth and Yield of Four Tomato Varieties," *Punjab Vegetable Growers*, Vol. 21, 1986, pp. 16-19.

[25] S. Kang, W. Shi and J. Zhang, "An Improved Water-Use Efficiency for Maize Grown under Regulated Deficit Irrogation," *Field Crops Research*, Vol. 67, No. 3, 2000, pp. 207-214. doi:10.1016/S0378-4290(00)00095-2

[26] E. Lahav and D. W. Turner, "Banana Nutrition," *International Potash Institute Bulletin*, Vol. 7, 1989, p. 33.

[27] K. S. Parthasarathi, M. Krishnappa, N. C. Gowda, S. Reddy and M. Anjanappa, "Growth and Yield of Certain Radish Varieties to Varying Level of Fertility," *Karnataka Journal of Agricultural Sciences*, Vol. 12, No. 1-4, 1999, pp. 148-153.

[28] M. H. Taha, "Chemical Fertilizers and Irrigation System in Egypt," *Proceedings of the FAO Regional Workshop on Guidelines for Efficient Fertilizers Use through Irrigation*, Cario, 14-16 December 1999.

5

Effect of Irrigation with Sea Water on Germination and Growth of Lentil (*Lens culinaris Medic*)

Basel Natsheh[1], Zaher Barghouthi[2], Sameer Amereih[1], Mazen Salman[1*]
[1]Palestine Technical University-Kadoorei, Tullkarm, Palestine
[2]Department of Natural Resources Research, National Agricultural Research Center (NARC), Jenin, Palestine
Email: *salman_mazen@daad-alumni.de

ABSTRACT

In an attempt to evaluate the efficiency of sea water irrigation on plant growth and germination, five cultivars of lentil ILL4400, 5582, 5845, 5883 and 8006 were grown in sandy soil and irrigated with sea water of different salinity levels (Ec 0.9-12). Percent of germination, seedlings lengths and mean germination time were recorded. The results showed that all cultivars were able to germinate at different salinity levels. The germination percent was increased. The percent of seed germination was significantly higher in cultivars ILL8006 and ILL5883. Sea water of salinities Ec3 and Ec6 was ideal for irrigation without negative impacts on lentil germination and growth. At higher salinity (Ec12) the germination rate was reduced and the mean germination time was greater than that in lentil seeds irrigated with sea water of Ec 0.9-9.

Keywords: Lentil; Germination; Seedling; Salinity; Sea Water

1. Introduction

The limited water resources in many parts of the world form a major constraint for agricultural and socio-economic development [1]. In arid and semi-arid areas, salinity is one of the most pronounced problems of agricultural irrigation. In these regions, there is an urgent need to use saline water in irrigation because of limited water resources. The success of using saline water for economic crop production can be achieved using best management practices to reduce the negative effects of salinity on crop productivity. In addition to that, introduction and cultivation of crops and varieties that tolerate salinity is highly required.

With increasing demand for irrigation water, alternative sources are being sought. Saline water was previously considered unusable for irrigation. However, this water can be used successfully to grow crops under certain conditions [2]. Saline water has been used in different crops including food, fuel and fodder crops [3]. However, seed germination is the major factor limiting the establishment of plants under saline conditions. Salinity may cause significant reductions in the rate and percentage of germination, which in turn may lead to uneven stand establishment and reduced crop yields [4]. Salt stress reduces the total dry matter, and relative water content, but increases proline accumulation, enzyme activities and electrolyte leakage [5,6]. The aim of this work was to characterize the effect of salinity on lentil germination. Addressing these issues may help generate knowledge valuable for production programs of lentil under saline conditions.

2. Materials and Methods

2.1. Plant and Growth Conditions

Lentil (*Lens culinaris*) cultivars ILL4400, 5582, 5845, 5883 and 8006 were used as the test plant species. Seeds were obtained from ICARDA, Aleppo, Syria. The greenhouse experiments were conducted at the Mediterranean Agronomic Institute, Valenzano, Bari, Italy South coast. All experiments were performed in greenhouse cabins at 20°C and 14 h light and at 18°C in the dark for 10 h/day. The plants were grown in pure sand in plastic pots (15 × 15 × 7 cm).

2.2. Irrigation Water

Saline water was prepared by mixing fresh water (0.9 ds/m) with sea water (48 ds/m) to achieve salinity levels of 3.0, 6.0, 9.0 and 12.0 ds/m. The chemical and physical compositions of irrigation water were determined at the laboratory (**Table 1**). The electrical conductivity (EC), soluble cations (Ca^{2+}, Mg^{2+}, Na^+ and K^+), and soluble anions (CO_3^-, HCO_3^-, Cl^- and SO_4^-) were tested. EC was determined using conductivity meter (Crison micro CM 2201) and expressed in ds/m. The pH was measured

---

*Corresponding author.

Table 1. Chemical characteristics of water used in irrigation.

EC (ds/m)	pH	soluble anions (meq/l)			soluble cations (meq/l)				SAR (meq)$^{1/2}$
		HCO$_3$	Cl	SO$_4^-$	Ca	Mg	Na	K	
0.9*	7.4	4.3	3.15	0.8	5.4	2.75	2.0	0.25	0.99
3.0	7.6	4.6	24.8	0.7	6.3	10.2	18.7	1.2	6.57
6.0	7.7	5.2	53.2	1.6	6.8	16.7	40.0	2.5	11.67
9.0	7.9	5.5	83.4	1.95	8.6	23.7	60.1	3.6	14.93
12.0	8.0	5.9	118	2.31	9.7	29.4	83.5	4.49	18.88

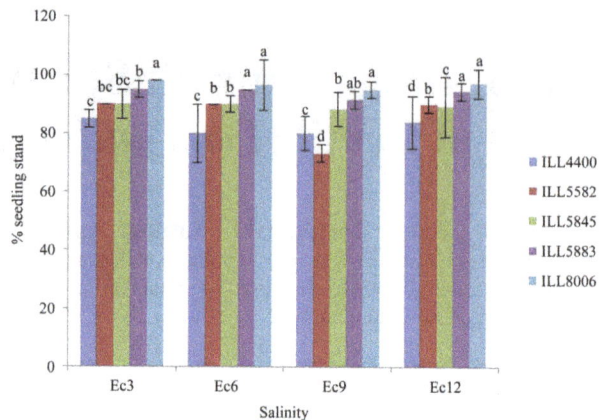

Figure 1. Germination percentage of lentil cultivars under irrigation with sea water of different salinity concentrations. Data of different letters are significantly different after ANOVA at P < 0.05. Bars represent standard errors.

using pH meter (CRISON Basic 20). Ca^{2+} and Mg^{2+} were determined by titration with EDTA (0.01N) according to versenate method [7], using ammonium purporate as an indicator for Ca and ECBT as indicator for Ca and Mg [7]. Na and K were measured photometricaly using a flame photometer (JENWAY PEP 7) as mentioned in Richards [7]. CO$_3$ and HCO$_3$ were estimated volumetrically by titration with a standard solution of sulphuric (0.01 N), using phenolphthalein and methyl-orange as indicators for CO$_3$ and HCO$_3$, respectively. Cl was determined by titration with a standard solution of silver nitrate (AgNO$_3$) in presence of K$_2$Cr$_2$O$_7$ (1%) an indicator. Sulfate was calculated by the difference between, cations and anions.

2.3. Experimental Design

The experiment was conducted in randomized block design. After sowing, seeds were irrigated with sea water for three weeks. After that, number of seedlings was counted and the percent of germination was calculated.In order to determine the mean germination time, seedlings stand was recorded every two days after sowing. Mean germination time (MGT) was calculated according to the equation of Ellis and Roberts [8].

The effect of salinity on lentil growth was also studied by measuring seedling length at the end of the experiment.

2.4. Statistical Analysis

For statistical analysis, XlStat program (Adinosoft) was used. Data on the percent of germination were Asin-transformed. All data were analyzed for variance by Analysis of Variance (ANOVA). Significant differences among treatments were computed after Tukey HSD test at $P < 0.05$.

3. Results and Discussion

3.1. Percent Germination

The effect of salinity on germination is presented at **Figure 1**. Rate of germination and final germination per-

centage differed significantly between cultivars after irrigation with sea water of different salinity levels. Our results showed that germination of lentil seeds was not affected due to increasing salinity concentrations. Seed germination was recorded 7 days after sowing (DAS) and continued every two day over a period of 11 days (**Figure 1**). The highest percent of germination (100%) was recorded in cultivar ILL 5883 irrigated with sea water of Ec 3 salinity. In cultivars ILL 5883 and ILL 8006 the percent of seed germination (100% and 99.15%, respectively) was significantly (P < 0.05) higher in seeds irrigated with fresh water and saline water of Ec 3. The lowest percent of germination was recorded in cultivar 5582 (89.5% and 90%) in seeds irrigated with water of Ec 0 and Ec 3, respectively.

Compared to the control treatments (*i.e.* fresh water irrigation), germination was enhanced after sea water application. Salinity is a major environmental stress factor that affects seed germination [9]. In contrast to what was achieved by Abazrian *et al.* [3], who reported a significant decrease in seed germination of different lentil cultivars with increasing salinity levels, our results showed an enhanced germination percentage after application of sea water.

Our results also showed that the different lentil cultivars were able to grow at different salinity concentration. The ranking of cultivars regarding the germination percent is presented in **Table 2**.

3.2. Seedlings Lengths

Germination and seedling stages are very sensitive to salinity. The failure development healthy seedlings will definitely lead to failure in yield production under saline irrigation practices. Seedlings should be well developed to sustain the harmful effects opposed due to continuous irrigation with saline water.

Table 2. Ranking for different lentil cultivars for the tolerance of salinity.

Varieties	Ec of Sea Water					
	0.9	3	6	9	12	Average
ILL4400	4	4	5	5	5	5
ILL5582	2	3	2	3	3	3
ILL5845	5	5	4	4	4	4
ILL5883	3	1	1	1	2	1
ILL8006	1	2	3	2	1	2

For the calculation of the mean ranking, scores were added and divided by 5.

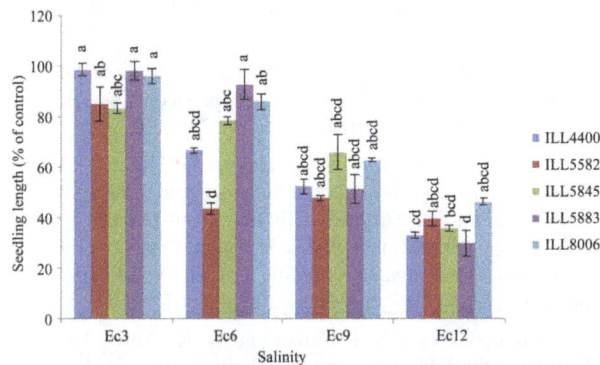

Figure 2. Seedlings lengths in (cm) for the investigated lentil cultivars under saline irrigation treatments. Figures represent the percent of seedlings lengths compared to the control. Data of different letters are significantly different after ANOVA at P < 0.05. Bars represent standard errors.

The interaction of cultivar × salinity stem length was significant different after ANOVA using Tukeys HSD test at P < 0.05. The seedlings lengths of the investigated cultivars are presented in (**Figure 2**). Results showed that salinity decreased plant height. Under fresh water irrigation, all cultivars showed higher stem lengths. However, no significant differences were found in stem lengths between cultivars irrigated with fresh water and cultivars irrigated with saline water of Ec 3. The highest stem-length (18.2 cm) was recorded in cultivar ILL 5883 irrigated with fresh water followed by cultivar ILL 4400 (17.7 cm).

3.3. Mean Germination Time

Germination is a critical phase in plant growth that determines plant establishment and final crop yield [2]. The presented data demonstrate that the gradual increments in Ec values of irrigation water did not stop the germination process of lentil seeds. However, seed germination was delayed in some cultivars after irrigation with sea water of higher EC values (**Figure 3**). Higher reduction of germination was recorded in seeds irrigated with sea water of Ec 12. Our results revealed that the rate of germination varied between different cultivars. In most culti-

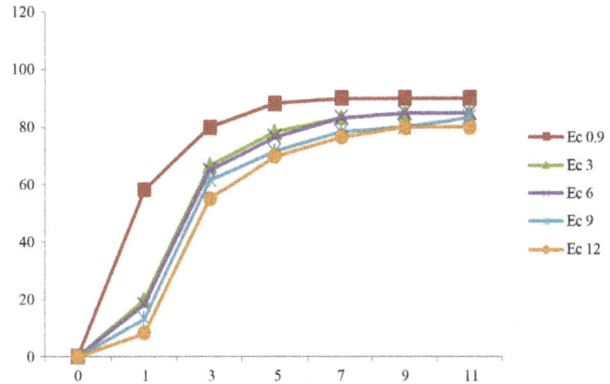

(a)

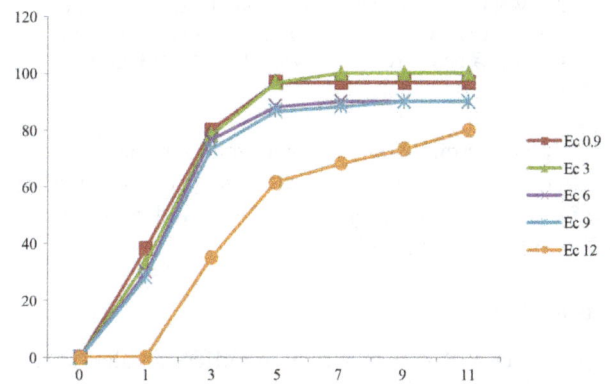

(b)

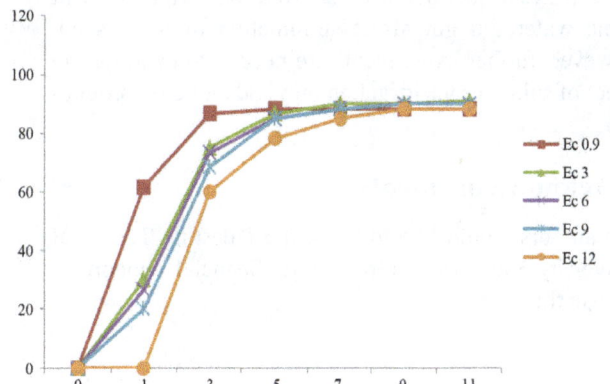

(c)

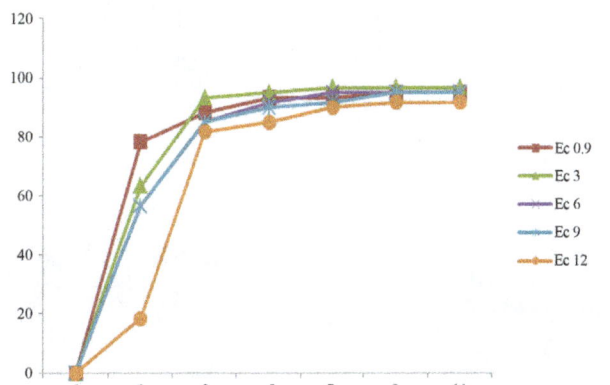

(d)

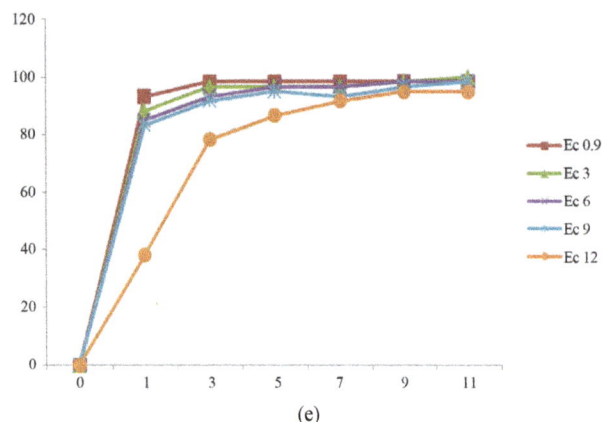

Figure 3. Mean gemrination times of lentile cultivars irrigated with sea water of different salinity levels. (a) ILL4400, (b) ILL5582, (c) ILL5845, (d) ILL5883 and (e) ILL8006.

vars complete germination was lower than 80% after 11 day after sowing. Seeds of cultivars ILL8006 and ILL5883 were the first to germinate followed by cultivars ILL5845, ILL5582 and ILL4400.

4. Conclusion

The use of sea water as in agriculture offers an alternative substitute to fresh water in areas suffering from water scarcity. In fact our work showed that irrigation with saline water did not affect germination of lentil seed. However, further experiments are needed to evaluate the effect of saline water irrigation on yield and crop production.

5. Acknowledgements

The authors would like to thank the Palestine Technical University-Kadoorie for providing financial support to publish this work.

REFERENCES

[1] A. Hamdy, "Saline Irrigation Management for a Sustainable Use," In: N. Katerji, A. Hamdy, I. W. Van Hoorn and M. Mastrorilli, Eds., *Mediterranean Crop Responses to Water and Soil Salinity*: *Eco-Physiological and Agronomic Analyses CIHEAM-IAMB*, Bari, 2002, pp. 185-229.

[2] I. M. Zeid, "Alleviation of Seawater Stress during Germination and Early Growth of Barley," *International Journal of Agriculture*: *Research and Review*, Vol. 1, No. 2, 2011, pp. 59-67.

[3] R. Abazarian, M. R. Yazdani, K. Khosroyar and P. Arvin, "Effects of Different Levels of Salinity on Germination of four Components of Lentil Cultivars," *African Journal of Agricultural Research*, Vol. 6, No. 12, 2011, pp. 2761-2766.

[4] M. R. Foolad and G. Y. Lin, "Genetic Potential for Salt Tolerance during Germination in Lycopersicon Species," *Horticulture Science*, Vol. 32, 1997, pp. 296-300.

[5] A. L. Tuna, C. Kaya, M. Dikilitas and D. Higgs, "The Combined Effects of Gibberellic Acid and Salinityon Some Antioxidant Enzyme Activities, Plant Growth Parameters and Nutritional Status in Maize Plants," *Environmental Experiments in Botany*, Vol. 62, No. 1, 2008, pp. 1-9. doi:10.1016/j.envexpbot.2007.06.007

[6] A. Nitika, B. Renu, S. Priyanka and H. K. Arora, "28-Homobrassinolide Alleviates Oxidative Stress in Salt-Treated Maize (*Zea mays* L.) Plants," *Brazian Journal of Plant Physiliolgy*, Vol. 20, 2008, pp. 153-157.

[7] L. A. Richards, "Agriculture Handbook No. 60," US Government Printing Office, Washington DC, 1954.

[8] R. A. Ellis and E. H. Roberts, "The Quantification of Ageing and Survival in Orthodox Seeds," *Seed Science and Technology*, Vol. 9, 1981, pp. 373-409.

[9] A. Khan and B. Gul, "Halophyte Seed Germination: Success and Pitfalls," In: M. A. Khan and D. J. Weber, Eds., *Ecophysiology of High Salinity Tolerant Plants*, Springer, Dordrecht, 2006, pp. 11-31. doi:10.1007/1-4020-4018-0_2

6

Irrigation Planning with Conjunctive Use of Surface and Groundwater Using Fuzzy Resources

D. G. Regulwar, V. S. Pradhan
Department of Civil Engineering, Government College of Engineering, Aurangabad, India
Email: regulwar@geca.ac.in, vizpradhan@gmail.com

ABSTRACT

Surface and groundwater are related systems. They can be used conjunctively to maximize the efficient use of available resources. Groundwater may be used to supplement surface water to cope with the irrigation demands to meet the deficits in low rainfall periods. The parameters involved in the present study are groundwater availability, surface water availability, water requirement of crops and crop area. The inclusion of such uncertain parameters leads to accept the decision making process beyond the consideration of economic benefits. In the present study, an irrigation planning model is formulated by considering the conjunctive use of surface and groundwater. The resources in the present model, *i.e.* the area, surface water and groundwater availability are represented by fuzzy set. The linear membership function is used to fuzzify the objective function and resources. The model is applied to a case study of Jayakwadi project and solved for maximization of the degree of satisfaction (λ) which is 0.546.

Keywords: Conjunctive Use; Irrigation Planning; Fuzzy Linear Programming (FLP); Uncertainties; Optimization

1. Introduction

In water management practices conjunctive use is considered important as surface and groundwater are related systems. Uncertainty in the parameters involved makes it difficult for the decision maker to derive the water use policy and estimate the returns from the system. Such uncertainties are efficiently tackled through fuzzy logic. Kashyap and Chandra [1] have developed a mathematical model for achieving an optimal conjunctive use policy incorporating spatially and temporally distributed groundwater withdrawals and spatially distributed cropping patterns. The groundwater withdrawals were constrained to keep the water table elevations within an appropriate range. Murthy [2] illustrated with a hypothetical example the case of conjunctive use of surface and groundwater to replace surface water. Onta *et al.* [3] have developed a stochastic dynamic programming model to derive the optimal operating policy and also a lumped simulation model is used to evaluate the alternative policies for each alternative and a multiple criteria decision making is used to select the most satisfactory alternative plan. Mohan and Jyothiprakash [4] have formulated a FLP model to derive optimal crop plans for an irrigation system for conjunctive use of surface and groundwater. The results of FLP model were compared with classical

linear programming model which showed that the fuzzy linear programming model maximized the degree of satisfaction. Vedula *et al.* [5] have developed a mathematical model to obtain an optimal conjunctive use policy for irrigation with the objective of maximizing the sum of relative yields of crops in a reservoir-canal-aquifer system. The crop water allocations are achieved by integration of the reservoir operation for canal release along with groundwater pumping. Srinivasulu and Satyanarayana [6] addressed the problem of a canal irrigated in saline groundwater areas by developing a linear programming model for allocation of land and water resources to different crops. A genetic algorithm model was developed by Nagesh Kumar *et al.* [7] to obtain an optimal operating policy and optimal crop water allocations with the objective of maximizing the relative yields from the crops in the study area. Manuel *et al.* [8] have presented an integrated non linear hydrologic economic modeling framework for optimizing conjunctive use of surface and groundwater for a river basin through capacity expansion. Khare *et al.* [9] have presented an economic optimization problem to explore the potential use of surface and groundwater resources using linear programming to arrive an optimal cropping pattern for maximization of net benefits. Kentel and Aral [10] have developed a simulation-optimization model with constraints on drawdown that have been used to optimize the additional groundwater withdrawal in a coastal aquifer. The results were used to determine the individual satisfaction degrees with respect to multiple objectives. Yang *et al.* [11] have developed a multi objective problem by integrating the multi-objective genetic algorithm, constrained differential dynamic programming and groundwater simulation model. These models are adopted to distribute the optimal releases. Regulwar and Anand Raj [12] have developed a Multi Objective Genetic Algorithm Fuzzy Optimization model which is applied to a multireservoir system. A 3-D optimal surface was evolved for deriving the optimal operation policies of the reservoir. Regulwar and Kamodkar [13] have developed a FLP model to evaluate the reservoir operation policy in which the resources were fuzzy in the first model; in the second model, the technological coefficients were fuzzified and in the third model both *i.e.* the resources and the technological coefficients were considered fuzzy. Gurav and Regulwar [14] have formulated an irrigation planning problem as (FLP) to tackle the uncertainties to derive the optimal cropping pattern with an objective of minimization of cost cultivation applied to a case study. Kamodkar and Regulwar [15] have developed a reservoir operation model using FLP which considers two objective functions. Fuzziness was considered in the objectives and constraints to derive the reservoir operation policy in a water year.

From the literature, it is evident that conjunctive use policy has been used to evolve the surface and groundwater resources allocation optimally and applied to an irrigation planning problem. Different strategies based on the status of the resources available with different alternatives of surface and groundwater utilization have been used for the allocation of resources utilized in irrigation planning on a linear programming framework. Other techniques such as non linear programming, stochastic dynamic programming and simulation appear in the literature for obtaining the solution of such problems. Linear programming based optimization methods are popularly used to derive the policies and are found as an effective tool in dealing with such problems. However it is observed that in a crisp optimization problem there is no flexibility in the constraints especially the resources. To incorporate relaxation in the constraints and look at the acceptability of the solution, the crisp optimization problem can be converted into fuzzy optimization problem.

The objective of the present study is to formulate a linear programming model to maximize the net benefits and to arrive at an optimal cropping pattern in a fuzzy environment. The groundwater availability of the study area is explored on the basis of Vertical Electrical Soundings (VES). The Language for Interactive General Optimization (LINGO) has been used to solve the linear optimization problem.

2. Methodology

Uncertainty is inevitable in irrigation planning due to random nature of events and fluctuating inputs. In the present study the applicability of irrigation planning model is improved by incorporating the uncertainties in the resources and representing them as fuzzy sets instead of crisp value. A FLP model is presented which considers uncertainty in the resources where as the other parameters are crisp in nature. The degree of satisfaction (λ) of a certain value of the resource within the fuzzy set is represented by membership function.

Model: The Fuzzy Linear Programming Problem with Fuzzy Resources

$$\max \sum_{j=1}^{n} c_j x_j$$
$$s.t. \sum_{j=1}^{n} a_{ij} x_j \leq \tilde{b}_i, \quad (i \in \mathbb{N}_m) \qquad (2.1)$$
$$x_j \geq 0. \qquad (j \in \mathbb{N}_n)$$

In this case, $\tilde{b}_i \ (i \in \mathbb{N}_m)$ is a fuzzy numbers with the following linear membership function

$$\mu b_i(x) = \begin{cases} 1 & \text{if } x \leq b_i, \\ \dfrac{b_i + p_i - x}{p_i} & \text{if } b_i < x < b_i + p_i, \quad (2.2) \\ 0 & \text{if } b_i + p_i \leq x. \end{cases}$$

Where $(x \in \mathbb{R})$ (**Figure 1**) for each vector

$$x = (x_1, x_2, \dots\dots, x_n),$$

we first calculate the degree, $D_i(x)$ to which x satisfies the *ith* constraint $(i \in \mathbb{N}_m)$ by the formula

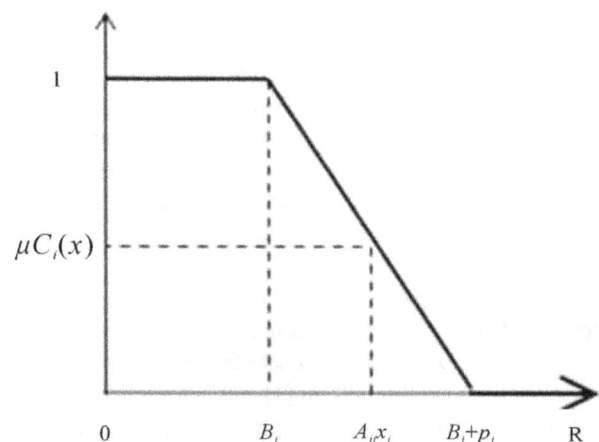

Figure 1. Linear membership function for resources $\tilde{(b_i)}$.

$$D_i(x) = B_i\left(\sum_{j=1}^{n} a_{ij}x_j\right). \qquad (2.3)$$

These degrees are fuzzy set on $\mathbb{R}^n$ and their intersection $\underset{i \neq}{\overset{m}{\cap}} D_i$ is a fuzzy feasible set. Next, we determine the fuzzy set of optimal values. This is done by calculating the lower and upper bounds of the optimal values first. The lower bound of optimal values z_l is obtained by solving the standard linear programming problem;

$$\max z = cx$$

$$s.t. \sum_{j=1}^{n} a_{ij}x_j \le b_i, \qquad (i \in \mathbb{N}_m) \qquad (2.4)$$

$$x_j \ge 0, \qquad (j \in \mathbb{N}_n)$$

The upper bound of the optimal values z_u is obtained by a similar linear programming problem in which each b_i is replaced with $b_i + p_i$:

$$b_i \max z = cx$$

$$s.t. \sum_{j=1}^{n} a_{ij}x_j \le b_i + p_i, \qquad (i \in \mathbb{N}_m) \qquad (2.5)$$

$$x_j \ge 0. \qquad (j \in \mathbb{N}_n) + p_i$$

Then the fuzzy set of optimal values, G which is fuzzy subset of $\mathbb{R}^n$ is defined by a linear membership function $G(x)$ for objective as,

$$G(x) = \begin{cases} 0 & if \ \sum_{j=1}^{n} c_j x_j < z_l, \\ \left(\sum_{j=1}^{n} c_j x_j - z_l\right)\Big/(z_u - z_l) & if \ z_l < \sum_{j=1}^{n} c_j x_j < z_u, \\ 1 & if \ \sum_{j=1}^{n} c_j x_j \ge z_u. \end{cases}$$

$$(2.6)$$

Graphical representation of linear membership function given by Equation (2.6) for fuzzy goals is shown in **Figure 2**.

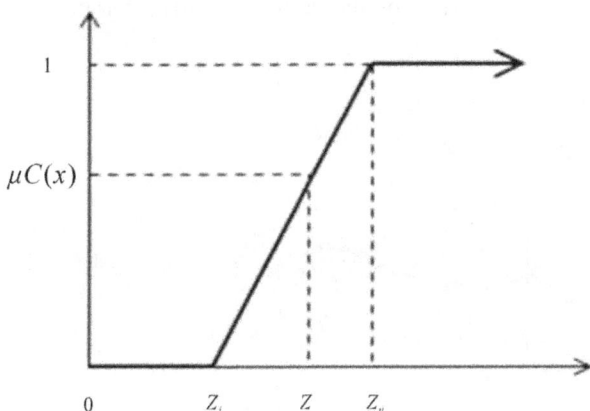

Figure 2. Linear membership function for goals.

The fuzzy set of the ith constraint, C_i which is subset of $\mathbb{R}^m$, is defined by

$$\mu, C_i(x) = \begin{cases} 1 & if \ \sum_{j=1}^{n} a_{ij}x_j < b_i, \\ \dfrac{\left(b_i + p_i - \sum_{j=1}^{n} a_{ij}x_j\right)}{p_i} & if \ b_i < \sum_{j=1}^{n} a_{ij}x_j \le b_i + p_i \\ 0 & if \ b_i + p_i < \sum_{j=1}^{n} a_{ij}x_j. \end{cases}$$

$$(2.7)$$

Graphical representation of linear membership function given by Equation (2.7) for fuzzy resources is shown in **Figure 1**.

By incorporating the above information (the problem given by Equation (2.1) becomes the following optimization problem

$$s.t. \ \mu G(x) \ge \lambda,$$
$$\mu C_i(x) \ge \lambda, \qquad 1 \le i \le m \qquad (2.8)$$
$$x \ge 0, \ 0 \le \lambda \le 1.$$

By using Equations (2.6) and (2.7), Equation (2.8) can be written as

$$s.t. \ \lambda(z_u - z_l) - cx \le -z_l,$$
$$\lambda p_i + \sum_{i=1}^{n} a_{ij}x_j \le b_i + p_i, \qquad (i \in \mathbb{N}_m) \qquad (2.9)$$
$$\lambda, x_j \ge 0. \qquad (j \in \mathbb{N}_n)$$

Where, λ is the level of satisfaction. The above problem is actually a problem of finding $x \in \mathbb{R}^n$ such that

$$\left[\left(\bigcap_{i=1}^{m} D_i\right) \cap G\right](x) \qquad (2.10)$$

reaches the maximum value. This is a problem of finding a point (λ) which satisfies the constraints and goal with the maximum degree. The method employed here is called as symmetric method (*i.e.* the constraints and the goal are treated symmetrically) [16].

3. Description of the Study Area

The Jayakwadi project is constructed across river Godavari which originates in the Nasik district of Maharashtra State, India. The project is a multipurpose project having a catchment area of 21,755 km^2, Gross storage capacity of 2909 Mm^3 and Live storage capacity of 2171 Mm^3. A system of canals i.e Left bank canal and Right bank canal serves a command area of 141640 ha and 41682 ha respectively. The entire command area lies within the Latitude: 18°46'N to 19°30'N and Longitude 75°20'E to

77°45'E. **Figure 3** shows the Index Map of Jayakwadi Project.

The cropping pattern of the command is determined on the considerations that there is a suitable combination of food crops and cash crops and minimizing delta to maximum possible extent consistent with the productivity of the project. The project is planned for 75% dependable yield and crops are grown throughout the year. The irrigation intensity adopted is 22% in kharif season, 45% in Rabi season, 28% in two seasonal, H.W. crop 3% and perennial 4.5% which gives a total irrigation intensity of 102%. A geoelectric field survey was carried out by applying the Vertical Electrical Sounding (VES) technique which measures the electrical resistivity variation with depth to locate the groundwater resources for irrigation in the canal command of Jayakwadi Project. The geoelectric field measurements are conducted using Computer Resistivity Monitor (CRM). The study was conducted at 10 locations of the area using Schlumberger configurations. Hence it is evident from the above studies that water is available between a depth of 9 - 12 m below ground level to a depth of 35 m and it is observed that this water level fluctuates to give a yield up to 69.56% based on the stage of groundwater development GEC 97 [17]. The monthly surface water availability cannot exceed the availability of water from the canal. The monthly optimal operating policy adopted for irrigation releases from the reservoir for a dependable yield of 75% based on the Preliminary Irrigation Program (PIP). The monthly irrigation requirement for all the crops is met by the releases from the reservoir and the water requirement is estimated by Modified Penman Method with an overall efficiency of 49% [18].

3.1. Model Formulation

3.1.1. Objective Function

The objective function consists of maximizing the benefits (Z) from the command area of Jayakwadi project. The net benefit is obtained by deducting the cost of production from the yield and income obtained from an individual crop on unit area basis and the cost of surface and groundwater.

$$\text{Maximize Net Benefits}(NB) = \sum_{i=1}^{n} (GB_i - IC_i) * A_i$$
$$-CST \sum_{j=1}^{t} SW_j - CGW \sum_{j=1}^{t} GW_j \tag{3.1}$$

where,

NB = Net benefit of ith crop in Million Rupees.
GB_i = Gross Benefit of ith crop per ha.
IC_i = Input Cost for i^{th} crop per ha.
A_i = Area under i^{th} crop (i = 1, 2, 3, ..., 10), 1 = Sugarcane, 2 = Banana, 3 = Chillies, 4 = L.S.Cotton, 5 =

Kharif Jowar, 6 = Paddy(Drilled), 7 = Rabi Jowar, 8 = Wheat, 9 = Gram, 10 = Groundnut (H.W).

Where GB = (Yield × Income)-Cost of Production excluding the cost of surface and groundwater (The Report of Commissionerate of Agriculture Maharashtra State 2010, Agricultural Statistical Information Maharshtra State, India).

CST is the cost of surface water and is estimated to be 11760 Rs./ha for a Benefit Cost Ratio of 1.5 for Jayakwadi Project (Jayakwadi Project Report).

CGW is the cost of groundwater which is estimated to be Rs 30,000 Rs/ha including the capital cost and charges been claimed by the local authorities (Maharashtra Jeevan Pradhikaran).

SW = Surface water allocation in j^{th} time period (1, 2, 3...12).

GW = Groundwater allocation in j^{th} time period (1, 2, 3...12).

3.1.2. Constraints

(1) Water Requirement Constraint

$$\sum_{i=1}^{n} WR_i * A_i = SW_j \tag{3.2}$$

Where, WR_i = Water requirement of ith crop in Mm^3
(2) Area Availability Constraint
The total area under each crop should not exceed the Culturable Command Area for all seasons

$$\sum_{i=1}^{n} A_i \leq CCA_i \tag{3.3}$$

CCA_i = Culturable command area for i^{th} crop in ha.
(3) Surface Water Availability Constraint
The monthly surface water availability cannot exceed the availability of water from the canal

$$\sum_{j=1}^{t} SW_j \leq \sum_{j=1}^{t} SWA_j \tag{3.4}$$

SWA_j = Surface water available in j^{th} time period (1, 2, 3... 12) in Mm^3 based on the Preliminary Irrigation Program for a water year.

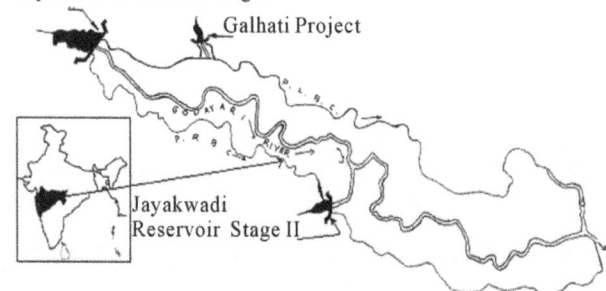

Figure 3. Index map of Jayakwadi project.

(4) Groundwater Availability Constraint

The total water pumped annually from the groundwater resources of the study area should not exceed the annual recharge.

$$\sum_{j=1}^{t} GW_j \leq \sum_{j=1}^{t} GWA_j \qquad (3.5)$$

GWA_j = Groundwater available in j^{th} time period (1, 2, 3…12) in Mm^3.

$$\text{Change in Storage} = h \times S_y \times A$$

where h = Fluctuation in water level in m, S_y = Specific Yield (for consolidated formations S_y = 0.02) and A = Area for computation of recharge.

4. Results and Discussions

In the present study a FLP model is applied to a case study of Jayakwadi Project which is a multipurpose project serving a command area of 141640 ha (PLBC) and 41682 ha (PRBC). The objective of the present model is to maximize the net benefits subject to constraints of area, surface water and groundwater availability. The model considers uncertainty involved in the resources $\left(\tilde{b} \right)_i$ i.e. area, surface water and groundwater availability. Whereas the other parameters of the model i.e. the technological coefficients $\left(\tilde{a}_i \right)$ and cost coefficient $\left(\tilde{c}_i \right)$ are crisp in nature. By using the methodology as explained earlier (Equations 2.4 and 2.5) the lower bound and upper bound of the optimal values are estimated first by solving the standard linear programming problem. As the canal command faces frequent water scarcity, 100 % availability of the resources is taken into consideration as the upper bound for crop area, surface water and groundwater. The lower bound for area is restricted to 77%, for surface water it is considered to be 78% as per the (PIP) and 69.56% for groundwater availability based on stage de-velopment (GEC 97). The fuzzy set of optimal values is defined by a linear membership function (Equation (2.6)). The values of upper and lower bounds of the objective function are 2308 Million Rupees and 1620 Million Rupees respectively. By incorporating the above information in the problem given by Equation (2.1) becomes the fuzzy optimization problem (Equation 2.8). Using Equations (2.6), (2.7) and (2.8), the maximum value which satisfies the constraints and goal with a maximum degree of satisfaction (λ) is achieved. Results for maximized Lamda (λ) for fuzzy resources are given in **Table 1**. It is observed that while solving the linear programming problem for upper bound and lower bound of the re-sources the area allocated for groundnut is the same but it shows a considerable increase in the area in the optimal cropping pattern. The irrigation intensities for upper bound, lower bound and for fuzzy optimization problem are observed to be 84.78%, 61.18% and 77.38% respec-tively from which it is evident that optimization of re-sources in a fuzzy environment gives results with a satis-faction level up to 0.546 taking care of the vagueness and uncertainties in the availability of resources. From the solution it is observed that a maximized benefit of 1994 Million Rupees is obtained for an irrigation intensity of 77.38%. **Figure 4** shows the graphical representation of the optimal cropping pattern.

5. Conclusions

The modeling procedure demonstrates the vagueness and uncertainties in the availability of resources which are tackled to evolve an optimal cropping plan. FLP model is formulated for maximization of net benefits applied to a case study of Jayakwadi Project. Following conclusions are drawn.

1) Uncertainty in the availability of resources makes it difficult for the decision maker to derive the water use

Table 1. Optimal cropping pattern for upper bound, lower bound and maximized lamda (λ).

Crop	Upper Bound for Fuzzy Resources Area (Ha)	Lower Bound for Fuzzy Resources Area (Ha)	Maximized Lamda Area (Ha) $(\lambda) = 0.546$
Sugarcane (A1)	3911.44	2975.243	3697.553
Banana (A2)	877.85	745.69	829.84
Chillies (A3)	5520	3864	4765.37
L.S. Cotton (A4)	46000	32200	39711.4
Kharif Jowar (A5)	774.23	657	731.89
Paddy (A6)	18400	12880	15884.57
Rabbi Jowar (A7)	27600	19320	23826.85
Wheat (A8)	42268	32200	39711.41
Gram (A9)	9200	6440	7942.283
Groundnut (A10)	874.91	874.91	4765
Total	155426.4	112156.8	141866.2

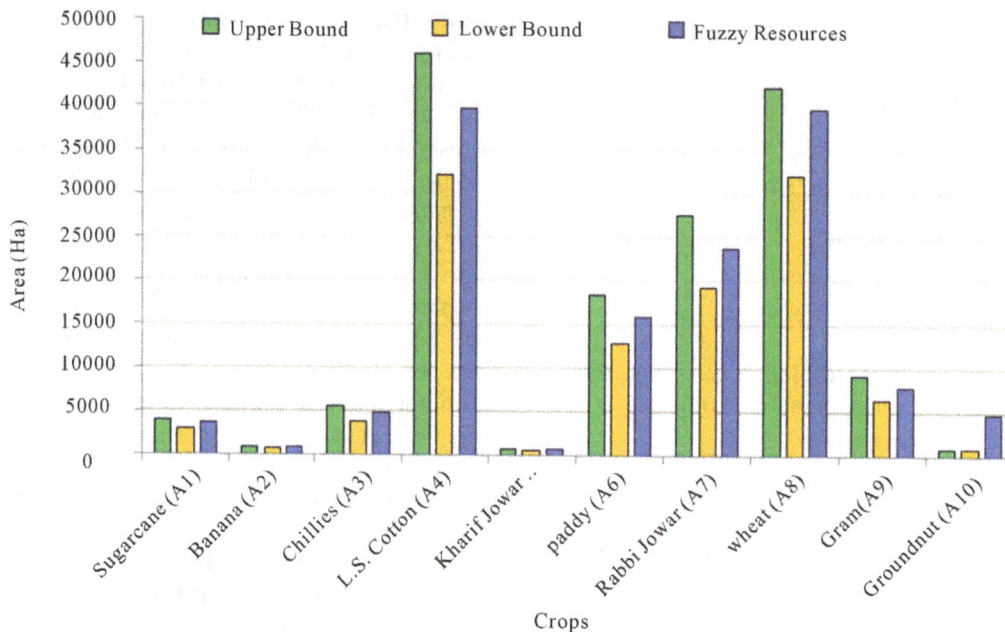

Figure 4. Optimal cropping pattern for upper bound, lower bound and fuzzy resources.

policy and estimate the returns from the system. Such uncertainties are efficiently tackled through fuzzy logic.

2) The irrigation intensities for upper bound, lower bound and for fuzzy optimization problem are observed to be 84.78%, 61.18% and 77.38%, respectively.

3) Optimization of resources in a fuzzy environment gives results with a satisfaction level up to 0.546 taking care of the vagueness and uncertainties in the availability of resources.

4) It is observed from the solution that the maximized benefit of 1994 Million Rupees is obtained for an irrigation intensity of 77.38%.

6. Acknowledgements

The authors are thankful to Command Area Development Authority, Aurangabad, Maharashtra State, India for providing necessary data for the analysis and Head, Geology Department, Government Institute of Science Aurangabad, for their support to carry out the field observations.

REFERENCES

[1] D. Kashyap and S. Chandra, "A Distributed Conjunctive Use Model for Optimal Cropping Pattern," *Proceedings of Symposium on Optimal Allocation of Resources*, IAHS, 1982.

[2] K. N. Murty, "Planning Conjunctive Groundwater and Surface Water Development in Public Irrigation Systems," *Journal of Groundwater Monitoring and Management*, Vol. 173, 1990, pp. 459-466.

[3] P. R. Onta and A. Das, "Multistep Planning Model for Conjunctive Use of Surface and Ground water Resources," *Journal of Water Resources Planning and Management*, Vol. 117, No. 6 , 1992, pp. 662-678. doi:10.1061/(ASCE)0733-9496(1991)117:6(662)

[4] S. Mohan and V. JyothiPrakash, "Fuzzy System Modeling for Optimal Crop Planning," *IE* (I) *Journal-CV*, Vol. 81, 2000, pp. 9-17.

[5] S. Vedula, P. P. Mujumdar and G. C. Sekhar, "Conjunctive Use Modeling for Multicrop Irrigation," *Agricultural Water Management*, Vol. 73, No. 3, 2005, pp. 193-221. doi:10.1016/j.agwat.2004.10.014

[6] A. Srinivasulu and T. V. Satyanarayana, "Development and Application of an LP Model for Conjunctive Use of Water Resources in Saline Groundwater Areas," *IE* (I) *Journal*, Agricultural Engineering Division, 2005, pp. 40-44.

[7] D. Nagesh Kumar, K. S. Raju and B. Ashok, "Optimal Reservoir Operation for Irrigation of Multiple Crops Using Genetic Algorithms," *Journal of Irrigation and Drainage Engineering*, Vol. 132, No. 2, 2006, pp. 123-129. doi:10.1061/(ASCE)0733-9437(2006)132:2(123)

[8] M. Pulido, Joaquin and Andres, "Economic Optimization of Conjunctive Use of Surface Water and Groundwater at the Basin Scale," *Journal of Water Resources Planning and Management*, Vol. 6, 2006, pp. 454-467.

[9] D. Khare, Jat and D. Sunder, "Assessment of Water Resources Allocation Options: Conjunctive Use Planning in a Link Canal Command," *Journal of Resources, Conservation and Recycling*, Vol. 51, 2007, pp. 487-506.

[10] E. Kentel and M. M. Aral, "Fuzzy Multiobjective Decision Making Approach for Groundwater Resources Management," *Journal of Hydraulic Engineering*, Vol. 12 No. 2, 2007, pp. 206-217. doi:10.1061/(ASCE)1084-0699(2007)12:2(206)

[11] C.-C. Yang, L.-C. Chang, C.-S. Chen and M.-S. Yeh, "Multi-Objective Planning for Conjunctive Use of Surface and Subsurface Water Using Genetic Algorithm and Dynamics Programming," *Journal of Water Resources Management*, Vol. 23, No. 3, 2009, pp. 417-437. doi:10.1007/s11269-008-9281-5

[12] D. G. Regulwar and P. Anand Raj, "Development of 3-D Optimal Surface for Operational Policies of a Multireservoir in Fuzzy Environment Using Genetic Algorithm for River Basin Development and Management," *Journal of Water Resources Management*, Vol. 22, No. 5, 2008, pp. 595-610. doi:10.1007/s11269-007-9180-1

[13] R. U. Kamodkar and D. G. Regulwar, "Derivation of Multipurpose Single Reservoir Release Policies with Fuzzy Constraints," *Journal of Water Resources and Protection*, Vol. 2, No. 12, 2010, pp. 1028-1039.

[14] D. G. Regulwar and J. B. Gurav, "Irrigation Planning under Uncertainty: A Multi Objective Fuzzy Linear Programming Approach," *Journal of Water Resources Management*, Vol. 25, No. 5, 2010, pp. 1387-1416. doi:10.1007/s11269-010-9750-5

[15] R. U. Kamodkar and D. G. Regulwar, "Multipurpose Single Reservoir Operation with Fuzzy Technological Constraint," *ISH Journal of Hydraulic Engineering*, Vol. 16 No. 1, 2010, pp. 49-62.

[16] G. J. Klir and B. Yuan, "Fuzzy Sets and Fuzzy Logic: Theory and Applications," Prentice Hall, 2000.

[17] National Water Commission, "Groundwater Resources Estimation Committee Report," National Water Commission, GEC, 1997.

[18] Government of Maharashtra, "Revised Project Report Jayakwadi Project Stage I," Irrigation Department, Government of Maharashtra, Maharashtra.

7

The Pollution Characteristic of Polycyclic Aromatic Hydrocarbons (PAHs) in Typical Sewage Irrigation Area in North of China

Jiale Li[1], Caixiang Zhang[1]*, Yihui Dong[1], Xiaoping Liao[1], Bin Du[2], Linlin Yao[1]

[1]State Key Lab of Biogeology and Environmental Geology, China University of Geosciences, Wuhan, China
[2]China National Administration of Coal Geology General Prospecting Institute, Beijing, China
Email: jllee.cug@gmail.com, *zeno448@163.com

ABSTRACT

This research aims to investigate the pollution characteristic of PAHs in Xiaodian sewage irrigation area. The result shows that the concentrations of Σ PAHs range from 47.94 to 46432.85 ng/g while that of the total components of the 16 kinds of PAHs are 5969.81 ng/g. PAHs with for rings and more than 4 rings are the main and important pollutants in topsoils of Xiaodian District. The main input of PAHs is combustion source, and the main pollution source in this area is fired coal. The topsoils in Xiaodian District are polluted by human activity in varying degrees. 23 of all 31 topsoil samples have been heavily polluted, especially those located nearby developed industrial townships and irrigation channels.

Keywords: Pollution Characteristic; Polycyclic Aromatic Hydrocarbons (PAHs); Typical Sewage Irrigation Area

1. Introduction

Due to the shortage of water resource for agriculture irrigation, sewage irrigation has become an effective approach to deal with this problem since 1950s [1,2]. However, sewage contains lots of contaminations such as heavy metals, soluble salt and other organic contaminations which are teratogenic and mutagenic such as polycyclic aromatic hydrocarbons [3].

Polycyclic aromatic hydrocarbons (PAHs) are compounds containing two or more fused benzene rings in linear, angular, and cluster like arrangements. They are mainly derived from incomplete fossil fuel combustion, volatilization of uncombusted petroleum, biomass burning and the early diagenesis of organic matter. PAHs exist in the environments ubiquitously and many of them are carcinogenic and mutagenic to human beings [4-6]. Due to their negative effects, U.S. Environmental Protection Agency (EPA) defined 16 kinds of PAHs as priority pollutants [5], they are naphthalene (Nap), acenaphthylene (Acy), acenaphthene (Ace), fluorene (Flo), phenanthrene (Phe), anthracene (Ant), fluoranthene (Fla), pyrene (Pyr), benz[a]anthracene (BaA), chrysene (Chry), benzo[b]fluoranthene (BbF), benzo[k]fluoranthene (BkF), benzo[a]pyrene (BaP), indeno[1,2,3-cd]pyrene (Ind), dibenzo[a,h]anthracene (DiA), and benzo[g,h,i]perylene (BghiP).

Many authors have reported the pollution characteristic of PAHs in soil. The background concentration of soils in village is 1 - 1300 ng/g, the concentration of agricultural soils is 5 - 900 ng/g while 145 - 166,000 ng/g in the polluted soils [7]. The concentration of polluted Σ 16PAHs in the surface sedimentary of river, estuary and sea reached 6 - 8399 ng/g-dw [8]. The concentration of PAHs in industrialized area, coal mining area and oil producing area in Welsh reached a high level of 54,500 ng/g while 2330 ng/g in the general industrial zone, the average range is 108 - 54,500 ng/g [9].

Xiaodian sewage irrigation area lies in the southeast of Taiyuan city, north of China. This area has a 30 years' history of sewage irrigation since 1970s. The seawage flowed through the area via East Main Channel which lies in the west of the area, irrigated the entire field and then flowed to Fenhe River through Beizhang Drainage and Taiyu Drainage. This study aimed to investigate the pollution characteristic of PAHs in typical sewage irrigation area—Xiaodian sewage irrigation area.

2. Materials and Methods

2.1. Sample Collection and Preparation

As shown in the **Figure 1**, the topsoils were collected with 2 km × 2 km grids in August, 2010. 31 samples

---

*Corresponding author.

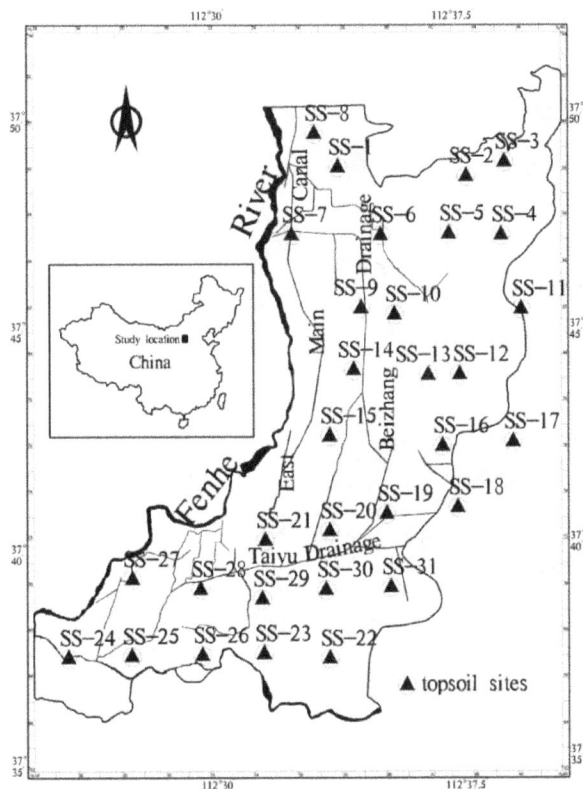

Figure 1. Location of Xiaodian sewage irrigation area and the sampling sites.

were positioned in aluminium boxes and marked with the sample number, sealed with parafilm, refrigerated in freezer at 0°C as soon as possible then air dried in sunless place before analyses.

2.2. Analytical Procedure

20 ng of recovery surrogate standards which were mixed by deuterated PAHs standards (naphthalene-d8, acenaphthene-d10, phenanthrene-d10, chrysene-d12, and perylene-d12) were added to 10 g of samples. Then, added 120 mL of Dichloromethane (DCM) to the samples and Soxhlet-extracted for 24 h. Concentrated the extracted liquid to 2 mL, filtered by alumina/silica gel (1:2 v:v) column with 30 mL of DCM-hexane at a radio of 2:3, concentrated to 0.2 mL, added internal standard (hexamethylbenzene, 1000 ng) before GC-MS analyses.

16 kinds of PAHs which were defined as priority pollutants by EPA were determined. 1 μl sample was injected in splitless/split mode. PAHs compounds were separated on a HP-5 capillary column (30 m × 0.25 mm i.d. ×0.25 film thickness) with nitrogen as carrier gas at a flow rate of 2.5 mL/min in a constant flow mode and determined by GC-MS (Agilent 6890N/5975MS). The GC operating conditions were: injectortemperature, 280 °C; ion source temperature, 180°C; temperature program: held at 60°C for 2 min, ramped to 290°C (3°C min^{-1}),

held for 30 min. The MSD was operated in the electron impact mode at 70 eV and the selectedion-monitoring mode. Quantitative determinations at the mass range m/z 50 - 500. Data acquisition and processing operated by a HP Chemstation software.

No detected target compounds were found in the daily method blanks. Surrogate standards were added to all the samples and analyzed for quality assurance and control. The mean recoveries for all the 31 samples were: 55% ± 15% for naphthalene-d8, 66% ± 12% for acenaphthane-d10, 74% ± 15% for phenanthrene-d10, 69% ± 8% for chrysene-d12, 84% ± 6% for perylene-d12.

3. Results and Discussion

As can be seen from **Table 1** and **Figure 2**, the average concentration of a sing component in 16 kinds of PAHs detected in topsoils of Xiaodian District ranges from 18.70 ng/g to 944.22 ng/g, while that of the total components of the 16 kinds of PAHs are 5969.81 ng/g. PAHs with two rings, accounting for 2.71% of the all PAHs, are Nap, Ace, Acy and Flo. Ace and Flo have low average concentrations of 34.0 ng/g and 12.4 ng/g, taking percentages of 0.57% and 0.21%, less than 1%. What' more,

Table 1. Distribution characteristics of PAHs in topsoils of Xiaodian sewage irrigation area.

	MAX	MIN	AVERAGE	Standard Deviation	Coefficient of Variance
	(ng/g)	(ng/g)	(ng/g)	(n = 31)	
Nap	394.57	4.07	83.93	102.57	1.22
Acy	377.65	0.46	34	80.62	2.37
Ace	57.78	ND	12.4	17.9	1.44
Flo	199.88	0.76	31.28	49.81	1.59
Phe	1940.88	7.19	247.63	405.85	1.64
Ant	410.09	7.32	67.55	82.07	1.21
Fla	3776.75	1.93	396.01	808.26	2.04
Pyr	2870.02	1.11	303.83	633.17	2.08
BaA	2301.95	1.67	200.96	480.37	2.39
Chry	5776.52	10.27	998.91	1579.81	1.58
BbF	4770.84	3.21	707.06	1224.03	1.73
BkF	8280.27	5.72	587.3	1525.72	2.6
BaP	7444.09	4.22	868.07	1718.06	1.98
InP	5371.45	ND	604.13	1129.16	1.87
DiA	1050.63	ND	141.55	233.34	1.65
BghiP	6918.69	ND	733.32	1491.51	2.03
ΣPAHs	46432.85	47.94	5969.81	10785.67	1.81

"ND" stands for this matter was not been found.

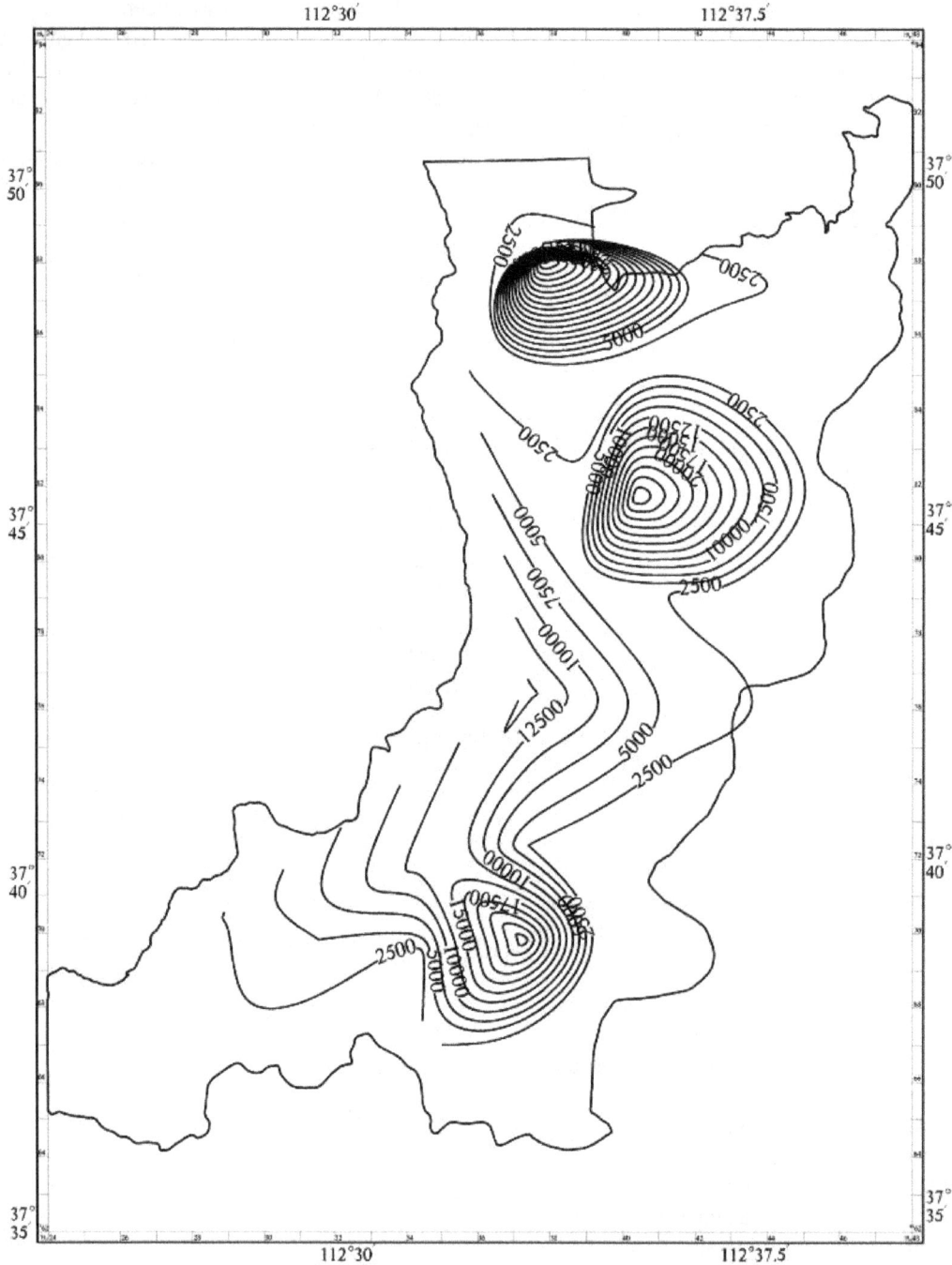

Figure 2. Concentration isoline map of ΣPAHs in topsoils of Xiaodian sewage irrigation area (unit: ng/g).

Ace is a detected object with a lowest concentration. PAHs with 3 rings are Phe, Ant and Fla, whose percentages are 4.1%, 1.13% and 6.63%. They account for 11.91% of the total PAHs. PAHs with for rings and more than 4 rings are the main and important pollutants in topsoils of Xiaodian District. Those with 4 rings take the largest percentage 46.87%. Chry and BbF are more than 11% of ΣPAHs separately. Chry, accounting for 16.73% of ΣPAHs in topsoils, is the most important pollutant

with a concentration of 998.91 ng/g, higher than any other kind of PAHs. The total percentage of PAHs with more than 5 rings is 38.51%. Except DiA with a little lower concentration, InP and BaP both have higher concentrations and have the percentages of more than 10% separately. BghiP, a kind of PAHs with 6 rings, accounts for 12.28% of all PAHs. The results showed that the main input of PAHs is combustion source, and the main pollution source in this area is fired coal.

The area where the concentration of ΣPAHs in topsoils of Xiaodian District is higher than other areas is located around Xiaodian Street Office. This is closely related with that area which is the economic and administrative center of Xiaodian District, a dense population, a developed industry and a frequent coal-fired use. In addition, this area has a relatively flat terrain and the surrounded places are open, easily to accept PAHs pollution carried by the atmosphere from northwest. Furthermore, the relationship between the concentration of ΣPAHs and TOC in topsoils showed a positive correlation.

The concentrations of ΣPAHs range from 47.94 to 46432.85 ng/g. Except that the sample SS-10 has the highest concentrations of Ace, InP and DiA and the sample SS-30 has that of Chry, all other highest concentrations of PAHs happened in the sample SS-1, as well the highest concentration of ΣPAHs. The samples in which the concentration of ΣPAHs are larger than 10,000 ng/g are SS-1, SS-10, SS-15, SS-21 and SS-30. The results showed above and the fact that all the 5 samples are located nearby the waste canal, reasonably explain the distribution of PAHs is closely related with sewage channel. The sample SS-1, located in the canal dam of Qinxian village, next to the sewage channel, is surrounded by residential buildings where there are some streets and dump pits. The lithologic-character of this topsoil sample is wet black dauk with many types of gravel. The topsoils in this place have been carried here to fill in the canal for over 10 years from an external place. The concentration of TOC in this sample reaches up to 7.87%, which is the highest in all the samples. Meanwhile, ΣPAHs here have a concentration of 46432.85 ng/g higher than any other samples.

SS-18, located in Xigia village, has the minimum concentration of ΣPAHs and that of all single components. The area within a hundred miles around the sample SS-18 is overgrown with weeds. A large number of silt sediments and saline-alkali soil can be seen there. It is estimated that the soil here was originally a field of swamp, rarely affected by human activity, fired-coal and factories. It is worth noting that this sample is located outside the irrigation area. The sample SS-11, located in the clean irrigation area, has a ΣPAHs concentration of 1743.71 ng/g, not very low, quite possibly affected by the nearby Wusu airport. Estimated to be impacted by atmospheric dry and wet deposition, the sample SS-3 located in Dong Mountain has a ΣPAHs concentration of 903.76 ng/g.

The PAHs produced by natural causes and plants' synthetic generally has a concentration between 1 - 10 ng/g. As a result, the topsoils in Xiaodian District are polluted by human activity in varying degrees.

Figure 3 shows the average percentage content of PAHs in topsoils of the research area. The abscissa on

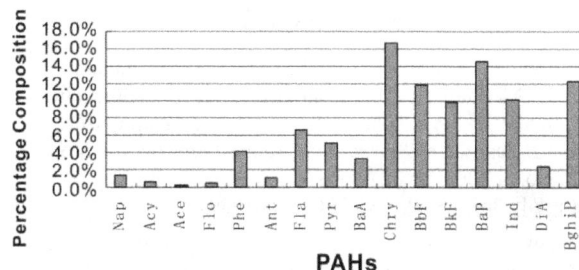

Figure 3. The average percentage composition of PAHs.

behalf of different kinds of PAHs, they are listed in order from left to right: Nap, Acy, Ace, Flo, Phe, Ant, Fla, Pyr, BaA, Chry, BbF, BkF, BaP, InP, DiA and BghiP. According to **Figure 3**, Chry, with an average concentration of 998.91 ng/g, higher than any other kind in the all 16 PAHs, accounting for 16.73% of all the PAHs pollutants, is the major pollutant in the study area. Other PAHs with more than 4 rings have a relatively higher concentration as well. This indicates that the main source of PAHs in the study area comes from the combustion.

Maliszewska Kordybach proposed that according to the concentration of ΣPAHs in soil, soil can be graded into 4 levels: when the concentration is less than 200 ng/g, the soil is considered to be clean, when it is between 200 - 600 ng/g, the soil is slightly polluted, when it ranges from 600 to 1000 ng/g, soil is polluted moderately, when it is larger than 1000 ng/g, the soil is heavily polluted. This standard is widely applied to distinguish whether soil is polluted in Europe [10]. Based on this standard, sample SS-18 is a clean soil, sample SS-31 is polluted slightly, and samples SS-4, SS-17, SS-19, SS-22, SS-24, SS-26 are polluted moderately, and the rest samples are all polluted heavily. 23 of all 31 topsoil samples have been heavily polluted, especially those located nearby developed industrial townships and irrigation channels. Consequently, the topsoils in Xiaodian District were polluted heavily and obviously effected by industrial townships and sewage irrigation. It is worth paying more attention to.

The concentrations of ΣPAHs in topsoils of the target area are higher than that in the industrial area of Linz, a industrial city in Australia (the average is 1450 ng/g) [11], even higher than that in the port city of Tallinn in Estonia (35.5 - 26,300 ng/g) [10] and in Dalian (6510 ± 5730 ng/g) [12]. This result illustrates that the topsoils are strongly influenced by coal burning.

4. Acknowledgements

This research was funded by National Natural Science Foundation of China (No. 40830748 and No.40972156). The authors would like to thank Zhao Xu, Xiang Qingqing, Li Feng, Liu Yuan, Liu Lian, Tao Zhihao for their help in sampling and sample treatment.

REFERENCES

[1] X. J. Wang, Y. Zheng, R. M. Liu, B. G. Li, J. Cao and S. Tao, "Medium Scale Spatial Structures of Polycyclic Aromatic Hydrocarbons in the Topsoil of Tianjin Area," *Part B Journal of Environment Science and Health*, Vol. 38, No. 3, 2003, pp.327-335. http://dx.doi.org/10.1081/PFC-120019899

[2] S. Tao, Y. H. Cui, F. L. Xu, B. G. Li, J. Cao, W. X. Liu, G. Schmitt, X. Wang, W. Shen, B. P. Qing and R. Sun, "Polycyclic Aromatic Hydrocarbons (PAHs) in Agricultural Soil and Vegetables from Tianjin," *Science of the Total Environment*, Vol. 320, No. 1, 2004, pp. 11-24. http://dx.doi.org/10.1016/S0048-9697(03)00453-4

[3] M. B. Yunker, R. W. Acdonald and R. Vingarzan, "PAHs in the Fraser River Basin: A Critical Appraisal of PAH Ratios as Indicators of PAH Source and Composition," *Organic Geochemistry*, Vol. 33, No. 4, 2002, pp.489-515. http://dx.doi.org/10.1016/S0146-6380(02)00002-5

[4] D. Broman, C. Naf, C. Roiff and Y. Zebuhr, "Occurrence and Dynamics of Polychlorinated Dibenzo-p-Dioxins and Dibenzofurans and Polycyclic Aromatic Hydrocarbons in the Mixed Surface Layer of Remote Coastal and Offshore Waters of the Baltic," *Environmental Science and Technology*, Vol. 25, No. 11, 1991, pp. 1850-1864. http://dx.doi.org/10.1021/es00023a002

[5] P. Patnaik, In: P. Patnaik, Ed., *Handbook of Environmental Analysis*, CRC Press, Boca Raton, 1997, p. 165. http://dx.doi.org/10.1201/9781420050608

[6] P. T. Williams, "Sampling and Analysis of Polycyclic Aromatic Compounds from Combustion Systems—A Review," *Journal of the Energy Institute*, Vol. 63, 1990, pp. 22-30.

[7] USDHHS, "Toxicological Profile for Polycyclic Aromatic Hydrocarbons," US Department of Health and Human Services, 1995, p. 482.

[8] Boonyatumanond, Ruchaya, Wattayakom. "Distribution and Origins of Polycyclic Aromatic Hydrocarbons (PAHs) in Riverine, Estuarine, and Marine Sediments in Thailand," *Manne Pollution Bulletin*, Vol. 52, No. 8, 2006, pp. 942-956. http://dx.doi.org/10.1016/j.marpolbul.2005.12.015

[9] K. C. Jones, J. A. Straford and K. S. Waterhouse, "Organiccontaminants in Welsh Soils: Polycyclic Aromatic Hydrocarbons," *Environmental Science & Technology*, Vol. 23, No. 5, 1989, pp. 540-550. http://dx.doi.org/10.1021/es00063a005

[10] M. Trapido, "Polycyclic Aromatic Hydrocarbons in Estonian Soil: Contamination and Profiles," *Environmental Pollution*, Vol. 105, No. 1, 1990, pp. 67-74. http://dx.doi.org/10.1016/S0269-7491(98)00207-3

[11] P. Weiss, A. Riss and E. Gschmeidler, "Investigation of Heavy Metal, PAH, PCB Patterns and PCDD/F Profiles of Soil Samples from an Industrialized Urban area (Linz, Upper Austria) with Multivariate Statistical Methods," *Chemosphere*, Vol. 29, No. 9-11, 1994, pp. 2223-2236. http://dx.doi.org/10.1016/0045-6535(94)90390-5

[12] Z. Wang, J. W. Chen and X. L. Qiao, "Distribution and Sources of Polycyclic Aromatic Hydrocarbons from Urban to Rural Soils: A Case Study in Dalian, China," *Chemosphere*, Vol. 68, No. 5, 2007, pp. 965-971. http://dx.doi.org/10.1016/j.chemosphere.2007.01.017

8

Agricultural Water Foot Print and Virtual Water Budget in Iran Related to the Consumption of Crop Products by Conserving Irrigation Efficiency

Azam Arabi[1*], Amin Alizadeh[1], Yaser Vahab Rajaee[2], Kazem Jam[3], Naser Niknia[4]

[1]Ferdowsi University of Mashhad, Mashhad, Iran
[2]Water Engineering Department, Azad University of Ferdows, Iran
[3]Regional Water Authority Company of Razavi Khorasan, Mashhad, Iran
[4]Water Engineering Department, Shiraz University, Shiraz, Iran
Email: *azamarabi@gmail.com

ABSTRACT

In this study we estimate agricultural water footprint and its components from consumption perspective in arid and semi-arid region like Iran. This study is based on blue water consumption in irrigated land. Iran has imported net virtual water about 11.64 billion cubic meters (bcm) as international crop trade in 2005-2006. Therefore, Iran has depended on virtual water imports. By conserving about 60% irrigation efficiency, the total water requirement to produce imported crops in Iran is nearly 20.78 billion cubic meters. It is nearly 9 percent of renewable water resources and 12.65% agricultural appropriated water which has added to internal water resources. Agricultural virtual water budget is about 112.78 Gm^3/yr. Agricultural water footprint is 110.2 Gm^3/yr. About 12.83% of agricultural water footprint of Iran is related to external water resources on the country boundaries. It means external water footprint. Water dependency, water self-sufficiency and water scarcity indexes in agricultural sector of Iran, are estimated 10.1%, 89.9% and 70.8%, respectively.

Keywords: Water Footprint; Virtual Water Trade; Blue Water Resource; Water Budget

1. Introduction

Water is an essential factor for producing goods and services. By increasing population, water resources have become scarce to sufficient production of food. Since the largest share of water consumption is used for food production, the relation between food and water is closed. Per person consumes about 4 liter water in different form, whereas for producing daily food per capita it needs about 2000 liter water [1]. It has estimated that population of Iran will increase to 100 million people in 2030 and we will need about 150 billion cubic meter of water per year to supply food security based on 2600 K Calories energy per capita [2]. Hoekstra and Hung discussed that water should be considered as an economic good. So, they defined three different levels at which decisions can be made and improvements be achieved; local water use efficiency, water allocation efficiency and global water use efficiency [3] and [4]. In addition, Allan introduced "virtual water" in relation to water, food and their trades as the total volume of water used in the production process of commodity [5-7].

There are some studies about virtual water trade that show water is a key factor to produce water-intensive products and countries can save their water resources due to using virtual water trade. Consequently, they will allocate water in other sections by virtual water import. This view can reform some water consumption pattern in future. Falkenmark [8], Savenije [9], Ringersma [10] and Obuobie et al. [11] mentioned green, blue and gray water distinction. Yang et al. [12] comprised green and blue water features. In addition, in 2002, Water footprint concept was introduced in order to have a consumption based indicator of water use that could provide useful information instead of the traditional production-sector-based indicators of water use [3]. The water footprint of a country is the volume of water needed for the production of the goods and services consumed by the inhabitants of the country. The water footprint of a nation is related to consumption pattern of nation [3] and [4]. Water footprint presents a wider perspective on how a consumer or producer relates to the use of water resources [13]. Hoekstra and Chapagain [14] defined water footprint components, internal water foot print and external water foot print. Van oel et al. [15] and Hoekstra et al. [16] classified water footprint to green, blue and gray.

In this research, we estimate Water footprint index on agricultural sector of Iran from the user perspective in 2006-2007 based on international virtual water flows related to crop product trades. Therefore, we subtract the virtual water export and add the virtual water import to total national water withdrawal to assess water footprint of Iran. In this study, we conserve irrigation efficiency related to international crop trade from consumption perspective. In Iran an agricultural sector annually uses over 93 percent of total renewable water resources (**Table 1**). The agricultural sector contributes about 27% to the Gross National product (GNP), 23% to employment and more than 80 % to food security of country.

Iran, with an area of 165 million hectares (Mha), is located in semi-arid region of the Middle East. Distribution of precipitation is uneven. The average amount of precipitation over the country is 252 mm/year, which is less than one-third of the world average [2]. **Table 1** shows water allocation and use in different categories in Iran.

2. Methods

In this study, virtual water of 31 important agricultural products (m^3/ton) is estimated based on crop water requirements and yields according to Hoekstra and Hung [3] These products are categorized in different groups such as cereals, fruits, industrial crops, oil seeds, summer crops, selected fodder crops. The selected crops which have more share in international trade. We have divided the whole irrigated area to 629 agricultural plains that have specific crop pattern (**Figure 1**). The computations of net water requirement per crop have been done based on the FAO Penman-Monteith equation in non standard condition. In addition, crop coefficient variation as a function of the plant growth stage was calculated [17]. We have considered typical irrigation methods and irrigation efficiency to calculate irrigation need per each crop. For water footprint calculation, the methodology described Hoekstra and Hung [3] and Van Oel et al. [15] have been used. In this approach, to assess water footprint (Top-Down method), we subtract virtual water export and add virtual water import to total national water withdrawal. National water scarcity, water dependency on foreign water resources and sustainability of national consumption (water self-sufficiency) index had calculated to explain relation between water scarcity and virtual water import of country according to Hoekstra and Hung [3].

The object of this study has been only focused on the consumption of agricultural products. Since based on Power ministry report, agricultural water allocation is about 94.25% of water resources (**Table 1**). The focus of this study will be on irrigated crops that is mentioned blue water use. In Iran more than 90% of agricultural

Table 1. Water availability and use in Iran, [2].

Component	Volume (bcm)	Percent of Total
Precipitation	413	100
Evaporation	283	70
Renewable water	130	30
Surface water	105	
Groundwater	25	
Total water use	87.5	100
Agriculture	82.0	94.25
Domestic	4.7	4.75
Industry (etc.)	0.8	1

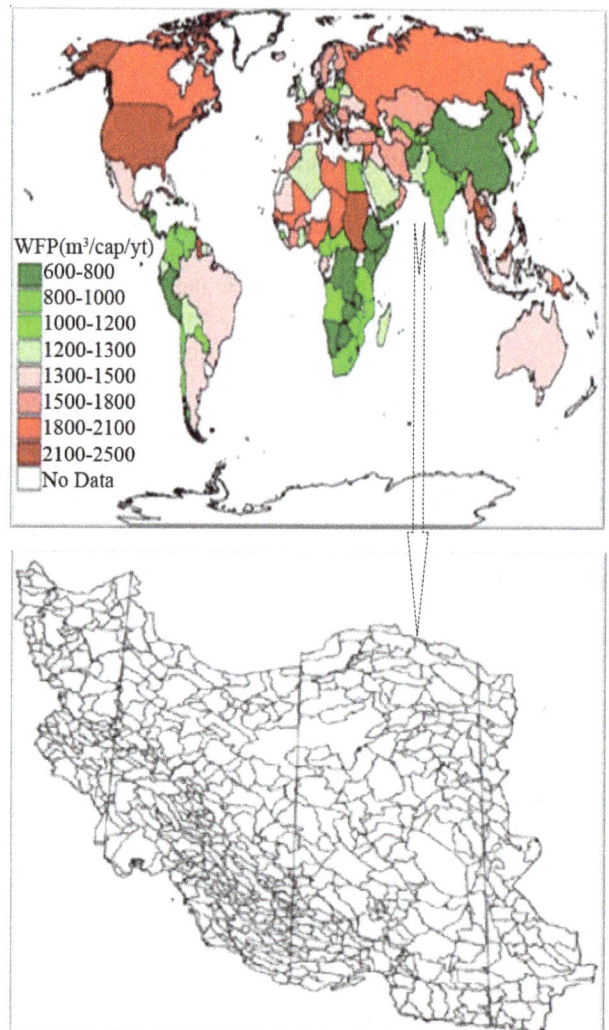

Figure 1. Agricultural plain distribution in Iran (source: Hoekstra and Chapagain, 2007; Agricultural ministry of Iran).

products have harvested from irrigated land due to climate condition. The blue water use is equal to the irrigation water requirement multiplied by the fraction of the total area of a crop that is irrigated [18]. Water saving due to agricultural virtual water trade by considering irrigation efficiency has estimated in 2006-2007.

Some information about crop pattern, crop productiveity, water resources and soil properties in each region of the country is derived from Power and Agricultural ministry [19]. Climatic data have been taken from Organization weather of Iran [20]. Climatic database have classified daily for 30-year period. In 2005 to 2006, total irrigated land was over 8.425 million hectares. **Figure 2** shows the proportion of major agricultural crops and garden products (fruits) from irrigated land area. Also total raw and dry agricultural products were about 77,411 million tons (statistics of agricultural ministry, 2006). Cereals such as wheat (10 million tons), barely (2 million tons), rice (2.5 million tons) were the great product category.

2.1. Calculation of International Virtual Water Trade

Virtual water content of product import is the amount of water that would have been required to produce the product at the place where the product is needed. This definition is particularly relevant how much water Iran needs to supply its agricultural water needs for food security. International crop trade statistics has derived from FAO [21]. The net virtual water import of a country is equal to the gross virtual water import minus the gross virtual water export as described [3]. This study is based on blue water resources (ground and surface water [10,11] as total volume of water resources consumed in agricultural sector in Iran [22,23].

2.2. Calculation of Agricultural Water Footprint

Calculating the water footprint of a nation requires quantifying the flows of virtual water leaving and entering the country. The sum of domestic water use and net virtual water import can be seen as a kind of "water footprint" of a country, [3,15]. The difference of import and export virtual water is the net virtual water content of a country. We have calculated weight average of irrigation efficiency in 629 agricultural plains. So, the real saving water due to the net virtual water import has predicted.

3. Results

Figure 3 shows virtual water content of different crops in Iran in 2005-2006. Virtual water content of each crop has been affected from water requirement and water productivity which are variable in each place. In Iran, some crops such as pistachio, almond, walnut and sunflower have high virtual water. In addition, agricultural water productivity is variable between 0.132 to 8.852 kg·m^{-3} in different crops without considering irrigation efficiency. In this way we need nearly 40 bcm to produce about 62 million tons of agricultural products. While, by considering irrigation efficiency about 60 %, water productivity average has decreased to 0.67 kg·m^{-3} at irrigated land. Water productivity for each crop varies due to agricultural practices and water management in each region.

Figure 4 represents the average of virtual water content

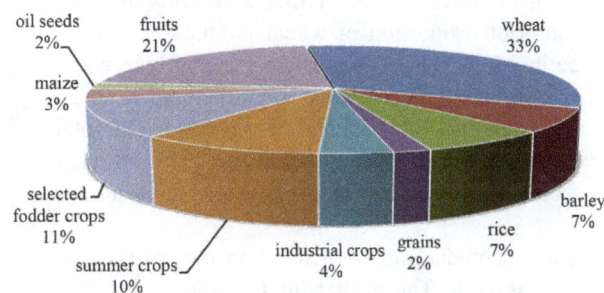

Figure 2. The share of major agricultural crops from irrigated land area in 2005-2006.

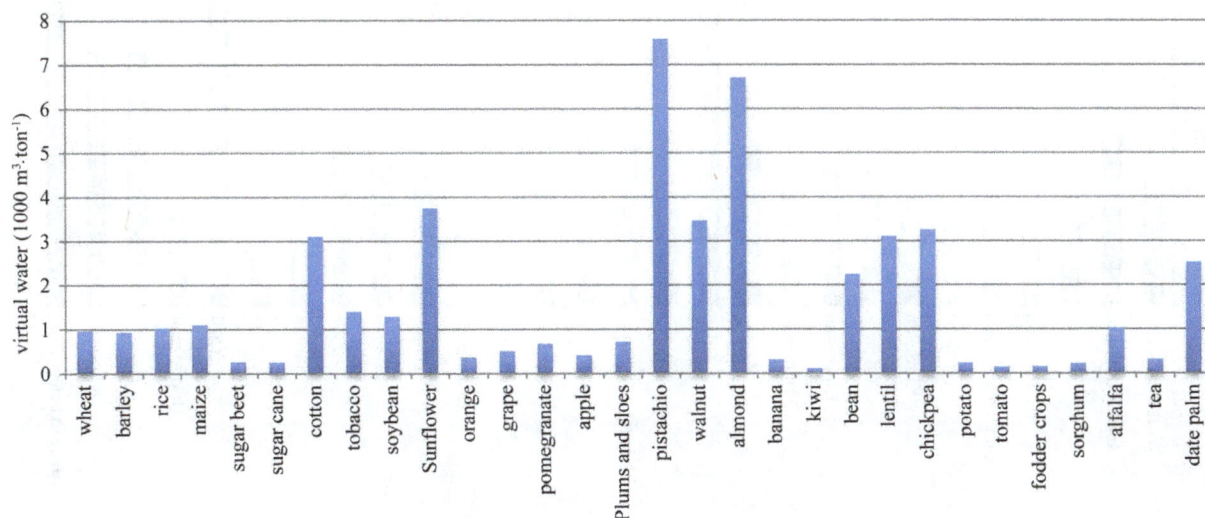

Figure 3. The average of virtual water content of agricultural crops in Iran (2006).

($m^3 \cdot ton^{-1}$) in separated categories of agricultural products in 2005-2006. Date palm, grains, cotton and oil seeds have virtual water content more than the virtual water standard limit as 1000 ($m^3 \cdot ton^{-1}$). So, they are included water-intensive products. Virtual water content of each type of crop depends on crop water requirements, climate conditions, crop productivity and irrigation efficiency. For instant, Wheat has variable virtual water content in some provinces. **Figure 5** illustrates virtual water content of wheat in different provinces of Iran in 2005-2006. Due to arid and semi-arid climate, it is found that northeastern and eastern provinces need more water to produce agricultural crops. It is noticeable that in Northern provinces like Gilan over than 94 percent of wheat production consumes green water resources, Consequently, blue water resources has consumed for crop production. In addition, in western provinces virtual water content of wheat is as weight average of virtual water in Iran (638 $m^3 \cdot ton^{-1}$). Since difference between virtual water content is related to climate condition, in this region, effective rainfall reduce irrigation water requirement. Rainfall amount in western provinces is variable between 300 to 900 mm per year, while it is about 225 mm per year in eastern provinces. So we can say virtual water content is affected on effective rainfall. **Table 2** presents the blue water footprint proportion of wheat production in each province in 2005-2006. The greatest share of blue water footprint is related to Fars, Great Khorasan and Khuzestan which is about 21.8, 17.6 and 11.4 percent. The total blue water footprint of wheat production is 9135 m^3/year that is about 10 percent of total blue water footprint per year.

Agricultural product import and export content are presented in **Figure 6**. The greater import product volumes are related in oil seeds (soybean), industrial crops (sugar)

and cereals [21]. Iran has imported about 7.80×10^8 tons soybeans and 7.5×10^8 raw soybean oil, so we transferred secondary products to primary products to calculate the volume of import and export products. The largest share of production export is related to fruits, especially pistachio and apple. On the other hand, **Figures 7** and **8** show the virtual water trade related to agricultural crop trade in import and export groups respectively. The production trade causes virtual water flow which import/export water in virtual form to/from nation. It is found the largest volume of products which are shared in import and export in 2006-2007.

Figure 9 presents net virtual water content in relation

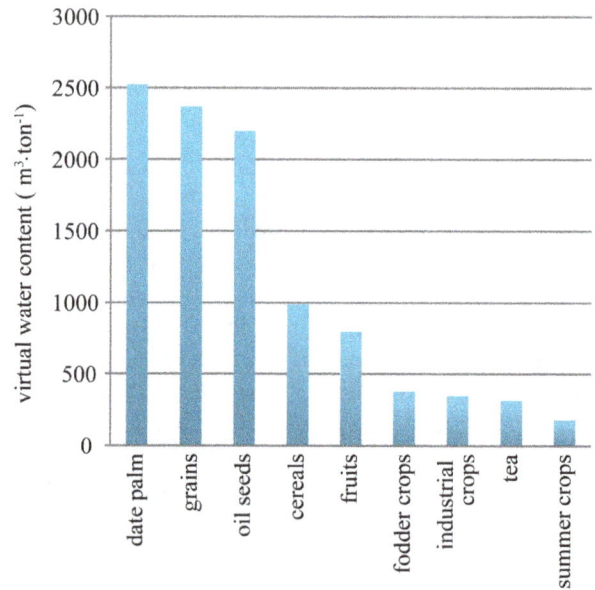

Figure 4. The average of virtual water content ($m^3 \cdot ton^{-1}$) in different groups of agricultural products in 2005-2006.

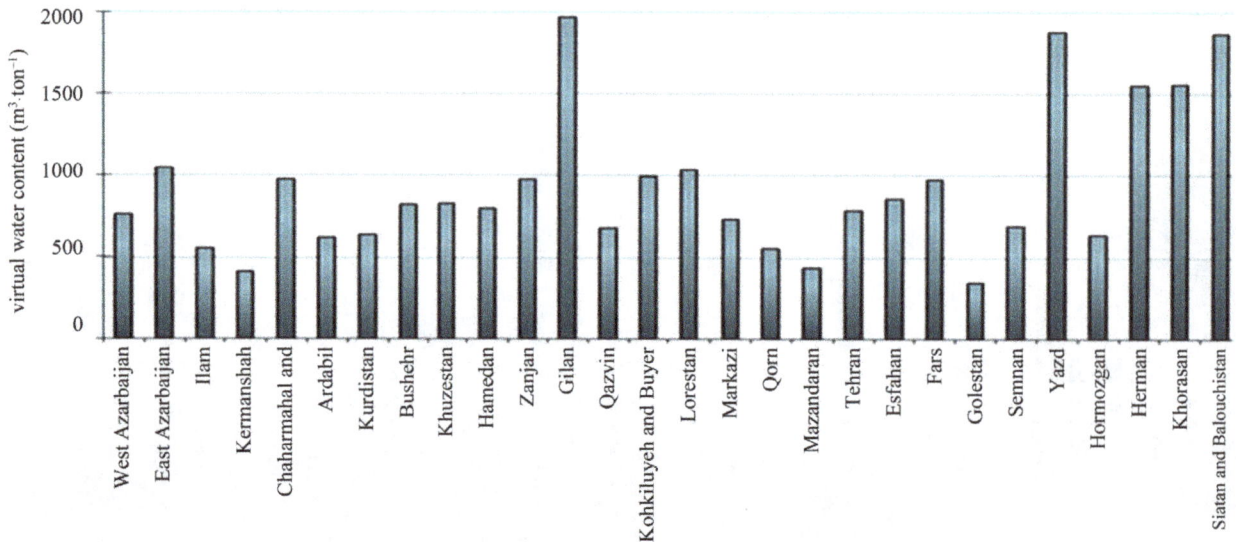

Figure 5. Virtual water content of wheat in different provinces of Iran in 2006.

Table 2. Proportion of provinces in blue water footprint of wheat production in Iran (2006-2007).

provinces of Iran	Yield (kg/ha)	production (ton)	Area (ha)	Blue water footprint of production (mcm/year)	Blue water footprint proportion (%)
Tehran	5359	370931.17	69217.5	290.35	3.2
Kermanshah	5342	517566.85	96886.5	213.03	2.3
Esfahan	5064	572628.27	113,088	489.37	5.4
Qazvin	4620	344936.21	74,667	234.72	2.6
Fars	4467	2044409.17	457,695	1989.13	21.8
Qom	4385	52158.33	11,894	28.76	0.3
Markazi	4326	317229.1	73,324	230.99	2.5
Kurdistan	4187	161441.85	38,554	102.51	1.1
Hormozgan	4179	56661.4	13,560	36.09	0.4
Ardabil	4014	305414.59	76,093	188.69	2.1
Semnan	3977	134622	33,853	92.52	1.0
Chaharmahal and Bakhtiari	3905	129458.86	33,156	126.19	1.4
Zanjan	3877	94406.95	24,348	92.21	1.0
Ilam	3813	154869.07	40,618	85.60	0.9
Hamedan	3641	381316.98	104,719	303.95	3.3
East Azarbaijan	3569	363356.56	101,809	380.82	4.2
Yazd	3413	89202.92	26,138	167.32	1.8
Khuzestan	3404	1260262.04	370,229	1040.12	11.4
Kerman	3394	219047.75	64544.6	340.03	3.7
West Azarbaijan	3211	372171.1	115,912	282.62	3.1
Golestan	3141	491029.78	156,335	169.08	1.9
Kohkiluyeh and Buyer Ahmad	3089	93987.78	30426	93.24	1.0
Lorestan	2967	303611.42	102,322	313.86	3.4
Great Khorasan	2863	1033284.83	360,945	1609.09	17.6
Mazandaran	2860	9336.03	3264	4.03	0.0
Bushehr	2706	55303	20,437	45.25	0.5
Sistan and Balouchistan	1872	99149.27	52,968	184.83	2.0
Gilan	1717	161.38	94	0.32	0.0

Source: agricultural ministry of Iran (2008).

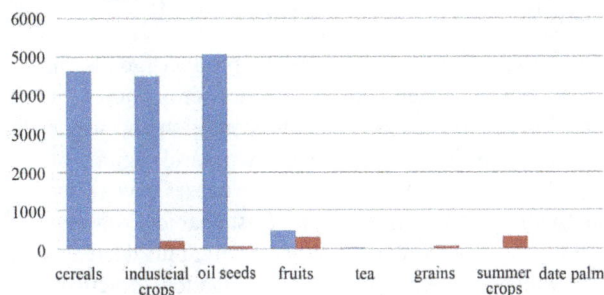

Figure 6. The import and export of agricultural products of Iran in 2006 [21].

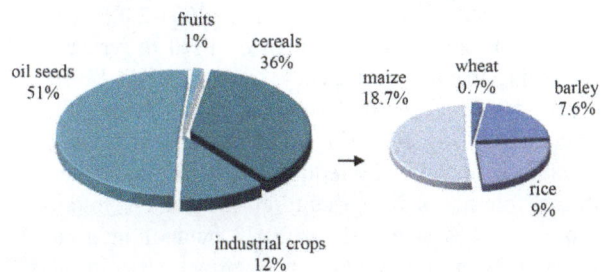

Figure 7. The largest categories proportion included in virtual water import in 2006-2007.

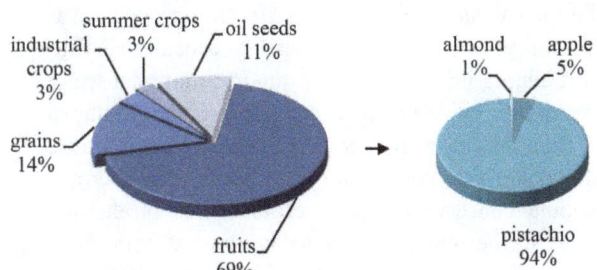

Figure 8. The share of the largest categories included in virtual water export in 2006-2007.

to import and export products. The positive amounts mean that these groups are more import than export virtual water and the negative amounts mean that these groups are more export than import virtual water. This graph shows Iran imported net virtual water about 13.3 bcm due to importing agricultural products and exported 1.6 bcm water in virtual form. The net virtual water of Iran in 2006 was about 11.64 billion cubic meters (bcm). Soybean and pistachio trade were the most important factor to import and export virtual water respectively.

Figure 10 shows the share of different categories of

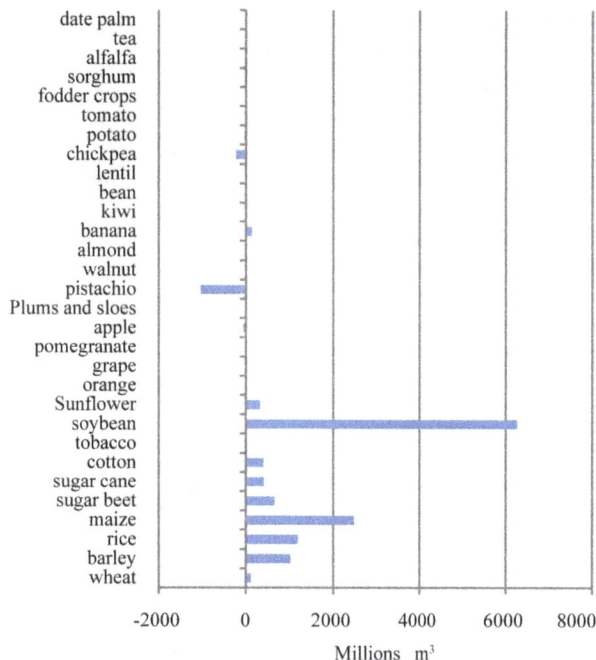

Figure 9. The net virtual water import (mcm) of different agricultural products in 2006.

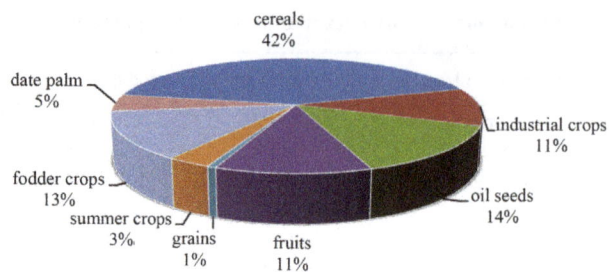

Figure 10. The proportion of agricultural crops groups in WF of Iran in 2005-2006.

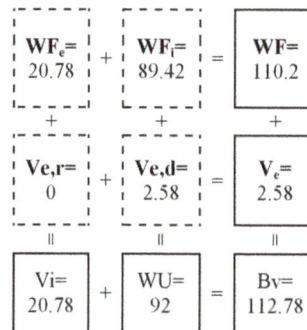

Figure 11. The budget of virtual water in Iran (2005-2006).

agricultural products in agricultural water footprint based on virtual water and production of crops in 2005-2006. Cereals has the largest proportion on agricultural water footprint about 42%, then relates 14% to oil seeds such as soybeans and sunflower seeds, 13% to fodder crops, 11% to industrial crops such as sugar cane, sugar beet, cotton and tobacco and 11% to fruits.

The total volume of water used for crop production is 110.2 Gm^3/yr or 1563.2 m^3/yr/cap. By considering **Figure 11**, adding net virtual water import to internal water use results virtual water budget [14,24]. If we produced production import from internal water resources, because internal water uses consist of gross water need for crops, we should conserve irrigation efficiency in production process. So the real virtual water import and total agricultural water footprint are about 20.78 and 110.2 bcm respectively.

This means that agricultural virtual water budget is about 112.78 Gm^3/yr. Hoekstra and Hung [3] estimated water foot print of Iran in period 1995 to 1999. Results are comparable in two researches. In that research, country population had predicted about 62,762,116 capita, gross virtual water export (import) were 803.41 × 10^6 (6623.1 × 10^6) m^3/yr respectively. So, net virtual water import had estimated about 5819.7 × 10^6 m^3/yr and agricultural water footprint was near 1457 m^3/yr/capita.

It is noted that water needs are affected not only on production volume but also their virtual water content. For instance, cereals include more than 50 percent of irrigated land in 2006, but due to their virtual water con-

tent, they consume about 42% of water resource appropriation in agricultural sector. Therefore, fruits, summer crops and grains have fewer shares in WF in comparison to appropriated land; however industrial crops and oil seeds due to high virtual water content consume more water as intensive water crops.

4. Conclusions

Virtual water is total water used to produce goods or services. Virtual water content of each crop varies in regions due to differences in climate condition and water productivity. For instant, in arid and semi arid regions of the country, high evaporative demand causes low water productivity to crop production. Sustainable cropping pattern can increase water use productivity, thus decrease virtual water content. Improving irrigation practices promote water use productivity. So, we can increase more irrigated land by significant appropriated water or decline pressure on water resources by saving water. Agricultural water footprint of Iran in 2005-2006 has estimated to be about 103.641 bcm or 1470 m^3/yr/cap. By conserving irrigation efficiency, this rose up to 110.2 Gm^3/yr or 1563.2 m^3/yr/cap. Iran has supplied about 87% of agricultural water footprint from national water resources and reminded water requirement by virtual water import as international agricultural production trade. Therefore Iran has saved about 20.78 bcm of internal water resources in 2005-2006. Although net agricultural water import was about 11.64 bcm, this is not consisting irrigation efficiency. In this research water footprint estimation has based only on agricultural crops consumption and total water footprint index includes comprehensive consumption pattern. Obviously water footprint of the whole goods and services is greater than agricultural water footprint. Since production process of each commodity requires water. Although industrial and withdrawal sectors consume less water than agricultural sector, their appropriation would be considered.

REFERENCES

[1] L. R. Brown, "Plan B2.0: Rescuing a Planet under Stress and a Civilization in Trouble," W. W. Norton, New York, 2006.

[2] A. Alizadeh and A. Keshavarz, "Status of Agricultural Water Use in Iran. Water Conservation, Reuse and Recycling," *Proceedings of an Iranian-American Workshop.* http//:www.nap.Edu/catalog/1124.html

[3] A. Y. Hoekstra and P. Q. Hung, "Virtual Water Trade: A Quantification of Virtual Water Flows between Nations in Relation to International Crop Trade," Value of the Water Research Report Series, No. 11, UNESCO-IHE, Delft, 2002. www.waterfootprint.org/Reports/Report11.pdf

[4] A. Y. Hoekstra and P. Q. Hung, "Globalization of Water Resources: International Virtual Water Flows in Relation to Crop Trade," *Global Environmental Change,* Vol. 15, No. 1, 2005, pp. 45-56. doi:10.1016/j.gloenvcha.2004.06.004

[5] J. A. Allan, "Fortunately There Are Substitutes for Water Otherwise Our Hydro-Political Futures Would Be Impossible," In: *ODA, Priorities for Water Resources Allocation and Management,* ODA, London, 1993, pp. 13-26.

[6] J. A. Allan, "Overall Perspectives on Countries and Regions," In: P. Rogers and P. Lyndon, Eds., *Water in the Arab World: Perspectives and Prognoses,* Harvard University Press, Cambridge, Massachusetts, 1994, pp. 65-100.

[7] J. A. Allan, "Virtual Water: A Long Term Solution for Water Short Middle Eastern Economies," Occasional Paper 3, School of Oriental and African Studies (SOAS), University of London, 1997.

[8] M. Falkenmark, "Coping with Water Scarcity under Rapid Population Growth," Conference of SADC Ministers, Pretoria, 1995, pp. 23-24.

[9] H. H. G. Savenije, "The Role of Green Water in Food Production in Sub-Saharan Africa," Article Prepared for FAO, IHE, Delft, The Netherlands, 2000.

[10] J. Ringersma, "Optimizing Green Water Use and Improved Crop Water Productivity under Rain Fed Agriculture in Sub-Sahara Africa," ISRIC Abstract of a Data Search and Literature Study, 2003. http://www.isric.nl/Greenwater/Green water ABSTRACT. doc

[11] E. Obuobie, P. M. Gachanja and A. C. Dorr, "The Role of Green Water in Food Trade," International Doctoral Studies, Center of Development Research University of Bonn, 2005.

[12] H. Yang, L. Wang, K. Abbaspour and A. J. B. Zehnder, "Virtual Water Highway Water Use Efficiency in Global Food Trade," *Hydrology and Earth System Sciences Discuss,* Vol. 3, 2006, pp. 1-26. doi:10.5194/hessd-3-1-2006 www.copernicus.org/EGU/hess/hessd/3/1

[13] A. Y. Hoekstra, W. Gerbens-Leenes and T. H. Van der Meer, "Water Footprint Accounting, Impact Assessment and Life-Cycle Assessment," *Proceedings of the National Academy of Sciences,* Vol. 106, No. 40, 2009, pp. 82-92. doi:10.1073/pnas.0909948106

[14] A. Y. Hoekstra and A. K. Chapagain. "Water Footprint of Nations, Water Use by People as a Function of Their Consumption Pattern," *Water Resource Management,* Vol. 21, No. 1, 2007, pp. 35-48. doi:10.1007/s11269-006-9039-x

[15] P. R. Van Oel, M. M. Mekonnen and A. A. Hoekstra, "The External Water Footprint of the Netherlands: Quantification and Impact Assessment," Value of Water Research Report Series No. 33, UNESCO-IHE, Delft, The Netherlands, 2008.

[16] A. Y. Hoekstra, A. K. Chapagain, M. M. Aldaya and M. M. Mekonnen, "Water Footprint Manual," State of the Art, 2009.

[17] M. M. Mekonnen and A. Y. Hoekstra, "A Global and High-Resolution Assessment of the Green, Blue and Grey Water Footprint of Wheat," *Hydrology and Earth System Sciences,* Vol. 14, 2010, pp. 1259-1276. doi:10.5194/hess-14-1259-2010

[18] F. Bulsink, A. Y. Hoekstra and M. J. Booij, "The Water Footprint of Indonesian Provinces Related to the Consumption of Crop Products," *Hydrology and Earth System Sciences,* Vol. 14, 2010, pp. 119-128. doi:10.5194/hess-14-119-2010

[19] Agricultural Scientific Information and Documentation Center (ASIDC), 2006. http://agrisis.areo.ir/

[20] Iran Meteorological Organization, 2006. http://www.irimo.ir/farsi/agro/index.asp

[21] Food and Agriculture Organization of United Nations, 2006. http://faostat.fao.org/site/342/default.aspx

[22] A. Arabi Yazdi, "Estimation of Agricultural Water Footprint in Iran and Assessment of Using Virtual Water Trade for Food Security Purposes," M.Sc. Dissertation, Water Department, Ferdowsi University of Mashhad, Iran, 2008.

[23] A. Arabi Yazdi, A. Alizadeh and F. Mohammadian, "Study on Ecological Water Footprint in Agricultural Section of Iran," *Journal of Water and Soil,* Vol. 23, No. 4, 2009, pp. 1-15.

[24] J. Ma, A. Y. Hoekstra and H. Wang, "Virtual versus Real Water Transfers within China," *Philosophical Transactions of the Royal Society,* Vol. 361, 2006, pp. 835-842. doi:10.1098/rstb.2005.1644

9

Determining Irrigators Preferences for Water Allocation Criteria Using Conjoint Analysis

Noor Ul Hassan Zardari[1*], Ian Cordery[2]

[1]Institute of Environmental and Water Resources Management (IPASA),
University of Technology Malaysia (UTM), Skudai, Malaysia
[2]School of Civil and Environmental Engineering, University of New South Wales, Sydney, Australia
Email: *noorulhassan@utm.my

ABSTRACT

Water allocation based on multiple criteria has the potential to maximize the total benefits to be gained from the use of a single unit of water. However most of the multi-criteria methods inherently include a considerable degree of subjectivity. In this study, we have attempted to reduce the subjectivity factor from water allocation decision-making process by introducing a conjoint analysis method. Opinions on the importance of a number of water allocation criteria were sought from a large number of irrigation farmers. The opinion survey data were then analyzed using the traditional conjoint analysis method which is widely used to analyze marketing surveys. The analysis allowed objective determination of the relative importance of five water allocation criteria (*i.e.* net farm income, percent of family working on the farm, amount paid to irrigation agency for canal water share). Each water allocation criteria was divided into three levels and utility values for each criteria level were estimated from the farmers' preferences on five water allocation criteria (attributes). The conjoint survey results revealed that the respondents prefer that "annual net farm income" be the most important attribute in water allocation decisions. As would be expected the vast majority of the respondents overwhelmingly placed the "water price" in the last position.

Keywords: Conjoint Analysis; Water Allocation; Pairwise Comparison; *Warabandi*

1. Introduction

Today because of changing climate, high population pressures, water scarcity and increased awareness of the long term implications of excessive use of water every effort should be made to use this resource optimally to enable more production from less water [1]. Previous studies conducted in Pakistan revealed that scarce irrigation water was being allocated to inefficient farmers under the *warabandi*[1] system, where value of a single unit of water was found very low [2,3].

The principles set in the 19th century for the design of the *warabandi* irrigation water delivery system in Pakistan, and notionally still being followed today are not in fact still being practiced. The farmers, with the assistance of irrigation officials have rearranged the irrigation water delivery system (*warabandi*) rules. The influential farmers have set their own rules to get extra water for meeting the crop-water demand of their farms [2,3]. These practiced rules favor the owners of the larger farms and those whose farms are at the upstream ends of the distributary watercourses [3]. Thus there is a need to develop a decision support system that can incorporate the needs of all farmers regarding canal water supplies, and to provide reliability and certainty of water delivery to give farmers confidence to invest in efficient water use practices. In this study a novel concept of developing improved water allocation based on multiple, farmer chosen criteria has been attempted. Given this, it is essential to know what factors or criteria the farmers consider should influence water allocation decisions so that an equitable and efficient system can be developed to improve productivity from the scarce water resource. Current productivity is very low. For example irrigated wheat production is about two tons per hectare whereas in developed countries dryland wheat cultivation typically averages more than this and irrigated yields are at least twice as high.

This paper focuses on the process of estimation of relativeness among the important water allocation criteria. This relativeness could be interpreted as a search for which criteria should be taken into consideration and which should be given little or no attention in water allocation decisions. Conjoint analysis (CA) is a technique

---

*Corresponding author.

[1]*Warabandi* is a rotational method for equitable distribution of the available water in an irrigation system by turns fixed according to a predetermined schedule specifying the day, time and duration of supply to each irrigator in proportion to the size of his landholding in the outlet command.

for establishing the relative importance of different attributes (in conjoint analysis, criteria/factors are called attributes) in the provision of a service [4]. It has its origins in market research where it has been used to establish what attributes influence the demand for different commodities, and thereby what combinations of such attributes will maximize the benefits of a service. It has also been widely used in transport. Journal of Transport Economics and Policy (JTEP) published a special issue on the application of conjoint analysis method in transport [5]. Conjoint analysis has also been applied for solving environmental problems [6,7]. However, to date its application in the area of water resources management is very limited. The next section describes the conjoint analysis method and the data collection process for this study. Following this, results are presented and discussed, and conclusions are drawn concerning the relative importance of water allocation attributes.

2. Designing Conjoint Study

Conjoint analysis is a multivariate technique developed specifically to understand how respondents develop preferences for any type of object (product or service or idea). It is based on the simple principle that respondents evaluate the worth of an object by combining the separate amounts of attribute values [8]. Individuals rarely express how they do this, but relative value judgments must be made and conjoint analysis attempts to emulate this process. Conjoint analysis is unique among multivariate methods in that the researcher first constructs a set of real or hypothetical objects/profiles by combining selected levels of each attribute. Conjoint analysis, compared to other multivariate techniques, has few statistical assumptions, and accordingly, it is based on logic and pragmatism when it concerns such issues as its design, estimation, and interpretation [9].

Designing attributes, assigning attribute levels, deciding which profiles should be presented to the respondents for preference elicitation, establishing the preferences, choosing a method of profile presentation, and selecting a method for estimating utility values are the six important stages in the design of a conjoint study. These conjoint design stages within the context of a water allocation study will now be explained.

2.1. Establishing the Attributes

Water resources planning and management objective is associated with many monetary and nonmonetary attributes or criteria [10]. Decision-making for managing scarce water resource by considering all important attributes could not be possible by merely applying customary cost-benefit analysis approach as this approach can only take monetary attributes into decision analysis. Therefore, a multi-criteria analysis (MCA) approach (e.g. conjoint

analysis) is required for making decisions on water resources management and planning.

It is important to say that the selection of attributes is a very important stage in a conjoint study as the final output entirely depends on the included attributes. In this study, initially ten water allocation attributes were discussed with a focus group of 20 people from an agricultural decision body in Sindh, Pakistan. These people were actively involved in farm and water management decisions as most of them were managing their own agricultural farms. Some of them were running their own agro-based businesses. The discussion with the focus group ended with the selection of the five attributes they thought to be the most important water allocation attributes for their region and these were included in this study. The attributes included were: percent of individual farmer's family working on the farm, the amount paid annually to the Provincial Irrigation Department (PID) for canal water share, the annual net farm income, water use efficiency—i.e. the proportion of received water effectively used, and the quality of groundwater beneath the farm.

2.2. Assigning Attribute Levels

Reference [4] suggested that the attribute levels should be plausible, actionable and capable of being traded-off. In this conjoint study, the attribute levels were decided from survey data gathered from 184 farms situated in Sanghar and Shaheed Benazir Abad (formerly known as Nawabshah) districts of Sindh, Pakistan. Three levels only, were adopted, to keep survey logistics within manageable bounds. Three levels determine there are 243 (3^5) possible combinations. Four levels would increase this number to 1024 (4^5). The first and the third levels of each attribute were decided as the minimum and maximum values of that attribute obtained from the survey. For example, on the average, the minimum and maximum amounts paid annually to the PID for canal water share were found to be USD 13 and USD 25 per hectare respectively. These figures were assigned Level-1 and Level-3 for that particular attribute. The average amount paid to PID was determined as about USD 18 per hectare per year. Thus, Level-2 of that particular attribute was decided as 13 - 25 USD/ha. Levels for the other attributes were determined in the same way. The attributes and levels included in the conjoint study are shown in **Table 1**.

2.3. Deciding Which Profiles to Present

Having established the attributes and their levels, hypothetical profiles with different combinations of attribute levels were presented to 62 individuals (farmers). The attributes and levels chosen in this study gave rise to 243 ($3 \times 3 \times 3 \times 3 \times 3$) possible profiles for the water allocation problem. Obviously, it would have been imprac-

Table 1. Attributes of water allocation and their levels.

Attributes	Attribute acronym used in SPSS	Units	Level-1	Level-2	Level-3
Percent of family working on the farm	FAMILY	%	<50	50 - 80	>80
Annual amount paid to PID for canal water share	PID	USD/ha	<13	13 - 25	>25
Water use efficiency (portion effectively used)	EFFICIENCY	%	<40	40 - 70	>70
Annual net farm income	INCOME	USD/ha	<500	500 - 1250	>1250
Groundwater quality beneath the farm	G_WATER	----	Fresh	Marginal	Saline

tical to ask individuals their preferences among so many profiles. Many methods exist to reduce the number of profiles to a manageable level. These include the use of fractional factorial designs; removing options that will dominate or be dominated by all other options; and dividing the possible options into blocks and establishing respondents' preferences for a block of possible profiles. It was decided to use a fractional factorial design using the statistical package Orthoplan provided in SPSS 11.5 [11]. The use of orthogonal main-effects design reduced 243 profiles to 16. In **Table 2** sixteen hypothetical profiles are shown in columns 1-6. In the last column of **Table 2** the average preferences from 62 respondents are shown.

2.4. Establishing Preferences

After selection of attributes, attribute levels, and profiles, the next step was to present the profiles to the participants and to ask for their preferences. The decision on the type of preference measure to be used must be based on practical as well as conceptual issues. Many researchers favor the rank-order measure because it depicts the underlying choice process inherent in conjoint analysis. From a practical perspective, however, the effort of ranking large numbers of profiles becomes overwhelming for respondents. On the other hand, the ratings measure has the inherent advantage of being easy to administer in any type of data collection context. Because of this characteristic, a rating preference measure was selected for this study to determine the respondents' preferences between nominated profiles. Each respondent was asked to rate their preference between two profiles on a scale from one to five (1 = no preference or rejection; 2 = weak preference; 3 = strong preference; 4 = very strong preference; 5 = absolute preference).

2.5. Choosing a Presentation Method

There are three methods by which the profiles could be presented to the respondents in a conjoint study. These presentation methods are: Trade-off, full-profile, and pairwise comparison [12]. In this study, a pairwise comparison was selected as a presentation method. But 16 profiles could produce 120 possible pairs of profiles.

In a conjoint survey, it was impractical to ask a respondent to show his preferences between 120 pairs.

Thus, these 16 profiles were randomly split into two groups (8 profiles in each group). Even 8 profiles in a pair-wise comparison could generate 28 pairs. It would have been difficult for a respondent to maintain concentration while showing his preferences to 28 profile pairs. Thus, one profile was randomly selected from each group and this quasi-profile was compared with each of the remaining 7 profiles of that specific group. This resulted in 7 pairs to be compared by each respondent and was thought to be practicable within the available time and finance limitations. These two profile groups formed the basis of two separate conjoint analysis questionnaires. Sixty-two subjects (31 for each group were randomly allocated between these two questionnaires. An example of one of the pair-wise choices is shown in **Figure 1**. The 1 - 5 rating scale was used to show the preference for one profile relative to the other.

2.6. Selecting Method for Utility Value Estimation

Estimating the utility value for each attribute level and the relative importance of the various attributes are the two main objectives of a conjoint study. To achieve those objectives, a relationship between the attributes and utility values needs to be specified. Generally there are two types of attributes, the benefit type and the cost type. The higher the benefit type value, the better it is, and for the cost type, the opposite is true. However, for some attributes (for example, "quality of groundwater"), it was difficult to assume a linear relationship between different attribute levels prior to the actual survey (*i.e.* whether respondents would prefer canal water to be supplied to farms underlain by fresh or saline groundwater). Because of this difficulty, a separate utility value relationship was assumed for each of the attributes used in this study and the methodology to do this is described next.

3. Collecting and Analyzing the Data

The results from the preliminary survey regarding the ranking of ten water allocation attributes from 20 decision makers described in Section 2.1, suggested that individuals understood the questionnaire and showed their preferences in a meaningful way. However, some respondents expressed difficulties in assigning absolute

Table 2. List of adopted hypothetical profiles and aggregate preferences.

Profile No.	Percent of family working on the farm (%)	Amount paid to PID annually for canal water share (USD/ha)	Water use efficiency (value of water) (%)	Annual net farm income (USD/ha)	Groundwater quality beneath the farm	Aggregate preferences (1-5 rating) (n = 31 for each group)
1	>80	<13	40 - 70	500 - 1250	Fresh	0.58
2	50 - 80	<13	<40	>1250	Marginal	2.32
3	50 - 80	13 - 25	40 - 70	<500	Fresh	1.61
4	<50	<13	>70	500 - 1250	Saline	2.15
5	>80	13 - 25	40 - 70	<500	Fresh	0.39
6	>80	>25	<40	>1250	Saline	2.68
7	50 - 80	13 - 25	>70	<500	Fresh	0.77
8	>80	<13	40 - 70	<500	Saline	0.90
9	>80	13 - 25	40 - 70	>1250	Fresh	2.21
10	<50	<13	>70	>1250	Marginal	3.16
11	50 - 80	13 - 25	<40	500 - 1250	Saline	2.10
12	<50	<13	<40	<500	Fresh	0.10
13	50 - 80	>25	40 - 70	500 - 1250	Fresh	1.84
14	>80	<13	<40	<500	Saline	0.52
15	<50	>25	40 - 70	<500	Fresh	0.23
16	50 - 80	13 - 25	40 - 70	<500	Marginal	1.71

Q: Considering only the attributes shown, which profile do you think should get greater priority in water allocations? Please rate the preferred profile using rating scale 1-5.

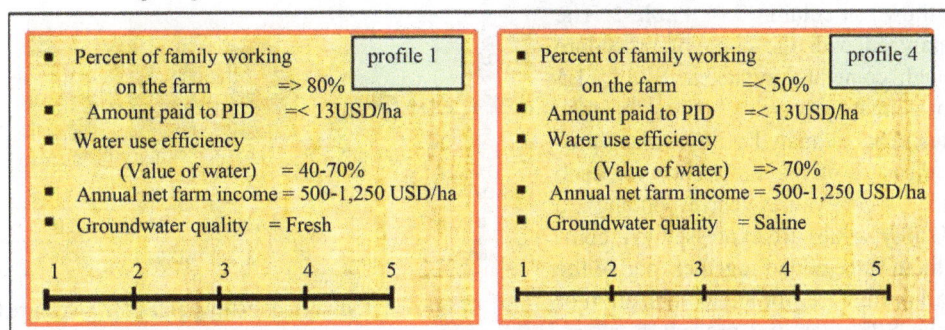

Figure 1. Example of pairwise choice to be made by participants.

rankings to the attributes and suggested that the preference scale should be flexible. Thus a rating scale of preferences was selected for the final conjoint questionnaire. Another problem faced by the survey participants was the large number of attributes they were asked to rank. In order to minimize that problem in the final conjoint questionnaire, the number of water allocation attributes included was reduced to five, as discussed above. The less important attributes were dropped from the questionnaire. The chosen water allocation attributes include: labor employed in the farming, farmers' income, revenue generated by PID, water use efficiency, and groundwater quality beneath the agricultural farm. A face-to-face survey was conducted with 62 decision makers in the Lower Indus River Basin of Pakistan (parts of districts Sanghar

and Shaheed Benazir Abad). In a pairwise comparison each respondent was asked to assign ratings to the profile he considered more deserving of receiving water. Each profile was displayed on a sample card, as shown in **Figure 1**, which contained one of the mixes of the levels for the five water allocation attributes shown in **Table 2**. Only one level of each attribute was presented in a single profile. An SPSS orthogonal sample design was used to select the particular levels to be included on each card to allow estimation over the entire range of profiles.

From the use of conjoint analysis, utility values for each of the attribute level were computed from the participants' preferences on five water allocation attributes. These utility values along with their rank order are shown in **Table 3**.

Table 3. Utility value and rank order of each attribute level.

Attribute	Attribute level	Utility value obtained from conjoint analysis	Percentage of participants' preference utility value within the attribute (%)	Percent difference between the highest and the lowest preferences within an attribute	Rank of attribute level
Percent of family working on the farm	<50%	1.41	32		9
	50% - 80%	1.73	40	12% (28 - 40)	4
	>80%	1.21	28		11
Amount paid to PID annually for canal water share	<13 USD/year	1.39	31		10
	13 - 25 USD/year	1.47	36	5% (31 - 36)	8
	>25 USD/year	1.58	33		6
Water use efficiency (value of water)	<40%	1.54	32		7
	40% - 70%	1.18	25	18% (25 - 43)	12
	>70%	2.03	43		3
Annual net farm income	<500 USD/ha	0.78	16		14
	500 - 1250 USD/ha	1.67	33	36% (16 - 51)	5
	>1250 USD/ha	2.60	51		1
Groundwater quality beneath the farm	Fresh	0.97	19		13
	Marginal	2.40	48	29% (19 - 48)	2
	Saline	1.67	33		5

3.1. Estimation and Interpretation of Utility Values

The averages of the preference scores assigned to the sixteen profiles are shown in column 7 of **Table 2**. The preferences were analyzed with the conjoint procedure (available only through command syntax in SPSS 11.5 standard version) to estimate the utility values for each level of the attributes. The estimated utility values provide a quantitative measure of the preference for each attribute level, with larger values corresponding to greater preference. Utility values are expressed in a common unit, allowing them to be added together to give the total utility, or overall preference, for any combination of attribute levels. In the SPSS conjoint procedure, all attributes were assumed to be discrete data. The estimated utility values for each attribute level along with rank orders are shown in **Table 3**. Other things being equal, a higher utility value for a particular attribute level indicates a higher influence of that attribute on the overall preference. **Figure 2** shows the trend of estimated utility values for each level of five water allocation attributes. The higher range of utility values for the "annual net farm income" attribute indicates that respondents consider this the most important attribute in water allocation decisions.

The highest utility value of 2.60 for Level-3 of "annual net farm income" attribute shown in **Table 3** means that a unit increase in annual net farm income (for instance from Level-1 to Level-3) will increase the preference or utility value of "annual net farm income" by 1.82 (2.60 – 0.78). On the other hand, almost identical utility values for Level-1 of "percent of family working on the farm"

Utility values for attribute levels

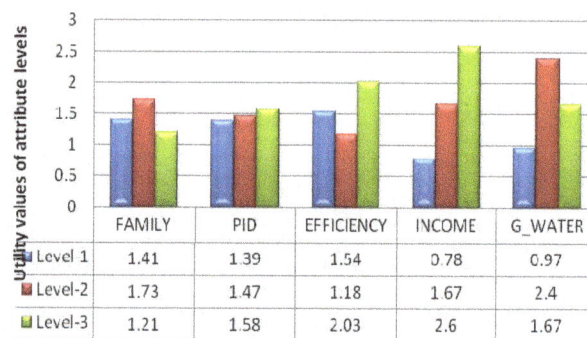

	FAMILY	PID	EFFICIENCY	INCOME	G_WATER
Level 1	1.41	1.39	1.54	0.78	0.97
Level-2	1.73	1.47	1.18	1.67	2.4
Level-3	1.21	1.58	2.03	2.6	1.67

Figure 2. Estimated utility values for attribute levels.

(*i.e.*, 1.41) and for Level-2 of "amount paid to PID for canal water share" (*i.e.*, 1.47) suggest that the respondents were indifferent in their choice between these two levels. The large preference for "marginal" in the groundwater attribute suggests farmers are of the view those who have access to fresh groundwater are less deserving of receiving canal water than those who have no alternative supply possibilities.

3.2. Relative Importance of Attributes

As the estimated utility values are on a common scale, so the relative importance of each attribute can be computed directly. The importance of an attribute is represented by the range of its levels (*i.e.*, the difference between the highest and lowest values of utility values) divided by the sum of the ranges across all attributes. This calculation indicates the relative impact or importance of each attribute based on the size of the range of its utility value estimates. Attributes with a larger range for their utility values have a greater impact on the calculated overall utility value and thus are deemed of greater importance. The relative importance weights across all attributes will total 100 percent. An example of how the relative importance weights for different attributes were determined is illustrated below, with reference to **Table 3**, for the attribute "percent of family working on the farm". The range of utility values for FAMILY attribute was determined as 0.52 (1.73 – 1.21). The sum of ranges of utility values for all five attributes was 4.81 [(1.73 – 1.21) + (1.58 – 1.39) + (2.03 – 1.18) + (2.60 – 0.78) + (2.40 – 0.97)]. The range of utility values for FAMILY attribute (0.52) divided by the sum of ranges of utility values for all attributes (4.81) gives the relative importance weight of 10.8% for this particular attribute. The relative importance weights for the remaining attributes were similarly determined and are plotted on **Figure 3**.

4. Overview of Conjoint Results

From the relative importance weights for each attribute, it can be seen that the respondents gave more importance (37.8%) to the "net farm income" attribute than to other water allocation attributes. This means the respondents preferred to allocate water to those who would make the highest income from it. The respondents considered "groundwater quality beneath the farm" the second most important water allocation attribute with relative importance of 29.7% followed by the "irrigation water use efficiency" (17.7%). The least important attribute with relative importance weight of merely 4.0% was "amount paid to PID". Though it appears individuals were more willing to improve water use efficiency than to engage more family members in the farming enterprise (**Figure 3**), there is some ambivalence as can be seen from the conflicting, or inconsistent utility values shown in **Table 3** for these attributes. Comparing the groundwater quality attribute with the remaining water allocation attributes, the respondents prefer water allocations to go to the less

efficient water users (1.18) and the districts where a very small charge was paid to PID for water supplies (1.39) rather than to the areas where "fresh" groundwater (0.97) was available. This preference clearly indicates that measurable economic or environmental benefits are not necessarily dominant in the thinking of these water users. Some compassion is apparently important, in this case for those who have no alternative means of obtaining crop water other than from the canal supply.

If the farmers were asked to choose one from two options of either: to raise their farm income from the existing income of <500 USD/ha to >1250 USD/ha or to increase the numbers of their family members working on the farm from <50% to >80%, the farmers would be about four times as attracted to raise the net farm income than to put more family members into farming—as the utility value difference between two extreme levels of net farm income was 1.82 (2.60 – 0.78) and the utility value difference between lowest and the highest levels of "percent of family working on farm" attribute was only 0.52 (1.73 – 1.21). The conjoint analysis findings indicate that when rating the alternative water allocation profiles, within attributes the respondents attached the highest value to the ">1250 USD/ha" level of farm income (utility value = 2.60), "marginal" quality groundwater (utility value = 2.40), ">70%" of water use efficiency (utility value = 2.03), "50% - 80%" proportion of family working on the farm (utility value = 1.73), and "saline" groundwater quality (utility value = 1.67). Thus, the total utility of an ideal agricultural farm would be: U = 1.73 + 1.58 + 2.03 + 2.60 + 2.40 = 10.34.

It is interesting to note in **Figure 2** and **Table 3** that for three of the five attributes respondents preferred the last level (level-3) rather than either Level-1 or Level-2, which would have indicated a trend and a preference for an increase or decrease for that attribute. On the other hand, farmers preferred Level-2 of two attributes (*i.e.* "percent of family working on the farm" and "farms with marginal quality groundwater"). It is suggested the preference for the middle level may indicate respondents considered the "extreme" levels, which were the maximum or minimum values obtained in an earlier, larger survey, to be generally unrealistic, or unattainable to them. So being realists they voted for what they considered attainable. In the case of INCOME everyone would like increased income and so the upper extreme was preferred—perhaps in an aspirational sense.

5. Concluding Comments

In water resources management, the application of conjoint analysis to determining the importance of different attributes in deciding priorities for allocation of irrigation water is an innovative approach. From a survey of 184 farms five water allocation attributes were determined to

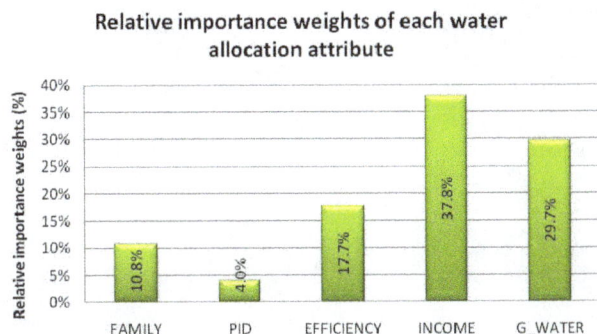

Figure 3. Relative importance weights for water allocation attributes.

be the most important in the thinking of the farmers. The levels or relative property characteristics of each of these five attributes were also decided from the large survey, which was conducted in the Lower Indus Valley of Pakistan. Sixty-two respondents participated in a conjoint analysis questionnaire and showed their preferences for particular farm scenarios based on nominated levels of the five water allocation attributes. The conjoint data analysis revealed that the respondents were more attracted to the "net farm income" attribute (relative importance weight = 37.8%) than any of other attributes. Quality of groundwater was second preference with relative importance weight of 29.7%. Here respondents preferred to provide canal water to those in marginal quality groundwater regions, presumably thinking those with no alternate irrigation water source were more deserving of an allocation of fresh canal water than those who had some access to an alternate (but much more expensive) freshwater supply. Not surprisingly "amount paid to the water provider" (considered as a tax) was assigned the lowest preference (merely 4.0% of relative importance weight).

While the study results cannot be considered to give definitive guidance on how water should be allocated in the Indus Valley, due to the limited range of attributes and numbers of participants included in the study, it does show that there could be an important place for conjoint analysis in objectively determining priorities in the redefining of water allocation guidelines and in the making of wider water resources decisions.

6. Acknowledgements

The authors would like to thank Ministry of Education, Government of Pakistan for the financial support provided to the first author. Partial funding was provided by Higher Education Commission (HEC) of Pakistan and the University of Technology Malaysia (under Vot No. 00L15). Special thanks are given to HEC and the UTM for their partial financial support.

REFERENCES

[1] R. Ali, J. Byrne and T. Slaven, "Modelling Irrigation and Salinity Management Strategies in the Ord Irrigation Area," *Natural Resources*, Vol. 1, 2010, pp. 34-56. doi:10.4236/nr.2010.11005

[2] D. J. Bandaragoda, "Institutional Conditions for Effective Water Delivery and Irrigation Scheduling in Large Gravity Systems: Evidence from Pakistan," *Proceedings of the ICID/FAO Workshop on Irrigation Scheduling*, Rome, Italy, 1996.

[3] N. H. Zardari and I. Cordery, "Water Productivity in a Rigid Irrigation Delivery System," *Water Resources Management*, Vol. 23, No. 6, 2009, pp. 1025-1040. doi:10.1007/s11269-008-9312-2

[4] M. V. Pol and M. Ryan, "Using Conjoint Analysis to Establish Consumer Preferences for Fruit and Vegetables," *British Food Journal*, Vol. 98, No. 8, 1996, pp. 5-12. doi:10.1108/00070709610150879

[5] JTEP, *Journal of Transport Economics and Policy*, Vol. 22, Special Issue, 1988.

[6] W. H. Desvousges, V. K. Smith and M. P. McGivney, "A Comparison of Alternative Approaches for Estimating Recreation and Related Benefits of Water Quality Improvements," Office of Policy Analysis, US Environmental Protection Agency, Washington DC, 1983.

[7] J. Opaluch, S. Swallow, T. Weaver, C. Wessels and D. Wichelns, "Evaluating Impacts from Noxious Facilities: Including Public Preferences in Current Sitting Mechanisms," *Journal of Environmental Economics and Management*, Vol. 24, No. 1, 1993, pp. 59-67. doi:10.1006/jeem.1993.1003

[8] J. Hair, R. Anderson, R. Tatham and W. Black, "Multivariate Data Analysis," 6th Edition, Pearson Prentice Hall, 2006.

[9] J. Hair, R. Anderson, R. Tatham and W. Black, "Multivariate Data Analysis with Readings," 5th Edition, Pearson Prentice Hall, 1998.

[10] D. G. Regulwar and J. B. Gurav, "Fuzzy Approach Based Management Model for Irrigation Planning," *Journal of Water Resource and Protection*, Vol. 2, No. 6, 2010, pp. 545-554. doi:10.4236/jwarp.2010.26062

[11] SPSS Inc., "SPSS 11.5 for Windows," Chicago, USA, 2002.

[12] B. K. Orme, "Getting Started with Conjoint Analysis: Strategies for Product Design and Pricing Research," Research Publishers, LLC, Madison, 2005.

10

Effect of Planting Density, Irrigation Regimes, and Maize Hybrids with Varying Ear Size on Yield, and Aflatoxin and Fumonisin Contamination Levels

Hamed K. Abbas[1*], **Henry J. Mascagni Jr.**[2], **H. Arnold Bruns**[3], **W. Thomas Shier**[4], **Kenneth E. Damann**[5]

[1]United States Department of Agriculture-Agricultural Research Service, Biological Control of Pests Research Unit, Stoneville, USA; [2]Northeast Research Center, Louisiana State University AgCenter, St. Joseph, USA; [3]United States Department of Agriculture-Agricultural Research Service, Crop Production Systems Research Unit, Stoneville, USA; [4]Department of Medicinal Chemistry, College of Pharmacy, University of Minnesota, Minneapolis, USA; [5]Department of Plant Pathology & Crop Physiology, Louisiana State University AgCenter, Baton Rouge, USA.
Email: *Hamed.Abbas@ars.usda.gov

ABSTRACT

Corn (maize, *Zea mays* L.) hybrids expressing the flexibility trait in ear size (number of kernels per ear) are marketed for ability to give higher yields under adverse conditions. Altered kernel number is associated with altered number of silk, a major route for infection of kernels by aflatoxin-producing fungi such as *Aspergillus flavus*. The effect of plant density and irrigation level on yield and accumulation of aflatoxins and fumonisins in harvested grain was compared in a fixed-ear hybrid (Pioneer 33K81), a semi-flexible ear hybrid (Pioneer 3223) and a flexible ear hybrid (Golden Acres 8460) over a range of seeding densities (49,400, 61,750, 74,700, 86,450, and 98,800 seeds·ha^{-1}) in non-irrigated, moderately-irrigated (6.4 cm soil water deficit) and well-irrigated plots (3.8 cm soil water deficit), during three years with variable rainfall. Irrigation increased yields in all hybrids, but in the absence of irrigation, yields were highest with the semi-flexible ear trait hybrid. In general, the hybrid with the flexible ear trait had lower optimal seeding densities than the other hybrids for each soil water regime. In general, kernel number was least affected by seeding density in the hybrid with fixed-ear trait compared to the semi- and flexible ear hybrids. The lowest levels of aflatoxin and of fumonisin contamination in harvested grain were associated with the flexible ear trait at all rainfall and irrigation levels, but there was no evidence that reducing stress by lowering seeding density reduced mycotoxin contamination. Inoculation with *A. flavus* resulted in much higher levels of aflatoxin and significantly higher levels of fumonisin contamination in grain of all hybrids under most conditions of rainfall and irrigation, suggesting that factors that promote *A. flavus* infection can affect production of both mycotoxins.

Keywords: Aflatoxin; Fumonisin; Mycotoxin; Corn; Maize; Environmental Manipulation; Irrigation; Flexible Kernel Number Trait

1. Introduction

Contamination by aflatoxin is a major determinant of crop quality in corn (maize, *Zea mays* L.) in the US, particularly in the Southern US, where the hot, dry conditions which favor *Aspergillus flavus* and other aflatoxin-producing fungi frequently occur [1-7]. Fungi cause a variety of root, stalk and ear rots in corn, all of which cause some loss of yield, but it is aflatoxin produced by *Aspergillus flavus* contaminating kernels that is responsible for the greatest economic losses [8]. *Aspergillus flavus* spores infect corn after over-wintering in reservoirs provided by

the soil and surface plant debris. The husk on the ear constitutes a major physical barrier to infection of kernels by *A. flavus*, but between silking (reproductive stage R1) and blacklayer (physiological maturity, reproductive stage R6) [9], the silks provide avenues for *A. flavus* to enter the ear and infect the kernels. The fungus may also infect the kernels by entering through exposed kernels, *i.e.* insect damage. Once inside the ear, a fungus may spread laterally and infect many kernels [10]. Contamination of harvested corn kernels by another class of mycotoxins, the fumonisins, is also a problem [1-3,7,11-13]. The fungus that produces fumonisins, *Fusarium verticillioides* (Sacc.) Nirenberg [syn: *Fusarium moniliforme*], is an endophyte present in the seed, which can infect the developing plant during germination. *F. verticillioides* is also present in the soil, where it is introduced by decaying corn plant material. *F. verticillioides* can also infect corn plants from the soil reservoir, but the relative contributions of the two infection sources to fumonisin production in kernels is not well established [14,15].

Fungi that produce mycotoxins are often found in corn kernels at harvest [16-21]. Aflatoxin and fumonisins are responsible for illnesses in humans and animals, and cause deterioration of the product resulting in economic losses to growers [8,22-25]. Thus, control of aflatoxin and fumonisins is a worldwide priority [26-28]. The United States Food and Drug Administration (FDA) has issued "action levels" for mycotoxins in foods and feeds in the US, which mandate the maximum level of aflatoxin at 20 ppb in foods for direct human consumption, but lower levels for some animal feed applications, whereas the action level of fumonisins is set at 2 ppm and is advisory [23,29,30]. Corn grain may be contaminated by both mycotoxins, which is expected to increase health risks. This regulatory environment has an important economic impact on the grain industry [1-3,31,32].

Kernel number is an important trait that affects a hybrid's ability to adapt to specific growing conditions. Corn hybrids adapted to the Northern and Eastern US, where growing conditions are almost ideal for corn production [33], usually have determinate or fixed kernel number, resulting in uniform cob length and girth even at high plant densities. In contrast, much of the corn production in the Southern US experiences drought stress at some time during the growing season. This is a particular problem on alluvial clay soils, which are more subject to drought stress due to a relatively shallow rooting zone and physical restrictions limiting plant-available water. Irrigation of corn is not consistently done in the lower Mississippi River Valley, particularly on the more slowly draining high clay soils. Many of the newer corn hybrids used in the Southern US have either the indeterminate (-flex) or partially indeterminate (semi-flex) kernel number trait [34]. Flex-ear hybrids respond to good growing conditions by producing an ear with more kernels than the plant would produce at high plant density. This trait may provide a competitive advantage when plant density is low. Examples of factors reducing plant density are soil surface residues which interfere with germination in no-till agriculture or use of lower seeding densities to minimize yield reduction associated with dry conditions. When a flex-ear corn plant responds to lower plant density by producing more kernels, it also produces a corresponding increased number of silks, which provides additional avenues for infection by *A. flavus*, as well as the desired increase in pollination potential. In principle, the

existence of additional routes for *A. flavus* infection provides the basis for a hypothesis that a flex-ear hybrid will have increased susceptibility to natural infection by *A. flavus* relative to a fixed ear hybrid under similar conditions. Increased infection by *A. flavus* should result in increased contamination of harvested kernels with aflatoxins, particularly at lower plant density. Differences in aflatoxin contamination in hybrids with different kernel numbers would not be expected when ears are artificially infected with *A. flavus* by a mechanism that circumvents the silks (e.g., a pin-bar applicator). The effect of silk/kernel number on fumonisin contamination is less predictable, because *F. verticillioides* is present as both endophyte and soil-derived pathogen [14,15].

Aflatoxin levels are believed to be affected by the level of stress in corn plants during the ear filling period after silking [7,35]. Among the types of plant stress believed to affect aflatoxin levels, high minimum (*i.e.*, nighttime) temperatures are believed to be very important. However, other types of plant stress including drought and population density may add to heat stress in enabling *A. flavus* infection, proliferation, and aflatoxin elaboration [35,36]. The ability to alter the number of kernels per ear has been introduced into corn cultivars in climates sub-optimal for corn production based on the rationale that these cultivars can minimize plant stress by altering ear size. If this is true, -flex and semi-flex hybrids may experience reduced aflatoxin contamination levels. A typical infection of corn by *A. flavus* involves the fungus entering the ear by growing down the silk, then spreading laterally in the cob. The larger the ear, the more silks and thus more potential avenues for infection and a greater likelihood of elevated aflatoxin levels. If this mechanism is important, full- and semi-flex corn hybrids would be expected to have higher aflatoxin levels at low seeding densities. The same plant stress factors are expected to affect levels of fumonisin contamination in a way similar to their affects on aflatoxin levels. However, the fumonisin-producing fungus, *Fusarium verticillioides*, is an endophyte in corn as well as a soil-derived contaminant, and the relative contributions of soil-derived *F. verticillioides* vs endophyte-derived *F. verticillioides* to production of the fumonisins that contaminate corn kernels under different sets of environmental conditions is poorly understood. This situation makes predicting how fumonisin levels in harvested corn kernels would be affected by additional kernels per ear and correspondingly larger numbers of silks more difficult than predicting effects on aflatoxin levels. A study was undertaken to determine the effects of seeding density and irrigation level on 1) Grain yield; 2) Number of kernels per ear and 3) Aflatoxin and fumonisin contamination in harvested corn kernels in hybrids with differing levels of ear development flexibility trait.

2. Materials and Methods

2.1. Experimental Site Characteristics

Field experiments were conducted in 2000 on the Louisiana Delta Plantation near Jonesville, LA, on an Alligator clay (very-fine, montmorillonitic, acid, thermic Vertic Haplaquepts) and 2001 and 2003 at the Northeast Research Station near St. Joseph, LA, on a Sharkey clay (very-fine, montmorillonitic, non-acid, thermic Vertic Haplaquepts) to evaluate the influence of seeding density on three hybrids differing in ear developmental traits on yield and aflatoxin and fumonisin accumulation in the grain. The following three soil water regimes were established using furrow irrigation: 1) No irrigation; 2) Moderate irrigation, in which crops were watered when the soil water deficit (SWD) reached 6.4 cm; and 3) Well-irrigated, in which crops were watered when the SWD reached 3.8 cm. The experimental design was a randomized complete block with five replications in 2000, three replications in 2001 and four replications in 2003 for each irrigation level. Timing of irrigations were determined using the Arkansas Irrigation Scheduler model [37,38]. In 2000, the well-irrigated regime was watered on June 2, 9, 14 and 26, and July 3, 7, 12, 18 and 25, and moderately irrigated regime was watered on June 2, 14 and 30, and July 11 and 22. In 2001 well-irrigated plots were watered on May 17 and 29, June 20 and 27, and July 3, 10 and 20, and moderately irrigated plots were watered on June 4 and July 2. In 2003 well-irrigated plots were watered on May 30, June 6, 23, and 30, and July 15, 21 and 29, and moderately irrigated plots were watered on June 3 and 27, and July 17 and 29.

2.2. Corn Hybrids

Hybrids were selected based on yield potential and differences in ear developmental traits reported by the seed companies. Hybrids evaluated were Pioneer brand (PB) 33K81 (a fixed-ear hybrid with a relative maturity of 112 days), PB 3223 (a semi-flex ear hybrid with a relative maturity of 118 days) and Golden Acres (GA) 8460 (a full-flex ear hybrid with a relative maturity of 120 days). However, a randomized block design was used in the study in which each hybrid was grown in a range of five population densities, which should allow valid comparisons of aflatoxin and fumonisin susceptibility to be made within the population density range for each hybrid. Further study will be required to determine the generality of any conclusions drawn from the study concerning relationships between ear developmental traits and mycotoxin susceptibility.

2.3. Plant Densities

Corn hybrids were planted with a plot planter at seeding densities of 49,400, 61,750, 74,700, 86,450, and 98,800 seeds·ha^{-1} on 11 April 2000, 23 March 2001, and 2 April 2003. Mid-silk dates occurred in early June. Plots consisted of four rows spaced 102 cm apart and 12.2 m long. Cultural practices for fertility and pest control recommended by the Louisiana State University AgCenter were followed. The two center rows of each plot were machine harvested, and yield reported at 15.5% grain moisture. Yield components of plant population, kernel weight, and kernels per ear were determined. Plant population was determined by counting the harvestable plants in each plot just before harvest. Mean ears per plant were also determined. Kernel weight was determined by averaging the weight of 100 kernels (g per 100 kernels). Kernels per ear was calculated using grain yield, ears·ha^{-1}, and kernel weight in the following formula: Kernels per ear = Yield/(ears^{-ha} × kernel weight).

2.4. Fungal Inoculation

In order to identify non-silk-related factors in the susceptibility of hybrids to aflatoxin contamination, pin-bar or needle inoculation of ears [39] was used to by-pass the silk-related parts of the infection process, and thereby produce controls for post-infection differences in aflatoxin between hybrids. Primary ears were inoculated with a suspension of about 1.0×10^6 spores·ml^{-1} of an aflatoxin-producing strain of *Aspergillus flavus*, F3W4 (NRRL 30796) about 20 days after anthesis (approximately mid-ear stage of development) using the pin-bar method [40] in 2000. In 2001 and 2003, needle inoculation was used. The needle was inserted under the husk and approximately 1.2 ml of suspension (1.0×10^6 spores·ml^{-1}) injected on each ear. In the case of fumonisins, in which kernel contamination can result from both soil-derived and endophyte *F. verticillioides*, the fungus is typically widespread in corn production areas so that artificial inoculation was not necessary.

2.5. Mycotoxin Determination

Aflatoxin and fumonisin were determined in naturally-infected corn from combine-harvested grain. *A. flavus*-inoculated ears were hand-harvested at approximately 15% grain moisture, shelled, dried at 50°C and thoroughly mixed on a sample splitter. Grain samples were ground using a Romer Mill (Romer Lab, Union, MO), and analyzed for aflatoxin and fumonisin at the USDA/-ARS laboratory in Stoneville, MS, using commercial ELISA kits (Neogen Corporation, Lansing, MI) as described previously [1-3]. Samples were analyzed in triplicate for both toxins.

2.6. Statistical Analyses

Statistical analyses were conducted using the Proc Mixed procedure of SAS 9.1 [41] to determine the effects of hybrid type and plant density at each irrigation level in each year, and Microsoft Excel 2010 for correlation analysis and Student's *t*-test. Fisher's Protected Least Significant Difference (LSD) was used to evaluate treatment differences when the F-test indicated a significance level (P ≤ 0.05).

3. Results and Discussion

3.1. Weather Conditions

Rainfall and temperatures in the Louisiana Delta Plantation (Jonesville) region and St. Joseph for the three years of the study are reported in **Tables 1** and **2**. Seasonal average temperatures were higher than the 40-year average for both maximum (daytime) and minimum (nighttime) temperatures for each of the three years studied. No single month was consistently responsible for elevated average temperatures. But in 2000, daily maximum temperatures from 14 July through 17 July and 19 July to 20 July were unusually high, ranging from 38.0°C to 39.8°C (**Table 2**). Rainfall varied substantially for the three years (**Table 1**). In 2000 rainfall was far below average in April and July, but at or above average in May and June, respectively. In 2001 rainfall was below average in April and May, while rainfall for June and July was above or near the long-term average, respectively. In 2003 rainfall for April, May, and July was below the long-term average, while June rainfall was well above normal.

Irrigation frequency was greatest in 2000, reflecting the large rainfall deficit that year (**Table 1**). Irrigation was triggered in the study by soil water deficit, which reflects the length of time between rainfall events as well as low total rainfall and the evapotranspiration rate, which is incorporated in the scheduling model. As a result, in 2000 there were five irrigations for plots in the moderate irrigation category (when the soil water deficit reached 6.4 cm) and nine irrigations for plots in the well-irrigated category (when the soil water deficit reached 3.8 cm). In 2001, a moderate drought year, there were two irrigations in the moderate irrigation category and seven irrigations in the well-irrigated category. In 2003, a relatively wet year by total rainfall, but with poor distribution, there were four irrigations in the moderate irrigation category and seven irrigations in the well-irrigated category. Soil water levels measured by Watermark sensors (Irrometer Co., Riverside, CA) confirmed that there was a consistent, expected difference in soil water between non-irrigated, moderately irrigated and well-irrigated plots (data not shown).

3.2. Effects of Planting Density (Seeding Rate) and Soil Water Deficit Kernel Number

If the hybrids in this study exhibited the kernel number variation expected on the basis of claims made by the

Table 1. Rainfall for the April through July growing season for the three locations in 2000, 2001 and 2003.

Month	Rainfall (mm/month)			
	2000	2001	2003	Long-term average[1]
April	23	53	69	127
May	137	69	107	135
June	114	155	206	99
July	19	107	84	104
Total	293	384	466	465

[1]Forty-year average for St. Joseph.

Table 2. Temperatures for the April through July growing season at the experimental sites in 2000, 2001 and 2003.

Month	Average Monthly Temperatures (°C)							
	2000		2001		2003		Long-term average[2]	
	Min	Max	Min	Max	Min	Max	Min	Max
April	12.3	24.6	14.6	26.9	12.3	26.3	12.7	25.3
May	19.6	31.9	17.4	30.8	19.0	30.2	17.4	29.3
June	21.3	33.0	20.7	31.4	21.3	32.5	21.1	32.6
July	22.4	36.4[1]	22.4	34.2	22.4	33.6	22.7	33.9
Season	18.9	31.5	18.8	30.8	18.8	30.6	18.5	30.3

[1]Temperatures for 6 days in mid-July ranged from 38.0°C - 39.8°C. [2]Forty-year average for St. Joseph.

marketing seed companies, then the number of kernels per ear would be expected to be affected by environmental factors that stress plants, including plant density and drought. A hybrid marketed as "flex" would be expected to yield ears with fewer kernels under stress conditions, whereas a hybrid marketed as "fixed-ear" would be expected to have yield with similar numbers of kernels under stress conditions. A hybrid marketed as "semi-flex" would be expected to exhibit intermediate characteristics. In the present study plant density correlated very closely with seeding density under all conditions encountered (R values were 0.989 ± 0.004). Hence, in the following discussion the term "seeding density" is used interchangeably with "plant density". In no treatment were there significantly more than one ear per plant. In 2000, non-irrigated plots showed no consistent effect of

seeding density on kernels per ear in the fixed-ear hybrid. In the semi-flex and flex hybrids, kernel number was reduced by an average of 171 and 199 kernels, respectively, over the range from the lowest to the highest seeding density used (**Table 3**). Similarly in 2001, non-irrigated plots showed no consistent effect of seeding density on kernels per ear in the fixed-ear hybrid, whereas kernels per ear was reduced by 141 kernels from the lowest to the highest seeding density in the semi-flex hybrid, and reduced by 151 kernels in the full-flex hybrid (**Table 4**). However, in 2003, kernels per ear in non-irrigated plots was progressively reduced by 125 - 191 kernels as the seeding density increased for all three hybrids (**Table 5**); indicating all three hybrids exhibited either a flex response to plant density, or a response to some unmeasured factor.

Table 3. Effect of hybrids with differing kernel number flexibility trait and seeding density on grain yield and yield characteristics in 2000[1].

Hybrid/Kernel number		Non-irrigated				Moderate irrigation (6.4 cm SWD)				Well-irrigated (3.8 cm SWD)			
Flexibility Trait	Seeding density (seeds/ha)	Grain yield (10^6g/ha)	Plant density (plts/ha)	Kernel weight (g/100)	Ear size (kernels per ear)	Grain yield (10^6g/ha)	Plant density (plts/ha)	Kernel weight (g/100)	Ear size (kernels per ear)	Grain yield (10^6g/ha)	Plant density (plts/ha)	Kernel weight (g/100)	Ear size (kernels per ear)
PB 33K81	49,400	4.3	43,500	22.0	464	5.0	44,260	23.3	497	5.8	48,930	25.0	481
Fixed-ear	61,750	5.1	54,760	19.9	478	5.7	58,240	23.3	423	6.7	60,560	24.2	464
	74,700	5.2	63,480	19.2	433	6.4	69,010	21.8	435	6.8	67,460	23.1	439
	86,450	5.2	73,190	19.1	375	6.1	74,740	21.4	388	7.8	75,630	23.0	452
	98,800	6.0	81,530	17.7	454	5.8	83,190	20.9	341	7.6	81,340	22.7	421
Average		5.2	63,290	19.6	441	5.8	65,890	22.1	417	6.9	66,780	23.6	451
PB 3223	49,400	6.0	51,050	24.0	496	6.8	50,760	28.8	474	7.2	54,270	31.7	417
Semi-flex	61,750	6.2	58,140	24.4	447	5.4	58,540	27.4	346	7.7	63,210	30.6	403
	74,700	5.7	67,280	22.4	390	6.6	69,110	26.9	361	8.1	70,570	29.4	399
	86,450	5.5	75,430	20.9	351	6.7	78,520	27.6	313	7.9	80,280	28.6	343
	98,800	5.8	84,840	21.5	325	5.9	87,170	27.2	267	8.2	91,540	27.7	328
Average		5.8	67,350	22.6	402	6.3	68,820	27.6	352	7.8	71,970	29.6	378
GA 8460	49,400	4.6	45,320	18.2	561	6.2	51,250	25.7	472	6.8	54,760	27.5	458
Full-flex	61,750	4.8	57,950	19.2	451	6.3	58,740	25.4	378	6.6	60,290	25.9	430
	74,700	4.7	62,320	17.1	455	5.6	65,830	24.1	366	7.3	72,520	24.8	406
	86,450	4.5	74,740	16.4	376	5.8	73,680	22.8	350	7.1	76,990	24.8	375
	98,800	4.6	83,580	15.6	362	5.6	84,450	23.8	282	6.8	86,700	24.2	325
Average		4.6	64,780	17.3	441	5.9	66,790	24.4	370	6.9	70,250	25.4	399
LSD (0.10):													
Ear flexibility (F)		0.4	NS	0.6	19	0.4	NS	0.6	19	0.2	2,180	0.5	18
Seeding density (D)		NS	3,850	0.8	24	NS	3,850	0.8	24	0.3	2,810	0.6	23
F × D		0.8	NS[1]	NS	NS	0.8	NS	NS	NS	0.4	NS	NS	NS

[1]Abbreviations: NS = Non-significant at the 0.10 probability level; LSD = Fisher's least significant difference test; flex = Flexible kernel number trait; SWD = Soil water deficit; plts = Plants.

Table 4. Effect of hybrids with differing kernel number flexibility trait and seeding density on grain yield and yield characteristics in 2001[1].

Hybrid/Kernel number		Non-irrigated				Moderate irrigation (6.4 cm SWD)				Well-irrigated (3.8 cm SWD)			
Flexibility Trait	Seeding density (seeds/ha)	Grain yield (10^6g/ha)	Plant density (plts/ha)	Kernel weight (g/100)	Ear size (kernels per ear)	Grain yield (10^6g/ha)	Plant density (plts/ha)	Kernel weight (g/100)	Ear size (kernels per ear)	Grain yield (10^6g/ha)	Plant density (plts/ha)	Kernel weight (g/100)	Ear size (kernels per ear)
PB 33K81	49,400	8.0	40,780	29.7	618	8.4	44,830	30.8	596	8.1	42,630	31.6	597
Fixed-ear	61,750	9.2	55,700	30.2	549	10.1	54,170	29.9	621	10.0	52,710	30.9	608
	74,700	9.8	57,060	28.3	606	10.9	63,530	29.5	572	11.9	61,870	32.3	583
	86,450	10.7	55,200	28.5	678	11.6	73,580	28.0	556	12.0	70,400	28.6	584
	98,800	11.3	71,930	25.4	619	11.8	77,110	26.3	586	12.3	74,350	26.9	607
Average		9.8	56,130	28.4	614	10.6	62,640	28.9	586	10.9	60,390	30.1	596
PB 3223	49,400	9.5	39,870	35.6	617	9.9	48,040	35.4	562	9.7	46,040	33.4	629
Semi-flex	61,750	10.6	52,390	34.2	578	11.2	55,110	34.3	571	11.3	55,450	36.4	549
	74,700	11.5	58,930	35.8	534	11.5	59,600	35.2	547	11.4	58,860	33.8	585
	86,450	11.3	63,840	33.7	508	13.0	77,010	34.2	485	13.2	71,460	35.1	517
	98,800	11.8	75,530	32.9	476	13.2	81,830	35.4	454	13.0	79,410	32.8	500
Average		10.9	58,110	34.4	543	11.8	64,320	34.9	524	11.7	62,240	34.3	556
GA 8460	49,400	9.7	45,990	30.0	703	10.5	50,390	29.8	680	10.5	50,460	30.6	673
Full-flex	61,750	10.7	54,880	28.5	679	11.5	60,560	28.7	690	11.1	57,080	29.5	653
	74,700	11.4	64,910	26.7	646	12.2	68,860	28.2	622	12.1	69,010	32.5	553
	86,450	11.0	72,450	26.3	575	12.1	75,710	25.9	610	11.7	74,670	27.4	568
	98,800	11.2	80,600	25.2	552	12.1	81,490	27.1	550	11.5	78,180	27.9	527
Average		10.8	63,770	27.3	631	11.7	67,400	27.9	630	11.4	65,880	29.6	595
LSD (0.10):													
Ear flexibility (F)		0.4	3,580	0.8	35	0.2	1,960	0.8	25	0.5	2,710	1.8	33
Seeding density (D)		0.6	4,620	1.0	45	0.3	2,530	1.0	32	0.5	3,500	2.3	42
F × D		NS	NS	NS	78	0.6	NS	1.7	NS	0.9	NS	NS	NS

[1]Abbreviations: NS = Non-significant at the 0.10 probability level; LSD = Fisher's least significant difference test; flex = Flexible kernel number trait; SWD = Soil water deficit; plts = Plants.

Table 5. Effect of hybrids with differing kernel number flexibility trait and seeding density on grain yield and yield characteristics in 2003[1].

Hybrid/Kernel Number		Non-irrigated				Moderate irrigation (6.4 cm SWD)				Well irrigated (3.8 cm SWD)			
Flexibility Trait	Seeding density (seeds/ha)	Grain yield (10^6g/ha)	Plant density (plts/ha)	Kernel weight (g/100)	Ear size (kernels per ear)	Grain Grain (10^6g/ha)	Plant density (plts/ha)	Kernel weight (g/100)	Ear size (kernels per ear)	Grain yield (10^6g/ha)	Plant density (plts/ha)	Kernel weight (g/100)	Ear size (kernels per ear)
PB 33K81	49,400	6.9	48,580	24.8	577	6.5	49,420	26.2	507	7.0	45,320	27.4	570
Fixed-ear	61,750	7.6	60,790	23.6	536	7.2	54,980	25.2	522	8.0	57,380	26.3	530
	74,700	8.3	67,060	23.2	536	8.4	66,440	25.6	495	9.2	63,850	26.2	547
	86,450	8.8	81,260	22.3	485	9.7	76,770	24.1	487	10.6	79,510	25.3	530
	98,800	8.4	89,240	22.4	424	9.5	85,220	23.3	478	10.3	83,440	23.9	518
Average		8.0	69,390	23.3	512	8.3	66,570	24.9	498	9.0	65,900	25.8	539
PB 3223	49,400	8.8	47,550	31.3	595	9.5	46,360	32.7	626	10.3	46,660	33.2	669
Semi-flex	61,750	9.6	56,070	31.2	549	10.1	59,060	32.1	532	11.1	57,970	33.6	573
	74,700	9.2	68,740	30.0	448	10.5	68,620	30.4	506	11.7	69,750	31.8	527
	86,450	9.9	77,360	29.9	427	11.0	79,340	31.4	445	12.2	75,210	32.1	508
	98,800	10.3	86,850	29.3	404	11.1	87,240	31.1	409	12.1	82,790	32.2	456
Average		9.6	67,310	30.3	485	10.4	68,120	31.5	504	11.5	66,480	32.6	547
GA 8460	49,400	7.8	42,310	28.2	658	8.6	48,840	27.6	713	9.0	43,250	29.1	722
Full-flex	61,750	8.6	51,670	26.7	626	9.6	51,850	27.4	676	10.2	52,610	28.6	681
	74,700	9.4	57,670	26.0	625	9.9	61,820	26.7	604	10.5	60,560	27.9	624
	86,450	8.7	67,280	25.4	517	10.2	72,620	26.0	544	11.2	68,300	27.4	602
	98,800	9.4	73,580	23.9	533	10.3	77,310	25.0	537	11.9	77,810	26.8	572
Average		8.8	58,500	26.0	592	9.7	62,490	26.5	615	10.6	60,510	28.0	640
LSD (0.10):													
Ear flexibility (F)		0.3	1,850	0.6	20	0.4	NS	0.6	26	0.3	2,800	0.4	22
Seeding density (D)		0.4	2,370	0.8	25	0.5	4,940	0.8	33	0.4	3,620	0.5	28
F × D		NS	4,120	NS	44	NS	NS	NS	57	0.7	NS	0.9	48

[1]Abbreviations: NS = Non-significant at the 0.10 probability level; LSD = Fisher's least significant difference test; flex = Flexible kernel number trait; SWD = Soil water deficit; plts = Plants.

The effect of drought on kernel number was less clear. In a comparison of average kernel number in non-irrigated plants between a drought year (2000) and a normal rainfall year (2003), all three hybrids had a lower kernels per ear under drought conditions. But the increase in kernels per ear in the normal rainfall year for the fixed ear hybrid (71 kernels) was less than for the semi-flex hybrid (83 kernels), and the largest increase was in the full-flex hybrid (151 kernels). However, the 2001 growing year was anomalous with a larger kernel number for all three hybrids, suggesting another unidentified factor was the primary determinant of kernel number that year. Inconsistent results were also obtained in a comparison of kernels per ear with increasing irrigation levels for the three growing years. Hybrids in the drought year (2000) would be expected to exhibit the largest kernel number flex response. As expected, the fixed ear hybrid changed kernel number little with increased irrigation, but the semi-flex and full-flex hybrids changed kernels per ear in unexpected directions, suggesting that the irrigation levels used were not sufficient to eliminate drought stress. In the semi-drought year (2001) there was insufficient kernels per ear change to allow conclusions to be drawn. In the normal rainfall year (2003) the fixed ear hybrid changed kernels per ear little with increased irrigation, as expected, while the semi-flex and full-flex hybrids increased in kernels per ear (by 64 and 74 kernels, respectively) with increasing irrigation levels as expected for effective relief of drought stress. In general, kernel number change in response to drought was only partially consistent with expectations for ear size change properties claimed by the seed companies that marketed the hybrids.

3.3. Factors Affecting Yield

3.3.1. Effect of Type of Hybrid on Yield

Hybrids with flexible kernel number are marketed for their purported ability to give higher yields under stress conditions than do fixed ear size hybrids, which have been selected for the ability to give high yields under optimal conditions. When not irrigated, only the main effect of hybrid type was significant. Among the three hybrids in this study, under non-irrigated conditions, the semi-flex hybrid gave the highest average grain yield in all three years, resulting in the average yield rank for the hybrids being semi-flex hybrid > fixed ear hybrid ≈ full-flex hybrid (**Table 3**). These data suggest that the full-flex hybrid may not be very tolerant of dry conditions. The semi-flex hybrid had the highest yield among hybrids regardless of the irrigation level. These higher yields for the semi-flex hybrid were primarily due to

higher kernel weights regardless of soil water levels. Hybrid and seeding density main effects affected kernels per ear, however, the interaction between hybrid and seeding density was not significant in either of the soil water treatments in 2000 (**Table 3**). The fixed ear hybrid exhibited less variation in ear size (kernels per ear) with increased seeding density than the semi-flex and full flex hybrids. In scatter plots of kernels per ear vs seeding density the data for the fixed ear hybrid with no irrigation correlated poorly ($R^2 = 0.235$) and the slope of the line fitted to the data = –0.1 kernels per ear change per 10^3 increase in seeds per ha. The corresponding values for moderate irrigation were $R^2 = 0.895$ and slope = –2.8, and for well irrigated were $R^2 = 0.832$ and slope = –1.1. For the semi-flex hybrid the corresponding values were $R^2 = 0.984$ and slope = –3.6 for no irrigation, $R^2 = 0.840$ and slope = –3.6 for moderate irrigation and $R^2 = 0.900$ and slope = –1.9 for well-irrigated conditions. For the full flex hybrid the corresponding values were $R^2 = 0.886$ and slope = –3.8 for no irrigation, $R^2 = 0.894$ and slope = –3.3 for moderate irrigation and $R^2 = 0.975$ and slope = –2.6 for well-irrigated conditions.

3.3.2. Effect of Seeding Density on Yield

Yields from hybrids with flexible kernel number are expected to correlate negatively with seeding density, particularly under drought conditions, because these hybrids can increase yields at low seed densities by increasing kernel number. In the drought year of 2000, yields from semi-flex and full flex hybrids correlated negatively with seeding density with no irrigation (R = –0.648 and –0.412, respectively) and with low irrigation (R = –0.128 and –0.817, respectively), whereas under well-irrigated conditions the correlations were positive (R = 0.863 and 0.295, respectively) (**Table 3**). In contrast, yields from the fixed kernel number hybrid were observed to correlate positively with seeding density with no irrigation (R = 0.919), moderate irrigation (R = 0.612) and well-irrigated (R = 0.929). These differences are reflected in a significant hybrid × seeding density interaction for yield in both moderately- and well-irrigated corn. In non-irrigated plots in 2000, yields for both the semi-flex hybrid and the full-flex hybrid were maximal at about 74,700 seeds·ha^{-1}, whereas yields for the fixed ear hybrid continued to increase as seeding density increased. Maximum yield for the semi-flex hybrid occurred at higher seeding densities than the other two hybrids in the well-irrigated plots, while yields for the fixed ear hybrid and the semi-flex hybrid increased with higher seeding densities in the moderately irrigated plots.

In years 2001 and 2003, when there was greater rainfall, yields from all hybrids correlated positively with seeding density with or without irrigation (R ranging from 0.722 to 0.984). Kernels per ear for the fixed ear hybrid remained relatively constant across seeding densities, while kernels per ear decreased with increasing plant populations for both the semi-flex hybrid and the full-flex hybrid. In 2001 the main effects of hybrid and seeding density were significant for kernels·ear^{-1} regardless of soil water level, and the interaction between hybrid and seeding density was significant for the non-irrigated plots. In 2003 there appeared to be a good relationship between optimum plant population and yield potential. Optimum seeding density was approximately 74,700 seed·ha^{-1} for non-irrigated plots and 86,450 seed·ha^{-1} for moderately irrigated plots regardless of hybrid. However, there was a significant hybrid × seeding density interaction for yield in well-irrigated plots. The optimum seeding density was 86,450 seed·ha^{-1} for the fixed ear hybrid and 74,700 seed·ha^{-1} for the semi-flex hybrid, while yield for the full-flex hybrid continued to increase with increased seeding density over the entire range studied. The semi-flex hybrid had the highest kernel weight in non-irrigated, moderately irrigated and well-irrigated plots (**Table 5**). Unlike the first two years, the hybrid × seeding density interaction for kernels·ear^{-1} was significant across all irrigation rates examined. The decrease in kernels per ear as plant population increased was less for the fixed ear hybrid than for the other two hybrids.

In general, the flex hybrid had lower optimum plant populations than the other two hybrids for each soil water regime. This was most obvious in both the non-irrigated and moderately irrigated plots, indicating that lower seeding densities could be used for this hybrid, particularly under drought conditions. However, the 2000 findings indicate that the -flex hybrid may not be as drought tolerant as the other hybrids in the study. For a grower, selecting an adapted drought-tolerant hybrid would be advisable in a dryland cropping system.

3.3.3. Effect of Moisture Stress on Yield

Reducing drought conditions by irrigation resulted in increased yields with all hybrids in all years and higher levels of irrigation resulted in higher yields (**Tables 3-5**). Averaged across the three hybrids and seeding densities, yields were 8.2 Mg·ha^{-1} for the non-irrigated corn, 8.9 Mg·ha^{-1} for moderately irrigated corn, and 9.6 Mg·ha^{-1} for well-irrigated corn, consistent with water stress being greatest in the non-irrigated plots, and least in well-irrigated plots. The largest increase in yield in response to irrigation was observed with the flex hybrid each year, although the relative sizes of yield increases were variable.

3.4. Mycotoxin Contamination in Hybrids Expressing Varying Levels of Kernel Number Flexibility Trait

3.4.1. Aflatoxin Contamination in Hybrids Expressing Varying Levels of Kernel Number Flexibility Trait

Aflatoxin levels in uninoculated corn averaged over the three years in the study were 33.5 ± 12.0 µg·kg^{-1} for non-irrigated plots, 29.3 ± 11.4 µg·kg^{-1} for moderately irrigated plots and 24.2 ± 8.5 µg·kg^{-1} for well-irrigated plots (**Table 7**). The effect of drought on aflatoxin levels can be examined by comparing average aflatoxin levels in corn grown without irrigation in the drought year of 2000, to corn grown in moderately irrigated plots and in well-irrigated plots. In 2000, corn inoculated with *A. flavus*, averaged aflatoxin levels of 827 ± 46 µg·kg^{-1} in non-irrigated plots, 783 ± 51 µg·kg^{-1} in moderately irrigated plots, and 877 ± 50 µg·kg^{-1} in well-irrigated plots. The only significant difference with differing irrigation levels was hybrid in both the moderately-irrigated and well-irrigated plots. The hybrid with the flexibile kernel number trait had the lowest aflatoxin levels at each of the irrigation levels. In 2001, average aflatoxin levels in corn inoculated with *A. flavus* were 1,008 ± 274, 660 ± 212 and 684 ± 145 for the non-irrigated, moderately-irrigated and well-irrigated plots, respectively. At each irrigation level, hybrid was significant for aflatoxin levels with the rank being flex ear > fixed ear = semi-flexible ear hybrid. In 2003 aflatoxin levels in corn inoculated with *A. flavus* were lower than the other two years, averaging 46.5 ± 12.2 µg·kg^{-1} for non-irrigated plots, 68.1 ± 16.7 µg·kg^{-1} for moderately irrigated plots, and 22.3 ± 5.5 µg·kg^{-1} for well-irrigated plots. The only significant effect was hybrid in the non-irrigated plots.

Very limited differences in aflatoxin levels in naturally infected corn (**Table 6**) or in corn inoculated with *A. flavus* (**Table 7**) were observed in hybrids with varying ear flex. In 2001, for moderate irrigation there was a significant hybrid type × seeding density interaction in which aflatoxin levels increased with increasing seeding density for all hybrids with *A. flavus* inoculation and for the flex hybrid with natural infection. For naturally-infected fixed ear or semi-flex hybrids there was insufficient aflatoxin contamination to evaluate the effect of seeding density in 2001 (**Table 6**). In 2003 the hybrid × seeding density interaction was significant only for the non-irrigated plots. There was considerable variation in aflatoxin levels among seeding densities for all hybrids. Effects of drought stress and seeding density alone do not fully explain the variation in aflatoxin levels, consistent with other factors playing a significant role.

Table 6. Effect of hybrids with differing kernel number flexibility trait and seeding density on aflatoxin levels in kernels harvested from corn naturally infected with *A. flavus* while growing at three soil water deficit levels in 2000, 2001, and 2003[1,2,3].

Hybrid/Kernel Number		Average aflatoxin levels ($\mu g \cdot kg^{-1}$)									
		Irrigation in 2000			Irrigation in 2001			Irrigation in 2003			Average
Flexibility Trait	Seeding density (seeds/ha)	None	Moderate	Well	None	Moderate	Well	None	Moderate	Well	
PB 33K81	49,400	99	36	19	14	0	0	1	2	13	20
Fixed-ear	61,750	41	178	30	0	0	0	1	2	30	31
	74,700	6	64	14	0	0	0	2	1	1	10
	86,450	289	141	27	0	0	0	1	66	1	58
	98,800	7	26	114	0	0	0	23	1	1	19
Average		88	89	41	3	0	0	6	14	9	28
PB 3223	49,400	34	109	31	0	0	0	62	1	23	29
Semi-flex	61,750	135	25	12	0	0	0	18	2	1	21
	74,700	41	63	34	0	0	0	1	2	1	16
	86,450	7	138	7	0	0	0	7	2	1	18
	98,800	20	42	147	11	0	0	1	1	1	25
Average		47	75	46	2	0	0	18	2	5	22
GA 8460	49,400	92	7	103	55	19	0	29	3	1	34
Full-flex	61,750	256	39	6	32	0	9	1	12	14	41
	74,700	25	23	10	31	26	113	2	2	2	26
	86,450	9	37	17	43	48	39	8	2	1	23
	98,800	85	63	83	16	136	180	3	1	1	63
Average		93	34	44	35	46	68	9	4	4	37
LSD (0.10):									—[2]		
Ear flexibility (F)		NS	NS	NS	21	22	42	NS		NS	
Seeding density (D)		NS	NS	59	NS	29	NS	NS		NS	
F × D		NS	NS	NS	NS	50	NS	NS	NS	NS	

[1]Abbreviations: NS = Non-significant at the 0.10 probability level; LSD = Fish's least significant difference test; flex = Flexible kernel number trait; SWD = Soil water deficit. [2]Some treatments contained only one replicate. [3]Moderate irrigation occurred at 6.4-cm SWD; well irrigated occurred at 3.8-cm SWD.

Table 7. Effect of hybrids with differing kernel number flexibility trait and seeding density on aflatoxin levels in kernels harvested from corn inoculated with *A. flavus* while growing at three soil water deficit levels in 2000, 2001, and 2003[1,2].

Hybrid/Kernel Number		Average aflatoxin levels ($\mu g \cdot kg^{-1}$)									
		Irrigation in 2000			Irrigation in 2001			Irrigation in 2003			Average
Flexibility Trait	Seeding density (seeds/ha)	None	Moderate	Well	None	Moderate	Well	None	Moderate	Well	
PB33K81	49,400	950	1,010	1,067	710	115	212	10	77	40	466
Fixed-ear	61,750	515	956	854	290	175	86	31	82	30	335
	74,700	779	547	771	50	67	36	34	8	61	261
	86,450	871	1,117	717	134	330	435	54	63	1	414
	98,800	688	693	673	260	237	363	41	76	11	338
Average		761	865	816	289	185	226	34	61	29	363
PB3223	49,400	839	921	1,053	475	180	89	20	90	50	413
Semi-flex	61,750	1,224	1,008	945	230	309	477	65	8	5	475
	74,700	728	937	1,249	207	508	689	11	1	15	483
	86,450	642	605	1,087	567	170	221	36	45	1	375
	98,800	867	849	1,036	620	282	678	36	36	47	491
Average		860	864	1074	420	290	431	34	36	24	447
GA8460	49,400	863	726	829	2,868	488	1,630	192	101	36	859
Full-flex	61,750	821	532	738	1,148	830	1,381	1	11	34	611
	74,700	1,048	628	633	3,152	1,345	1,373	41	181	1	934
	86,450	636	670	581	1,948	3,008	1,370	105	18	1	926
	98,800	936	542	906	2,458	1,855	1,218	21	225	1	907
Average		861	620	737	2315	1505	1394	72	107	15	847
LSD (0.10):											
Ear flexibility (F)		NS	231	251	679	431	398	29	NS	NS	
Seeding density (D)		NS	NS	NS	NS	557	NS	NS	NS	NS	
F × D		NS	NS	NS	NS	964	NS	65	NS	NS	

[1]Abbreviations: NS = Non-significant at the 0.10 probability level; LSD = Fisher's least significant difference test; flex = Flexible kernel number trait; SWD = Soil water deficit. [2]Moderate irrigation occurred at 6.4-cm SWD; well irrigated occurred at 3.8-cm SWD.

3.4.2. Fumonisin Contamination in Hybrids Expressing Varying Ear Flex

Fumonisin contamination levels in harvested grain were significantly higher in the full- and semi-flex hybrids than in the fixed kernel number hybrid with (**Table 8**) or without (**Table 9**) inoculation with *A. flavus*, under most (six of nine) experimental conditions of rainfall and irrigation level. In hybrids not inoculated with *A. flavus* (**Table 8**), fumonisin levels in harvested grain averaged over the three years were significantly lower (analysis of variance, P < 0.05) for the fixed ear hybrid (1.32 ± 0.08 mg kg^{-1}) than for the semi-flex hybrid (2.30 ± 0.22 mg·kg^{-1}) and the flex hybrid (3.02 ± 0.19 mg·kg^{-1}). In hybrids that were inoculated with *A. flavus* (**Table 9**), fumonisin levels in harvested grain averaged over the three years were significantly lower (ANOVA, P < 0.01) for the fixed ear hybrid (1.94 ± 0.17 mg·kg^{-1}) than for the semi-flex hybrid (2.84 ± 0.19 mg·kg^{-1}) and the flex hybrid (3.80 ± 0.10 mg·kg^{-1}). Fumonisin levels in corn grain not inoculated with *A. flavus* averaged over the three years in the study was 1.94 ± 0.20 mg·kg^{-1} for non-irrigated plots, 2.18 ± 0.36 mg·kg^{-1} for moderately irrigated plots, and 2.54 ± 0.35 mg·kg^{-1} for well-irrigated plots (**Table 8**). Inoculation with *A. flavus* is not expected to affect fumonisin levels unless the two fungi are mutually antagonistic or synergistic during fungal growth and mycotoxin production. Fumonisin levels in corn grain inoculated with *A. flavus* averaged over the three years in the study were 2.13 ± 0.19 mg·kg^{-1} for non-irrigated plots, 3.11 ± 0.39 mg·kg^{-1} for moderately irrigated plots, and 3.31 ± 0.46 mg·kg^{-1} for well-irrigated plots (**Table 9**), which was significantly higher (P < 0.05, Student's *t* test) than observed for corn in plots not inoculated with *A. flavus*.

Fumonisin levels in naturally infected corn (**Table 8**) were significantly affected by hybrid in 2000 and 2003 with higher fumonisin levels being associated with hybrids expressing increased ear size flexibility. In corn inoculated with *A. flavus* (**Table 9**) fumonisin levels were significantly affected by hybrid in 2000 in moderate and well irrigated corn and in 2003 with all irrigation levels, also with higher fumonisin levels being associated with hybrids expressing increased ear size flexibility. Overall average fumonisin levels were significantly higher in hydrids expressing increasing ear size flexibility in naturally infected corn (ANOVA, P < 0.001) and in corn inoculated with *A. flavus* (ANOVA, P < 0.001). No consistent effects of irrigation level or seeding density on fumonisin levels were observed in either naturally infected corn (**Table 8**) or corn inoculated with *A. flavus* (**Table 9**).

Table 8. Effect of hybrids with differing kernel number flexibility trait and seeding density on average fumonisin levels in kernels harvested from corn naturally infected with *A. flavus* while growing at three soil water deficit levels in 2000, 2001, and 2003[1,2,3].

Hybrid/Kernel number		Average aflatoxin levels (µg·kg^{-1})									
		Irrigation in 2000			Irrigation in 2001			Irrigation in 2003			Average
Flexibility Trait	Seeding density (seeds/ha)	None	Moderate	Well	None	Moderate	Well	None	Moderate	Well	
PB33K81	49,400	0.3	1.6	2.1	1.2	2.6	0.4	1.2	0	0.8	1.1
Fixed-ear	61,750	0.6	0.4	1.4	1.9	3.8	0.7	0.7	0	1.3	1.2
	74,700	0.3	1.8	2.7	1.1	1.7	1.1	1.9	0	0.9	1.3
	86,450	0.7	1.7	2.4	1.8	2.9	1.5	1.6	0.1	1.2	1.5
	98,800	0.4	2.6	2.3	1.7	1.8	0.4	1.9	0.1	2	1.5
Average		0.5	1.6	2.2	1.5	2.6	0.8	1.5	0	1.2	1.3
PB3223	49,400	0.8	3.8	4.9	1.8	1.1	0.4	2.2	0	1	1.8
Semi-flex	61,750	1.3	2.2	5.1	1.5	1.8	0.8	1.7	0	1.4	1.8
	74,700	0.9	3.3	5.5	1.9	2.1	1.3	3.8	3.9	1.1	2.6
	86,450	0.7	6.3	6.7	3.2	0.9	0.5	2.3	0	0.8	2.4
	98,800	1.5	4.3	7	4.7	1	0.5	5.5	0	1.7	2.9
Average		1	4	5.8	2.6	1.4	0.7	3.1	0.8	1.2	2.3
GA8460	49,400	0.8	6	6	3.4	2.5	0.7	3.5	0.9	4	3.1
Full-flex	61,750	1.7	6.1	6.8	2	1.1	1.2	1.4	0	1.2	2.4
	74,700	0.5	5.1	5.2	4	3.2	2.7	2.2	0	2.4	2.8
	86,450	1.4	9.2	8.3	2.4	1.4	1	6.7	0	1.3	3.5
	98,800	1.3	4.8	9.4	2.2	3.3	1.4	2.7	2	2.8	3.3
Average		1.1	6.2	7.1	2.8	2.3	1.4	3.3	0.6	2.3	3
LSD (0.10):											
Ear flexibility (F)		0.4	1.2	1.4	NS	NS	0.6	1.3	−2	0.7	
Seeding density (D)		NS	1.6	NS	0.9	NS	0.7	NS	-	NS	
F × D		NS	NS	NS	NS	NS	NS	NS	-	NS	

[1]Abbreviations: NS = Non-significant at the 0.10 probability level; LSD = Fisher's least significant difference test; flex = Flexible kernel number trait; SWD = Soil water deficit. [2]Some treatments contained only one replicate. [3]Moderate irrigation occurred at 6.4-cm SWD; well irrigated occurred at 3.8-cm SWD.

Table 9. Effect of hybrids with differing kernel number flexibility trait and seeding density on average fumonisin levels in kernels harvested from corn inoculated with *A. flavus* while growing at three soil water deficit levels in 2000, 2001, and 2003[1,2].

Hybrid/Kernel number		Average aflatoxin levels ($\mu g \cdot kg^{-1}$)									
		Irrigation in 2000			Irrigation in 2001			Irrigation in 2003			Average
Flexibility Trait	Seeding density (seeds/ha)	None	Moderate	Well	None	Moderate	Well	None	Moderate	Well	
PB33K81	49,400	2.1	3.4	7.5	0.4	4.3	1	1.5	0.8	1.2	2.5
Fixed-ear	61,750	3.4	3	3.1	0.6	3.1	0.8	1.4	0.6	0.4	1.8
	74,700	1.6	2.5	4.2	0.9	0.9	0.7	1.2	0.4	0.7	1.5
	86,450	1.6	2.1	5.6	1.4	2.9	0.7	0.6	0.7	0.6	1.8
	98,800	2.4	7.3	3.7	1.4	0.9	0.5	1.6	0.4	1	2.1
Average		2.2	3.7	4.8	0.9	2.4	0.7	1.3	0.6	0.8	1.9
PB3223	49,400	3.2	7.3	8.9	2.2	1	0.3	1.4	1.9	1.5	3.1
Semi-flex	61,750	3.7	5.2	5.7	0.4	0.2	0.4	1.7	0.9	1.4	2.2
	74,700	4.9	3.3	6.3	1.6	0.2	0.5	4.5	1.1	1.5	2.7
	86,450	2.2	6.6	5.5	1.3	1.2	0.5	3.5	3.3	1.9	2.9
	98,800	2.5	7.6	7.6	0.7	1.2	1.3	2.3	2.2	4	3.3
Average		3.3	6	6.8	1.2	0.8	0.6	2.7	1.9	2.1	2.8
GA8460	49,400	2.5	8.7	6	1.8	3.1	1.5	5.8	2.3	4.5	4
Full-flex	61,750	3.1	6.9	7.9	0.8	3.7	1.5	2.7	2.1	3.8	3.6
	74,700	1.7	8	8.5	1.3	1.7	0.9	3.9	3.5	5.3	3.9
	86,450	1.9	7.9	10.4	0.5	1.3	0.7	3.9	1.2	3.5	3.5
	98,800	3.9	9.5	11.8	1.8	2	0.4	2	1.7	3.1	4
Overall average		2.9	6	6.8	1.1	1.9	0.8	2.5	1.5	2.3	
LSD (0.10):											
Ear flexibility (F)		NS	2.5	2.3	NS	0.9	NS	0.8	0.7	1.2	
Seeding density (D)		NS	NS	NS	NS	1.2	NS	NS	NS	NS	
F × D		NS	NS	NS	NS	2	NS	19	1.5	NS	

[1]Abbreviations: NS = Non-significant at the 0.10 probability level; LSD = Fisher's least significant difference test; flex = Flexible kernel number trait; SWD = Soil water deficit. [2]Some treatments contained only one replicate. [2]Moderate irrigation occurred at 6.4-cm SWD; well irrigated occurred at 3.8-cm SWD.

3.5. Effect of Irrigation Level on Aflatoxin and Fumonisin Levels

In the drought year of 2000, aflatoxin levels were significantly higher (P < 0.001, Student's *t*-test) than in years with moderate (2001) or good (2003) rainfall at all levels of irrigation with or without inoculation with *A. flavus* (**Tables 8** and **9**). There were no consistent significant differences in aflatoxin levels among irrigation regimes in the drought year of 2000 or in years with moderate (2001) or good (2003) rainfall, suggesting that factors other than drought were primarily responsible for determining aflatoxin levels in harvested grain.

In the drought year of 2000, fumonisin levels were significantly lower (P < 0.001, Student's *t*-test) in non-irrigated corn than in moderately or well-irrigated corn with or without inoculation with *A. flavus* (**Tables 8** and **9**). There were no consistent significant differences in fumonisin levels among irrigation regimes in years with moderate (2001) or good (2003) rainfall, suggesting that factors other than drought were primarily responsible for determining fumonisin levels in harvested grain.

3.6. Effect of Seeding Rate on Aflatoxin and Fumonisin Levels

Seeding rate and the resulting plant density is one of several crop management techniques which have been extensively studied for their effects on aflatoxin and fumonisin contamination in harvested corn kernels based on the model that mycotoxin production is associated with factors that increase plant stress [42]. Effects of plant density have become of increased concern as plant population recommendations for maize production have increased to the point at which they are now double those recommended in the 1950's. For example, Rodriguez-del-Bosque [43] observed in Mexico that the two factors most associated with enhanced aflatoxin contamination were late planting and ear insect damage, whereas plant density did not significantly affect aflatoxin contamination. However, Alvarado-Carrillo et al. [44] observed that high plant densities resulted in lower grain yield and increased aflatoxin content and charcoal rot disease in corn in Mexico. In Italy Blandino et al. [45] reported significantly higher levels of fumonisins in harvested corn using a combination of agronomic techniques including higher seeding density.

In this study, no consistent significant correlations were observed between seeding density and aflatoxin levels in either naturally infected corn (R ranged from –0.73 to +0.71) (**Table 6**) or corn inoculated with *A. flavus* (R ranged from –0.94 to +0.84) **Table 7**). In hybrids not inoculated with *A. flavus* (**Table 8**), the average fumonisin levels across the three years correlated well with seeding density for the fixed kernel number hybrid (R = 0.97) and less well for the semi-flexible kernel number hybrid (R = 0.91) and the fully flexible kernel number hybrid (R = 0.54). All significant correlations with seeding density for individual years and irrigation levels in the study were positive, but average fumonisin levels did not correlate (R < 0.5) consistently with seeding density for hybrids (R ranged from –0.47 to +0.96). In hybrids inoculated with *A. flavus* (**Table 9**) there were no consistent significant correlations between fumonisin levels and seeding density (R ranged from –0.96 to +0.99).

4. Conclusion

In this study, characterization of the effects of irrigation and population density (seeding density) on kernel number in corn hybrids with different ear size flexibility traits generally reflected expectations based on seed company claims. Under drought conditions, kernels per ear for the fixed-ear hybrid remained relatively constant across the range of seeding densities studied, whereas kernels per ear decreased with increasing plant population for both the semi-flexible and flexible kernel number hybrids. However, drought alone resulted in smaller numbers of kernels per ear for all hybrids. Irrigation increased yields in all hybrids, but in the absence of irrigation, yields were highest for the hybrid with the semi-flexible kernel number trait. Under irrigation yields of all hybrids correlated with seeding density. Under drought conditions, yields were maximal at 74,700 seeds/ha for semi-flexible and flexible ear hybrids, but yields for the fixed ear hybrid continued to increase with seeding density evaluated. The lowest levels of aflatoxins and fumonisins in harvested grain were observed for the hybrid with the flex ear hybrid at all rainfall and irrigation levels, but there was no evidence that reducing stress by lowering planting density played any role in achieving low mycotoxin contamination. Inoculation with *A. flavus* resulted in much higher levels of aflatoxins and significantly higher levels of fumonisins in all hybrids under most conditions of rainfall and irrigation, suggesting that factors promoting successful infection by *A. flavus* can have a major effect on production of both toxins. Further studies with larger numbers of hybrids will be needed to determine more specific roles played by the ear flex trait

5. Acknowledgements

We thank Ms. Bobbie J. Johnson for her technical assistance in this project. Disclaimer: The mention of trade names or commercial products in this report is solely for the purpose of providing specific information and does not imply recommendation or endorsement by the US Department of Agriculture.

REFERENCES

[1] H. K. Abbas, W. P. Williams, G. L. Windham, J. C. Pringle, W. Xie Jr. and W. T. Shier, "Aflatoxin and Fumonisin Contamination of Commercial Corn (*Zea mays*) Hybrids in Mississippi," *Journal of Agricultural and Food Chemistry*, Vol. 50, No. 18, 2002, pp. 5246-5254.

[2] H. K. Abbas, R. D. Cartwright, W. Xie and W. T. Shier, "Aflatoxin and Fumonisin Contamination of Corn (Maize, *Zea mays*) Hybrids in Arkansas," *Crop Protection*, Vol. 25, No. 1, 2006, pp. 1-9. doi:10.1016/j.cropro.2005.02.009

[3] H. K. Abbas, W. T. Shier and R. D. Cartwright, "Effects of Temperature, Rainfall and Planting Date on Aflatoxin and Fumonisin Contamination in Commercial Bt and Non-Bt Corn Hybrids in Arkansas," *Phytoprotection*, Vol. 88, No. 2, 2007, pp. 41-50.

[4] H. A. Bruns and H. K. Abbas, "Responses of Short-Season Corn Hybrids to a Humid Subtropical Environment," *Agronomy Journal*, Vol. 97, No. 2, 2005, pp. 446-451. doi:10.2134/agronj2005.0446

[5] R. K. Jones, H. E. Duncan and P. B. Hamilton, "Planting Date, Harvest Date, and Irrigation Effects on Infection and Aflatoxin Production by *Aspergillus flavus* in Field Corn," *Phytopathology*, Vol. 71, 1981, pp. 810-816. doi:10.1094/Phyto-71-810

[6] E. B. Lillehoj, W. F. Kwolek, A. Manwiller, J. A. Durant, J. C. LaPrade, E. S. Horner, J. Reid and M. S. Zuber, "Aflatoxin Production in Several Corn Hybrids Grown in South Carolina and Florida," *Crop Science*, Vol. 16, 1976, pp. 483-485.

[7] G. A. Payne, "Aflatoxins in Maize," *Critical Review in Plant Science*, Vol. 10, No. 5, 1992, pp. 423-440. doi:10.1080/07352689209382320

[8] J. Robens and K. Cardwell, "The Costs of Mycotoxin Management to the USA: Management of Aflatoxins in the United States," *Journal of Toxicology—Toxin Reviews*, Vol. 22, No. 2-3, 2003, pp. 139-152.

[9] C. S. T. Daughtry, J. C. Cochran and S. E. Hollinger, "Estimating Silking and Maturity Dates of Corn for Large Area," *Agronomy Journal*, Vol. 76, 1984, pp. 415-420.

[10] H. K. Abbas, J. Wilkinson, R. M. Zablotowicz, C. Accinelli, C. A. Abel, H. A. Bruns and M. A. Weaver, "Ecology of *Aspergillus flavus*, Regulation of Aflatoxin Production and Management Strategies to Reduce Aflatoxin Contamination of Corn," *Toxin Reviews*, Vol. 28, No. 2-3, 2009, pp. 142-153. doi:10.1080/15569540903081590

[11] W. J. Chamberlain, C. W. Bacon, W. P. Norred and K. A.

Voss, "Levels of Fumonisin B1 in Corn Naturally Contaminated with Aflatoxin," *Food Chemistry and Toxicology*, Vol. 31, No. 12, 1993, pp. 995-998. doi:10.1016/0278-6915(93)90009-N

[12] S. N. Chulze, M. L. Ramirez, M. Farnochi, M. Pascale, A. Visconti and G. March, "*Fusarium* and Fumonisins Occurrence in Argentinian Corn at Different Ear Maturity Stage," *Journal of Agricultural and Food Chemistry*, Vol. 44, No. 9, 1996, pp. 2797-2801. doi:10.1021/jf950381d

[13] J. P. Rheeder, W. F. O. Marasas and H. F. Vismer, "Production of Fumonisin Analogs by *Fusarium* Species," *Applied and Environmental Microbiology*, Vol. 68, No. 5, 2002, pp. 2101-2105. doi:10.1128/AEM.68.5.2101-2105.2002

[14] C. W. Bacon, A. E. Glenn and I. E. Yates, "*Fusarium verticilloides*: Managing the Endophytic Association with Maize for Reduced Fumonisins Accumulation," *Toxin Reviews*, Vol. 27, No. 3-4, 2008, pp. 411-446. doi:10.1080/15569540802497889

[15] G. P. Munkvold, D. C. McGee and W. M. Cariton, "Importance of Different Pathways for Maize Kernel Infection by *Fusarium moniliforme*," *Phytopathology*, Vol. 87, No. 2, 1997, pp. 209-217. doi:10.1094/PHYTO.1997.87.2.209

[16] K. F. Cardwell, J. G. King, B. Maziya-Dixon and N. A. Bosque-Perez, "Interactions between *Fusarium verticallioides*, *Aspergillus flavus*, and Insect Infestation in Four Maize Genotypes in Lowland Africa," *Phytopathology*, Vol. 90, No. 3, 2000, pp. 276-284. doi:10.1094/PHYTO.2000.90.3.276

[17] H. K. Abbas, "Special Issue on Aflatoxin and Food Safety, Part II," *Journal of Toxicolog—Toxin Reviews*, Vol. 23, No. 2-3, 2003, pp. 153-154.

[18] Council for Agriculture Science and Technology, "Mycotoxins: Risks in Plant, Animal, and Human Systems," IA Task Force Report, CAST, Ames, 2003, p. 139.

[19] D. C. McGee, O. M. Olanya and G. M. Hoyos, "Populations of *Aspergillus flavus* in the Iowa Cornfield Ecosystem in Years Not Favorable for Aflatoxin Contamination Corn Grain," *Plant Disease*, Vol. 80, 1996, pp. 742-746. doi:10.1094/PD-80-0742

[20] H. H. L. Gonzalez, S. L. Resnik, R. T. Boca and W. F. O. Marasas, "Mycoflora of Argentinian Corn Harvested in the Main Production Area in 1990," *Mycopathologia*, Vol. 130, No. 1, 1995, pp. 29-36. doi:10.1007/BF01104346

[21] N. Zummo and G. E. Scott, "Interaction of *Fusarium moniliforme* and *Aspergillus flavus* on Kernel Infection and Aflatoxin Contamination in Maize Ears," *Plant Disease*, Vol. 76, 1992, pp. 771-773. doi:10.1094/PD-76-0771

[22] F. S. Dowling, "Fumonisin and Its Toxic Effects," *Cereal Foods World*, Vol. 42, No. 1, 1997, pp. 13-15.

[23] National Toxicology Program, "Toxicology and Carcinogenesis Studies on Fumonisin B_1 in F344/N Rats and B6CF1 Mice (Feed Studies)," Technical Report Series, No. 496. NIH Publication No. 99-3955, US Department of Health and Human Services, National Institutes of Health, Research Triangle Park, 1999.

[24] C. Vardon, C. McLaughlin and C. Nardinelli, "Potential Economic Costs of Mycotoxins in the United States," Council for Agricultural Science and Technology Task Force Report No. 139, Council for Agricultural Science and Technology, Ames, 2003.

[25] N. W. Widstrom, "The Aflatoxin Problem with Corn Grain," *Advances in Agronomy*, Vol. 56, 1996, pp. 219-279. doi:10.1016/S0065-2113(08)60183-2

[26] H. K. Abbas and W. T. Shier, "Mycotoxin Contamination of Agricultural Products in the Southern United States and Approaches to Reducing It from Pre-Harvest to Final Food Products," In: M. Appell, M. Kendra and D. Trucksess, Eds., *Mycotoxin Prevention and Control in Agriculture. American Chemical Society Symposium Series*, Oxford University Press, Oxford, 2009, pp. 37-58.

[27] H. K. Abbas, R. M. Zablotowicz, B. W. Horn, N. A. Phillips, B. J. Johnson, X. Jin and C. A. Abel, "Comparison of Major Biocontrol Strains of Non-Aflatoxigenic *Aspergillus flavus* for the Reduction of Aflatoxins and Cyclopiazonic Acid in Maize," *Food Additives and Contaminants*, Vol. 28, 2011, pp. 198-208. doi:10.1080/19440049.2010.544680

[28] J. F. Robens and R. L. Brown, "Aflatoxin/Fumonisin Elimination and Fungal Genomics Workshops," San Antonio, 23-26 October 2002, pp. 393-505.

[29] H. P. van Egmond, R. C. Schothorst and M. A. Jonker, "Regulations Relating to Mycotoxins in Food Perspectives in a Global and European Context," *Annals of Bioanalytical Chemistry*, Vol. 389, No. 1, 2007, pp. 147-157.

[30] US Food and Drug Administration, "Guidance for Industry: Fumonisin Levels in Human Foods and Animal Feeds," Washington DC, 20 September 2005. http://www. cfsan.fda. gov/~dms/fumongu2. html

[31] J. W. Park, E. K. Kim, D. H. Shon and Y. B. Kim, "Natural Co-Occurrence of Aflatoxin B1, Fumonisin B1 and Ochratoxin A in Barley and Corn Foods from Korea," *Food Additives and Contaminants*, Vol. 19, No. 11, 2002, pp. 1073-1080. doi:10.1080/02652030210151840

[32] M. Picco, A. Nesci, G. Barros, L. Cavaglieri and M. Etcheverry, "Aflatoxin B1 and Fumonisin B1 in Mixed Cultures of *Aspergillus flavus* and *Fusarium proliferatum* on Maize," *Natural Toxins*, Vol. 7, No. 6, 1999, pp. 331-336.

[33] L. J. Abendroth, R. W. Elmore, M. J. Boyer and S. K. Marlay, "Corn Growth and Development," Iowa State University of Science and Technology Cooperative Extension Service, Ames, 2011.

[34] P. R. Thomison and D. M. Jordan, "Plant Population Effects on Corn Hybrids Differing in Ear Growth Habit and Prolificacy," *Journal of Production Agriculture*, Vol. 8, No. 3, 1995, pp. 394-400.

[35] K. M. Tubajika, H. J. Mascagni, K. E. Damann and J. S. Russin, "Aflatoxin Production in Corn by *Aspergillus flavus* Relative to Inoculation, Planting Date, and Harvest Moisture in Louisiana," LSU AgCenter Research Report Number 102, Louisiana State University, Baton Rouge, 2000.

[36] G. A. Payne, "Process of Contamination by Aflatoxin-Producing Fungi and Their Impact on Crops," In: K. K.

Sinha and D. Bhatnagar, Eds., *Mycotoxins in Agriculture and Food Safety*, Marcel Dekker, New York, 1998, pp. 279-306.

[37] J. Cahoon, J. Ferguson, D. Edwards and P. Tacker, "A Microcomputer-Based Irrigation Scheduler for the Humid Mid-South Region," *Applied Engineering in Agriculture*, Vol. 6, No. 3, 1990, pp. 289-295.

[38] E. D. Vories, R. Hogan, P. L. Tacker, R. E. Glover and S. W. Lancaster, "Estimating the Impact of Delaying Irrigation for Midsouth Cotton on Clay Soil," *Transactions of the American Society of Agricultural and Biological Engineers*, Vol. 50, 2007, pp. 929-937.

[39] G. L. Windham, W. P. Williams, P. M. Buckley and H. K. Abbas, "Inoculation Techniques Used to Quantify Aflatoxin Resistance in Corn," *Journal of Toxicolog—Toxin Reviews*, Vol. 22, No. 2, 2003, pp. 313-325.

[40] H. K. Abbas, R. M. Zablotowicz and H. A. Bruns, "Modeling the Colonization of Maize by Toxigenic and Non-Toxigenic *Aspergillus flavus* Strains: Implications for Biological Control," *World Mycotoxin Journal*, Vol. 1, No. 3, 2008, pp. 333-340. doi:10.3920/WMJ2008.x036

[41] SAS, "SAS® Proprietary Software Release 9.1," SAS Institute Inc., Cary, 2003.

[42] H. A. Bruns, "Controlling Aflatoxin and Fumonisin in Maize by Crop Management," *Journal of Toxicolog—Toxin Reviews*, Vol. 22, No. 2, 2003, pp. 153-173.

[43] L. A. Rodriguez-del-Bosque, "Impact of Agronomic Factors on Aflatoxin Contamination in Preharvest Field Corn in Northeastern Mexico," *Plant Disease*, Vol. 80, 1996, pp. 988-993. doi:10.1094/PD-80-0988

[44] M. Alvarado-Carrillo, A. Diaz-Franco, E. Delgado-Aguirre and N. Montes-Garcia, "Impact of Agronomic Management on Aflatoxin (*Aspergillus flavus*) Contamination and Charcoal Stalk Rot (*Macrophomina phaseolina*) Incidence," *Tropical and Subtropical Agrosystems*, Vol. 12, No. 3, 2010, pp. 575-582.

[45] M. Blandino, A. Reyneri and F. Vanara, "Strategy for Fumonisin Reduction in Maize Kernel in Italy," *Mycotoxines fusariennes* des Cereals, Arcachon, 11-13 September 2007. http://www.symposcience.org

11

Tillage and Rice-Wheat Cropping Sequence Influences on Some Soil Physical Properties and Wheat Yield under Water Deficit Conditions

Sandeep Kumar[1*], Pradeep K. Sharma[2], Stephen H. Anderson[3], Kapil Saroch[4]

[1]Carbon Management & Sequestration Center, School of Environment & Natural Resources, Ohio State University, Columbus, USA; [2]Department of Soil Science, CSK HPKV, Palampur, India; [3]Department of Soil, Environmental and Atmospheric Sciences, University of Missouri, Columbia, USA; [4]Department of Agronomy, Forage and Grassland Management, CSK HPKV, Palampur, India.
Email: [*]kumar.278@osu.edu

ABSTRACT

Adopting a better tillage system not only improves the soil health and crop productivity but also improves the environment. A field experiment was conducted to investigate the effects of tillage and irrigation management on wheat (*Triticum aestivum* L.) production in a post-rice (*Oryza sativa* L.) management system on silty clay loam soil (acidic Alfisol) for 2003-2006. Four irrigation levels (RF: rainfed; I_1: irrigation at crown root initiation (CRI); I_2: irrigation at CRI + flowering; I_3: irrigation at CRI + tillering + flowering), and two tillage systems (ZT: zero tillage and CT: conventional tillage) were tested. Zero tillage compared to CT, resulted in higher bulk density (1.34 vs 1.23 $Mg \cdot m^{-3}$), lower total porosity (48.7% vs 52.9%), higher penetration resistance (1.51 vs 1.37 MPa), lower saturated hydraulic conductivity (1.60 vs 92.0 $mm \cdot h^{-1}$), lower infiltration rate (9.40 vs 36.6 $mm \cdot h^{-1}$) and higher volumetric available water capacity (7.9% vs 7.5%) in the surface 0.15 m soil layer. Irrigation levels significantly affected crop water use, wheat yield, and water use efficiency (WUE). Average total water use was 461, 491, 534 and 580 mm under RF, I_1, I_2 and I_3 treatments, respectively. Grain and straw yield of wheat were statistically the same under ZT and CT during 2003-2004; the values, averaged over four irrigation levels were 2.10 and 2.38 $Mg \cdot ha^{-1}$ for grain, and 3.46 and 3.67 $Mg \cdot ha^{-1}$ for straw, respectively. Grain yield declined by 22%, 11% and 8% of I_3 (2.32 $Mg \cdot ha^{-1}$) with RF, I_1 and I_2 treatments, respectively, under ZT; and by 13%, 8% and 5% of I_3 (2.61 $Mg \cdot ha^{-1}$) with RF, I_1 and I_2 treatments under CT. Average values of WUE were 4.33 $kg \cdot ha^{-1} \cdot mm^{-1}$ and 2.35 $m^3 \cdot kg^{-1}$ grain for the ZT and CT treatments. Wheat yield increased with increased irrigation levels for all the cropping seasons. Results from this study concluded that ZT system was better compared to the CT system even with lower yields due to lower input costs for this treatment.

Keywords: Conventional Tillage; Soil Physical Properties; Infiltration; Water Retention; Water Use Efficiency; Zero Tillage

1. Introduction

Rice and wheat in sequence are cultivated in two contrasting soil environments. Rice requires soft, puddled and water-saturated soil conditions, while wheat requires well aggregated and well aerated soil with fine tilth. Puddling (wet tillage) is the most common technique of land preparation for rice in South Asian countries. Puddling creates soil conditions ideal for rice cultivation, but unsuitable for upland crops which follow rice [1,2]. After rice harvest, puddled soils, upon drying shrink, become compact and hard, and develop surface cracks of varying sizes and shapes. The draught power requirement for tilling such soils is very high, sometimes beyond the reach of local ploughs and small tractors. Nevertheless,

when tilled, these soils often break into larger clods, having high breaking energy [3]. In spite of spending significant time and energy, it is often difficult to obtain seedbeds with the desired tilth for sowing wheat. Wheat planted in seedbeds with coarse tilth, due mainly to poor seed-soil contact, results in poor seedling emergence and unsatisfactory crop stands. This lowers wheat productivity.

Frequent stirring opens the soil, breaks soil clods and aggregates, and enhances the oxidation of soil organic matter [4,5]. The loose soils especially on sloping landscapes and in high rainfall areas, are excessively prone to soil erosion. Thus, this system enhances land degradation and results in a decline in soil quality.

To achieve satisfactory soil tilth, soils must be tilled at optimum moisture content. The optimum water content

---

*Corresponding author.

range in puddled soils is generally narrow [1], and many times difficult for farmers to observe. Further, puddled soils may take from several weeks to months to dry and reach a moisture content optimum for tillage. This increases the lag time between rice harvest and wheat planting. The delayed sowing of wheat is another cause of low productivity in post-rice management. An estimate suggests each day delay in planting after 15th of November lowers wheat yield by about 0.04 Mg·ha^{-1} [6,7].

Conventional cultivation of wheat involves several repetitions (3 to 7) of ploughing and planking with animal-drawn local ploughs and wooden planks. The idea of repeated tillage is to create a seedbed with fine tilth and create a dust mulch to conserve soil moisture in the seedbed. In high rainfall areas (annual rainfall varying between 1500 - 2500 mm), these soils suffer from severe soil erosion. Hence, a conservation tillage system is required which is less intensive, is better for the environment (reduces carbon emissions), and enhances soil structural stability and helps to conserve soil by reducing erosion risks [8].

Field experiments with zero tillage in wheat at several locations in the Indo-Gangetic plains have shown encouraging results [9-11]. Farmers have found direct drilling of wheat into post-rice systems without tillage feasible and beneficial at several locations. Wheat yields with zero tillage are either equal or even better than those obtained with conventional tillage because of timely planting of wheat, efficient use of fertilizers and weed control. In addition, zero tillage is fuel and energy efficient but also reduces greenhouse gas emissions [12]. Zero tillage systems conserve the land resource and are cost effective and efficient. Moreover, this tillage system also avoids challenges with clod formation.

Benefits of zero-till planting have been reported under irrigated conditions. Whether similar benefits can be obtained under deficit water conditions still remains a question. The issue is more relevant to hilly areas, as in the Himachal Pradesh (HP) state of India, where wheat is principally a rainfed crop. Irrigated areas under wheat in HP, India are less than 18%. Deficit irrigation systems also need to be evaluated relative to performance with zero-till planting. The objectives of the current study were to 1) compare soil physical properties under zero and conventional tillage systems for a rice-wheat cropping system, and 2) compare wheat yields, and wheat water use efficiency (WUE) for zero and conventional tillage systems.

2. Materials and Methods

2.1. Experimental Site and Management

The experiment was conducted at the Experimental Farm of the Department of Soil Science, CSK Himachal Pradesh Krishi Vishvavidyalaya, Palampur, Himachal Pradesh (HP). The experimental site was situated at 32°6'N latitude and 76°3'E longitude at an elevation of about 1290 m above mean sea level. The area lies in the Palam Valley of Kangra district in the foothills of Dhauladhar Range and represents the high rainfall, mid-elevation, wet-temperate zone of Himachal Pradesh in the Northwest Himalayas.

The climate of the study area is wet temperate, characterized by severe winters and mild summers. The average annual rainfall of the study area in the last 10 years was 2058 mm and pan evaporation was 1215 mm. The annual maximum temperature is 22.4°C and the minimum temperature is 13.3°C. Annual mean temperature varies from 8.2°C in January to around 28.0°C during the months of May-June. The soil temperature drops as low as 2°C during the winters and frost incidences are common.

The soils of the region are classified as Gray Brown Podzols, as per the Genetic System of Classification. Taxonomically, these soils fall under the order of Alfisols [13]. These soils owe their origin to the fluvio-glacial parent materials developed from rocks like slate, phyllites, quartzites, schists and gneisses. The soils are acidic (pH 5.2 to 6.3). The experimental soils belonged to subgroup Typic Hapludalfs.

2.2. Treatment Details

A field experiment was conducted in 2003-2006. Soil physical properties were measured for 2003-2004 year, and wheat yield and water use efficiency were also compared for three cropping seasons from 2003 to 2006 for different treatments. Two tillage systems (ZT: zero tillage and CT: conventional tillage) and four irrigation levels (RF: rainfed; I_1: irrigation at CRI; I_2: irrigation at CRI + flowering; I_3: irrigation at CRI + tillering + flowering) were tested. The two tillage systems included 1) zero tillage: wheat sown in lines 0.20 m apart by opening narrow slits with a hand plough in untilled plots (ZT), and 2) conventional tillage: wheat was sown in lines 0.20 m apart with the help of a hand plough in well pulverized plots (CT). The four irrigation management systems included 1) rainfed (RF), 2) irrigation at CRI (CRI), 3) irrigation at CRI and flowering stage (CRI + F), and 4) irrigation at CRI, active tillering and flowering stages (CRI + T + F). It is noted that each irrigation of about 5 cm was applied as surface flooding per treatment. One pre-sowing irrigation was applied to all plots.

The total number of treatment combinations was eight with three replications for 24 total number of plots (9 m^2 areas). The treatment effect was investigated for wheat crop (Surbhi, HPW-89) and the experimental design was a randomized complete block. Land preparation was done with the help of a power tiller (Model CT-85, V.S.T. Tillers Tractors Ltd., Bangalore, India). Different field operations and irrigation scheduling during the experiment are summarized in **Tables 1** and **2**.

Each plot received a uniform application of 120 N kg·ha^{-1}, 60 kg·ha^{-1} P$_2$O$_5$ and 40 kg·ha^{-1} K$_2$O as urea, single super phosphate and muriate of potash during the growing season. For the rainfed treatment (no irrigation applications), all fertilizers were band-placed at the time of sowing in November. For the irrigation treatment at crown root initiation (CRI; I$_1$), 50% of the N and all of the P and K were band-placed at the time of sowing with 50% of N broadcast applied at CRI stage during November and December. For the irrigation treatments at CRI + flowering (I$_2$) and irrigation at CRI + active tillering + flowering (I$_3$), 50% of the N and all of the P and K were band-placed at sowing with 25% of N broadcast applied, each at CRI and flowering stage.

2.3. Soil Physical Properties

Particle size analysis of surface and sub-surface soil samples (0 - 0.15 and 0.15 - 0.30 m) was done using the pipette method [14]. Particle density of the soil was determined by the pycnometer method [15]. The soil textural class was silty clay loam and the particle density of the soil was 2.60 and 2.61 Mg·cm^{-3}, respectively for 0 - 0.15 and 0.15 - 0.30 m soil depths. The soil bulk density (ρ_b) was determined before land preparation and 30-days after sowing of wheat by the core sampler method [16], using metallic cores having 0.138 m length and 0.103 m internal diameter. Undisturbed soil cores were collected from the 0 - 0.60 m depth at 0.15 m depth intervals in all plots. Four soil cores were removed at each depth and the moist mass was recorded. Gravimetric moisture content was determined in a sub-sample of each soil core and was used to determine the dry mass of soil in each core.

The total porosity (f) of the 0 - 0.15 m soil layer was determined at 30-days after seeding from data on particle density and ρ_b, using the following relationship:

$$f = (1 - \rho_b/\rho_s) \times 100 \tag{1}$$

where, f is the total porosity (%), ρ_b the bulk density (Mg·m^{-3}) and ρ_s the particle density (Mg·m^{-3}).

Table 1. Summary of field operations performed during the study for the 2003-2004 cropping season. The same operations were performed for the 2004-2005 and 2005-2006 cropping years within a couple days of the dates for the 2003-2004 season.

Date	Field operation	Remarks
Nov. 6	Pre-plant irrigation	Flood-irrigation method.
Nov. 9	Land preparation in conventionally-tilled plots	Land prepared using a power tiller.
Nov. 11	Planting of wheat	Wheat sown in rows at 0.20 m spacing using hand plough at 100 kg·ha^{-1} seed rate.
Dec. 27	Hand weeding and hoeing	Weeding in all plots. Hoeing only in conventionally-tilled plots.
Jan. 28	Hand weeding and hoeing	Weeding in all the plots. Hoeing only in conventionally-tilled plots.
May 6	Crop harvesting	Wheat harvested in 5.76 m^2 area in centre of each plot.

Table 2. Irrigation timing and amount of water applied to the zero tillage (ZT) and conventional tillage (CT) treatments for the irrigation regime treatments for the 2003-2004 cropping year. The same amounts were applied within a couple of days of the dates for the 2003-2004 cropping years for the 2004-2005 and 2005-2006 cropping years.

Irrigation Regime Treatment	Tillage Treatment	Amount of Irrigation Water Applied (mm)				Total water applied
		Irrigation Application Dates				
		Nov. 6	Dec. 12	Jan. 11	March 3	
RF	ZT	50	-	-	-	50
	CT	50	-	-	-	50
I$_1$	ZT	50	50	-	-	100
	CT	50	50	-	-	100
I$_2$	ZT	50	50	-	50	150
	CT	50	50	-	50	150
I$_3$	ZT	50	50	50	50	200
	CT	50	50	50	50	200

RF = Rainfed; I$_1$ = Irrigation at crown root initiation (CRI); I$_2$ = Irrigation at CRI and flowering stage; I$_3$ = Irrigation at CRI and active tillering and flowering stage.

Undisturbed soil cores were collected from the 3 replicates in metal cores of 0.11 m length and 0.081 m diameter in the 0 - 0.15 m soil depth in both zero-till and conventionally-tilled plots at crop harvest. The saturated hydraulic conductivity (K_{sat}) was determined by the constant head method [17].

2.4. Soil Water Content

Soil water content was determined gravimetrically in the 0 to 0.60 m profile at 0.15 m depth intervals at sowing, one-day before and after each irrigation and at crop harvest. The mass wetness was converted into volume wetness for each soil layer using the ρ_b of each respective soil layer.

2.5. Soil Penetration Resistance

Soil penetrometer resistance (SPR) refers to the resistance offered by the soil to a metal probe (representing a plant root) pushed into soil. The SPR at field moisture content was determined in the 0 - 0.03 and 0.10 - 0.13 m soil depths at tillering stage. A Proctor penetrometer having a 0.18 m long probe with a flat tip of 6 mm diameter was used for SPR determination. About seven observations were made per plot at each depth for computing the average SPR. After recording the SPR value, soil samples from the layer of the same depth thickness were collected with the help of a tube auger for determining gravimetric moisture content.

2.6. Soil Water Retention

Undisturbed soil core samples, 0.03 m long and 0.054 m diameter were collected from each replication in the middle of the 0 - 0.075 m soil layer with metal cores at the flowering stage of wheat. Moisture content at −33, and −1500 kPa matric potential was determined with a pressure plate apparatus (Soil Moisture Equipment Co., Santa Barbara, California, USA). Soil samples were saturated for 24 hours on the porous plate and then equilibrated to the applied pressures.

Plant-available water capacity (PAWC) was determined for each treatment at flowering stage of wheat as follows:

$$PAWC = FC - PWP \qquad (2)$$

where FC is the moisture retained at −33 kPa matric potential, and PWP (permanent wilting point) is the moisture retained at −1500 kPa matric potential.

2.7. Infiltration Measurements

The infiltration behavior of the soil under zero and conventional tillage treatments was studied at the time of wheat harvest using double ring infiltrometers. The infil-

trometers were pushed into the ground to a depth of 0.15 m. Care was taken to avoid formation of cracks at the soil surface while the infiltrometers were driven into the soil. Water was filled almost to the same level in the inner and outer rings with 0.25 and 0.30 m of inner and outer diameter, respectively. The volume of water which infiltrated into the soil as a function of time was measured. The depth of water infiltrated was computed by dividing the volume by the cross-sectional area of the inner infiltrometer. Regular determinations were made at periodical intervals until a steady water flux was reached. The water intake rate (i) and the cumulative intake (I) were plotted on a simple scale as a function of time. The Gree-Ampt model (1911) was used to fit the infiltration data.

The Green-Ampt [18] infiltration equation was modified by Philip [19] for time (t) vs. cumulative infiltration (I), as follows:

$$t = \frac{I}{K_s} - \frac{\left[S^2 \ln\left(1 + \frac{2IK_s}{S^2}\right)\right]}{2K_s^2} \qquad (3)$$

where t (T) is time (h), I (L) is the cumulative infiltration (mm), S (L·T$^{-0.5}$) is the sorptivity (mm·h$^{-0.5}$), and K_s (L·T^{-1}) is the saturated hydraulic conductivity (mm·h^{-1}). For estimating the S and K_s parameters, the method proposed by Clothier et al. [20] was used.

The method to estimate field saturated hydraulic conductivity (K_{fs}) suggested by Reynolds et al. [17] was used for estimating this parameter. It assumes one-dimensional water flow in the infiltration ring, and uses the following equation:

$$K_{fs} = \frac{q_s}{\left(\dfrac{H}{C_1 d + C_2 a}\right) + \left\{\dfrac{1}{\left[\alpha^* \left(C_1 d + C_2 a\right)\right]}\right\} + 1} \qquad (4)$$

where K_{fs} is the field-saturated hydraulic conductivity (mm·h^{-1}), q_s is the quasi-steady infiltration rate (mm·hr^{-1}), a is the radius of the infiltration ring (mm), H is the hydraulic head of ponded water in the ring (mm), d is the depth of ring insertion into the soil (mm), C_1 and C_2 are dimensionless quasi-empirical constants (C_1 = 0.993 and C_2 = 0.578 for this infiltrometer), and α^* is the soil macroscopic capillary length, assumed to be equal to 0.0036 mm^{-1} for the conventional tillage, and 0.012 mm^{-1} for the zero tillage treatment [17].

2.8. Wheat Yield and Water Use Efficiency

The wheat crop was harvested from ground level from the net plot area of 5.76 m^2, centered in the plot, with the help of sickles, and was left in the respective plots for sun-drying for 2 - 3 days. When most of the straw in a handful bundle broke up on folding, total produce was weighed and recorded as biological yield. The produce was then threshed with thresher and grains were separated out. The grains thus collected were weighed and the yield was recorded as $Mg \cdot ha^{-1}$.

The water use efficiency (WUE) was computed as: 1) WUE ($kg \cdot grains \cdot ha^{-1} \cdot mm^{-1}$) = Grain yield ($kg \cdot ha^{-1}$)/Total water use (mm), and 2) WUE ($m^3 \cdot kg^{-1}$) = Total water use (m^3)/Wheat grains (kg).

2.9. Statistical Analysis

The statistical design for the study was a factorial experiment with two levels of tillage and four levels of irrigation arranged in a randomized complete block design with three replicates. Some parameters were only evaluated for tillage plots; these were sampled from the rainfed (RF) irrigation treatment. Statistical differences were declared significant at the $\alpha = 0.05$ level. The statistical analysis was conducted with SAS software [21].

3. Results and Discussion

3.1. Soil Penetration Resistance

Soil penetration resistance (SPR) values, determined immediately before the application of irrigation at the tillering stage of wheat in the 0.10 - 0.13 m soil layer, are shown in **Table 3**. The SPR was significantly affected by tillage, but the effect of irrigation treatments for the CT tillage system on SPR was non-significant. SPR values varied between 1.40 and 1.61 MPa with a mean value of 1.51 MPa under ZT, and between 1.31 and 1.35 MPa with a mean value of 1.34 MPa under CT. The SPR was significantly higher under ZT than CT for all irrigation

levels. The gravimetric soil moisture content was 17.6% - 20.0% under ZT, and 19.1% - 21.8% under CT (**Table 3**).

Soil penetration resistance (SPR) values averaged over four irrigation levels, determined at 0 - 0.03 m and 0.10 - 0.13 m soil depths at crop harvest are shown in **Table 4**. The SPR at field moisture content (9.3% - 12.2%) was higher in the ZT system than the CT system with a magnitude of about 4 times in the 0 - 0.03 m layer, and about 2.5 times in 0.10 - 0.13 m layer. Higher SPR in the ZT plots were found due to the higher soil ρ_b value (1.34 $Mg \cdot m^{-3}$) in ZT plots compared to CT plots (1.23 $Mg \cdot m^{-3}$; **Table 5**).

3.2. Soil Bulk Density, Saturated Hydraulic Conductivity and Porosity

The ρ_b was about 8.9% higher in ZT compared to CT plots (**Table 5**). The soil K_{sat} (0 - 0.15 m depth), determined at wheat harvest, was 57 times higher under CT than ZT (**Table 5**). Infiltration rate and K_{sat} are both functions of pore size distribution. Both of these processes increase with an increase in soil macroporosity. Conventional tillage caused loosening of the surface soil layer thereby increasing the macroporosity and hence increasing the infiltration rate and K_{sat}.

Singh *et al.* [22] also observed an increase in K_{sat} in a post-rice soil after tillage. Higher values of infiltration as well as K_{sat} under CT than ZT were also reported by Barzegar [23]. The situation, however, may be different under continuous zero till systems than in rice-wheat system. A soil, continuously under zero till management, especially when crop residues are left on the soil surface, may show higher infiltration rates and K_{sat} values due to root channels formed in soil and enhanced earthworm

Table 3. Soil penetration resistance (SPR), and soil water content in the subsurface soil layer (0.10 - 0.13 m) at tillering stage of wheat under different tillage and irrigation treatments measured during the 2003-2004 cropping season.

Tillage Treatment	Irrigation Regime Treatment	Gravimetric soil water content (g/g %)	SPR (MPa)
ZT	RF	17.9[de]	1.61[a]
	I_1	17.6[e]	1.55[b]
	I_2	20.0[bc]	1.44[c]
	I_3	19.5[c]	1.40[d]
CT	RF	19.8[bc]	1.35[e]
	I_1	21.8[a]	1.31[f]
	I_2	19.1[dc]	1.33[ef]
	I_3	21.0[ab]	1.35[e]
Analysis of variance $P > F$			
Treatment		<0.01	<0.01

ZT = Zero tillage; CT = Conventional tillage; RF = Rainfed; I_1 = Irrigation at CRI stage; I_2 = Irrigation at CRI and flowering stage; I_3 = Irrigation at CRI, tillering and flowering stage; Note: The SPR values have been averaged over different irrigation treatments because of small differences in soil moisture content.

Table 4. Soil penetration resistance (SPR) and soil water content for two soil depths at crop harvest in the zero tillage (ZT) and conventionally-tilled (CT) treatments measured at crop harvest during the 2003-2004 cropping season.

Tillage	0 - 0.03 m soil depth		0.10 - 0.13 m soil depth	
	Water content	SPR	Water content	SPR
	(g/g %)	(MPa)	(g/g %)	(MPa)
ZT	12.2[a†]	5.16[a]	9.3[a]	8.36[a]
CT	11.7[b]	1.28[b]	11.4[b]	3.39[b]
Treatment	<0.01	<0.01	<0.01	<0.01

[†]Means with different letters are significantly different at the 0.05 probability level.

Table 5. Saturated hydraulic conductivity (K_{sat}), bulk density, and total porosity of the surface soil layer (0 - 0.15 m) under zero tillage (ZT) and conventional tillage (CT) treatments at 30 days after seeding of wheat for the 2003-2004 cropping season.

Tillage	K_{sat} (mm·h^{-1})	Bulk density (Mg·m^{-3})	Total porosity (%)
ZT	1.62[a†]	1.34[a]	49.3[a]
CT	92.0[b]	1.23[b]	53.7[b]
	Analysis of variance $P > F$		
Treatment	<0.01	<0.01	<0.01

[†]Means with different letters are significantly different at the 0.05 probability level.

activity as was observed by Barnes and Ellis [24]. Several other workers reported higher infiltration rate under ZT system due to the formation of continuous soil biopores [25-27]. Loch and Coughlan [28] reported higher deep drainage under ZT than CT due to the presence of continuous macropores under ZT.

3.3. Soil Water Retention

Water retention of the surface 0.075 m soil layer on a mass basis at –33 kPa and –1500 kPa soil water pressure was always higher in CT than in ZT plots. The differences, however, narrowed with the decrease in water potential. Water content on a volume basis was higher in the ZT system than the CT system at –33 and –1500 kPa pressure due to differences in ρ_b (**Table 6**). The plant available water capacity (PAWC) on a volume basis was lower for the CT (7.5%) than ZT (7.9%) treatment (**Table 6**).

Soil water retention (0 - 0.75 m soil layer) at –33 and –1500 kPa water pressures varied with tillage system. These differences could be explained with differences in pore size distribution since the water retention of soils depends primarily on 1) the number and size distribution of soil pores and 2) the specific surface area of soils. Pore size distribution affects water retention mainly at higher water potentials, such as those at saturation and field capacity, where the water retention is a function of soil structure. At lower water potentials, close to the permanent wilting point, the water retention is a function of soil texture, and also depends on the specific surface area of soil particles [29]. Tillage modified the soil structure

thereby affecting water retention at –33 kPa water potential; however, tillage did not affect soil texture, hence differences in water retention between CT and ZT narrowed at –1500 kPa water potential.

The water retention on a volume basis at –33 and –1500 kPa pressure was higher under ZT than CT (**Table 6**). This occurred in part because of the higher ρ_b under ZT than CT. Although differences in PAWC between ZT and CT were not very large, the soil water retention under ZT was slightly better than under CT.

3.4. Infiltration Rate

The steady-state infiltration rate in conventionally-tilled plots (32.6 mm·h^{-1}) was more than 4 times higher than that of zero tilled plots (7.2 mm·h^{-1}). The cumulative infiltration was also higher in CT (665 mm) than in ZT plots (278 mm). The steady-state infiltration in both cases was achieved in about an 11-hour period (**Figure 1**).

The Green-Ampt (GA) model was fitted to measured infiltration data. The GA model appeared to fit the measured infiltration data (**Figure 1**). The K_s parameter was significantly different, whereas, the S parameter was not significantly different between the tillage systems (**Table 7**). The K_s parameter was 47 times higher for the conventional tillage compared to zero tillage treatment (**Table 7**).

The K_{fs} and q_s parameters were significantly different between the CT and ZT systems (**Table 8**). The K_{fs} was about 5.4 times higher for CT compared to ZT (**Table 8**); whereas the q_s was about 4.5 times higher for this system

Table 6. Effect of tillage treatment on soil water retention at selected water potentials (0 - 0.075 m soil layer) for the 2003-2004 cropping season.

Water potential (kPa)	Volume wetness, (m^3/m^3 %)	
	ZT	CT
−33	36.5[a]	33.5[b]
−1500	28.6[a]	26.0[b]
PAWC	7.9	7.5

[†]Means with different letters are significantly different at the 0.05 probability level. The comparisons were made between ZT and CT at respective pressures; Note: average bulk density values of soil cores were 1.34 and 1.23 $Mg \cdot m^{-3}$ in ZT and CT plots, respectively; ZT = Zero tillage; CT = Conventional tillage; PAWC: plant available water capacity.

Table 7. Means of saturated hydraulic conductivity (K_s) and sorptivity (S) fitted parameters estimated with the Green-Ampt model for the zero and conventional tillage treatments under a rice-wheat cropping system along with analysis of variance for the 2003-2004 cropping season.

Tillage	K_s Mean ($mm \cdot h^{-1}$)	S Mean ($mm \cdot h^{-0.5}$)
ZT	0.63[a†]	62.55[a]
CT	29.88[b]	61.28[a]
	Analysis of variance $P > F$	
Treatment	<0.01	0.95

[†]Means with different letters are significantly different at the 0.05 probability level.

Table 8. Means of quasi-steady state infiltration rate (q_s) and field-saturated hydraulic conductivity (K_{fs}) for the zero and conventional tillage systems along with analysis of variance for the 2003-2004 cropping season.

Tillage	q_s Mean ($mm \cdot h^{-1}$)	K_{fs} Mean ($mm \cdot h^{-1}$)
ZT	7.19[a†]	4.49[a]
CT	32.59[b]	24.1[b]
	Analysis of variance $P > F$	
Treatment	<0.01	<0.01

[†]Means with different letters are significantly different at the 0.05 probability level.

compared with the ZT treatment.

3.5. Crop Yield

The wheat yield was observed for three cropping seasons during 2003-2004, 2004-2005 and 2005-2006. The yield was not consistent for all the three cropping seasons. Grain yield was affected significantly by irrigation levels but not by tillage treatments during 2003-2004 and 2005-2006 (**Table 9**).

During 2003-2004, grain yield decreased progressively

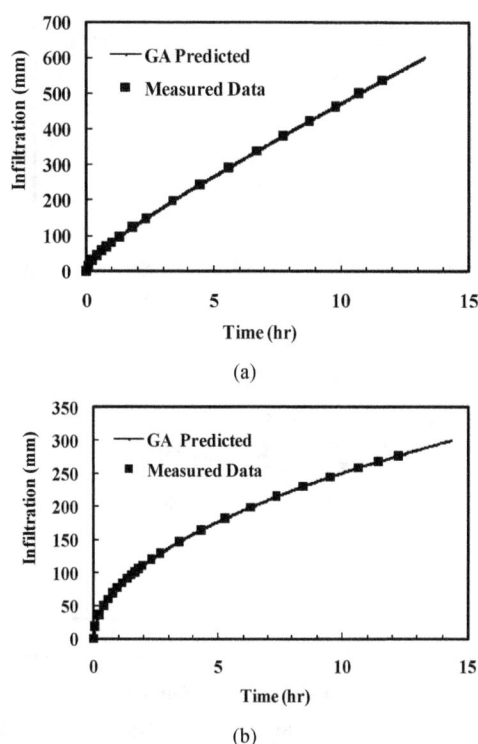

Figure 1. The Green-Ampt (GA) model fitted to measured ponded infiltration data for typical replicates under (a) conventional and (b) zero tillage for a rice-wheat cropping system.

with the reduction in irrigation levels under both ZT and CT systems, but yield differences were significant between RF and I_2 in ZT, and RF and I_3 in CT; I_1, I_2 and I_3 were statistically at par for grain yield under both ZT and CT. Grain yields with RF, I_1 and I_2 were about 80%, 89% and 92% of I_3 under ZT, and about 77%, 92% and 95% of I_3 under CT, respectively (**Table 9**). Thus grain yield declined by about 8%, 11% and 20% under ZT, and 5%, 8% and 23% under CT, respectively, with reduction in irrigation levels from I_3 to I_2, I_1 and RF. According to these data, ZT was relatively more sensitive to moisture stress than CT for RF and I_1 levels. At each irrigation level, CT numerically produced more grain than ZT. Averaged over four irrigation levels, grain yield with CT was about 13% higher than ZT. The grain yield showed a significant linear relationship with total water use with r^2 ranging from 0.36 to 0.69 for CT and ZT, respectively (**Figure 2**).

Tillage system had a significant effect on wheat yield; however, results were different between the 2003-2004 and 2004-2005 seasons. During 2004-2005, average yield across four irrigation levels was higher with ZT (2.56 $Mg \cdot ha^{-1}$) by about 22 percent relative to CT (2.10 $Mg \cdot ha^{-1}$). This difference was attributed to good precipitation distribution during the 2004-2005 cropping year;

(a)

(b)

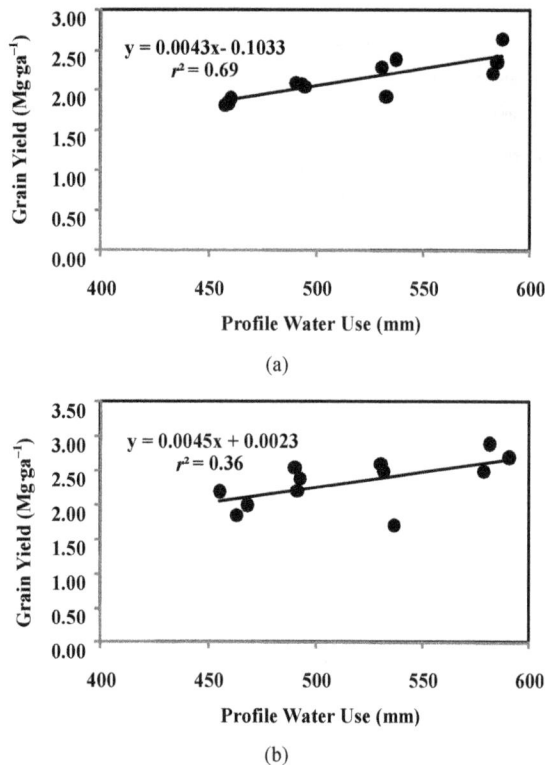

Figure 2. Relationship between wheat yield and total water use for zero tillage (a) and conventional tillage (b) for the 2003-2004 cropping season.

no irrigation (except, pre-sown irrigation) was needed during the entire cropping season. During the 2005-2006 cropping year, wheat grain yield increased significantly over the RF treatment by 18, 46 and 52 per cent with the I_1, I_2 and I_3 irrigation treatments, respectively (**Figure 3**). According to these data (2005-2006), wheat yield under both tillage systems (zero and conventional) responded similarly to deficit irrigation.

Higher SPR in ZT plots probably resulted in higher root resistance and less root growth and less wheat yield during the 2003-2004 cropping year. In contrast, the lower ρ_b, higher K_{sat} and infiltration rate in CT plots probably resulted in less runoff and better plant growth and wheat yield compared to ZT plots.

Yield data for tillage treatments showed that in 2003-2004, CT performed better; for 2004-2005, ZT produced higher yield; and for 2005-2006, no differences occurred in the wheat yield between tillage treatments. Compared to irrigation management, grain yield increased with increased number of irrigations; except for 2004-2005 since during this year rainfall distribution was enough to meet the irrigation requirement for the wheat crop. Results from this study conclude that irrespective of irrigation levels; there were not large differences in yield between the ZT and CT systems; however, ZT was more

economical compared to CT system because of low input cost. When compared to rainfed treatments, ZT performed better compared to CT, hence farmers can make more profit by relying on the ZT treatment rather than on the CT system.

3.6. Water Use Efficiency

Tillage treatments as well as irrigation levels showed a significant effect on water use efficiency (WUE) during the 2003-2004 cropping year (**Table 9**). Numerically, the highest WUE of 4.93 kg·ha^{-1}·mm^{-1} was found under CT with I_1 treatment and the lowest of 3.99 kg·ha^{-1}·mm^{-1} under ZT with I_3. Conversely, the amount of water used (m^3) to produce 1 kg of wheat grain varied between 2.03 m^3·kg^{-1} in I_1 (CT) and 2.51 m^3·kg^{-1} in I_3 (ZT) treatments. Similar to grain yield data, the WUE for 2003-2004 and 2004-2005 cropping seasons were different. During 2004-2005, the highest WUE (3.47 kg·ha^{-1}·mm^{-1}) was obtained with the I_1 and ZT treatment, and the lowest (2.48 kg·ha^{-1}·mm^{-1}) with the RF and CT treatment (**Table 10**). During 2005-2006, WUE was statistically the same with zero and conventional tillage (data not shown). The WUE increased progressively with the level of irrigation from rainfed through the three irrigation treatments (**Figure 3**).

Averaged over two tillage and four irrigation levels, WUE values were 4.33 kg·grain·ha^{-1}·mm^{-1} and 2.32 m^3·kg^{-1} grain. The interaction of tillage × irrigation was also significant (P < 0.01) for WUE for the 2003-2004 cropping season (**Table 9**). It was concluded that more water was needed for less grain production with ZT compared to CT plots.

4. Conclusions

The irrigation treatments did not affect soil physical properties but tillage systems did affect these properties. The ρ_b and SPR values of CT plots were 8.2% and 13%, respectively, lower compared to ZT plots which increased the porosity (8.9%), K_{sat} (57 times) and the steady infiltration rate (4.5 times) under CT plots.

The rice crop management, rice crop was grown previous to the wheat crop, created adverse soil conditions which partially caused the lower values of infiltra- tion rate and K_{sat} in wheat plots which followed rice. When comparing tillage systems, the soil was loosened with a plough for CT which decreased the ρ_b, increased soil porosity and also increased K_{sat} as well as the steady infiltration rate compared to zero-tilled wheat at the time of crop harvest. The improved soil properties under CT systems improved the wheat yield; however the yield differences were not significant between tillage treatments during the 2003-2004 cropping year. The results were

(a)

(b)

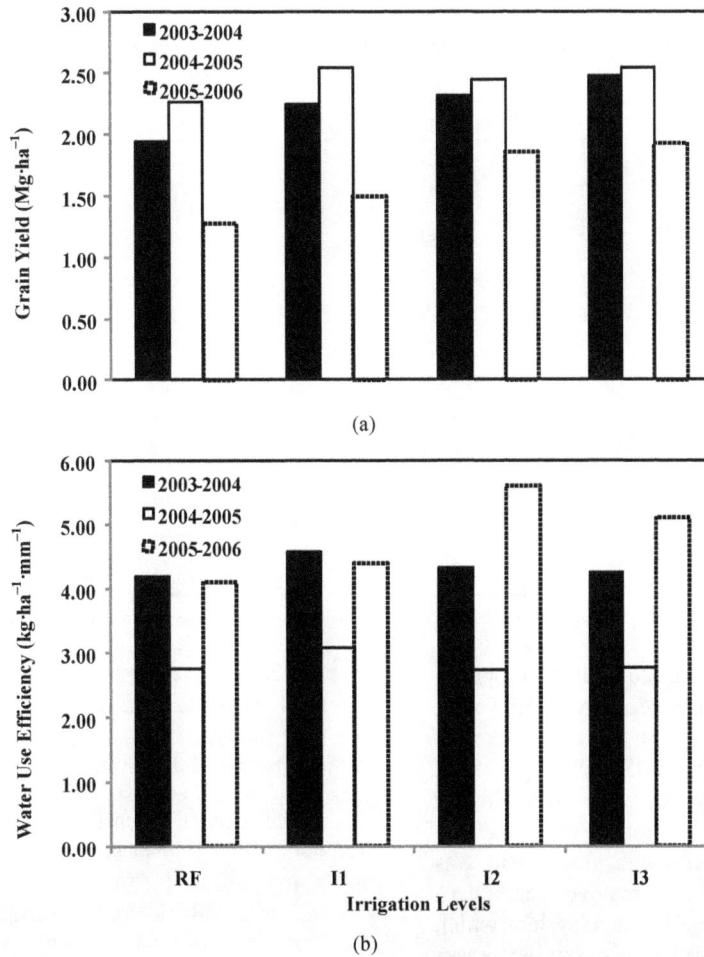

Figure 3. Wheat yield (a) and water use efficiency (b) for RF (rainfed), I_1 (irrigation at CRI), I_2 (irrigation at CRI and flowering), and I_3 (irrigation at tillering, CRI and flowering stages) irrigation levels during the 2003-2004, 2004-2005 and 2005-2006 cropping seasons.

Table 9. Effect of different irrigation regime treatments on total water use, grain yield, and water use efficiency (WUE) for wheat under zero tillage (ZT) and conventional tillage (CT) treatments for the 2003-2004 cropping season.

Irrigation regime Treatment	Tillage Treatment	Total water use (mm)	Grain yield (Mg·ha⁻¹)	WUE (kg·grain·ha⁻¹·mm⁻¹)
RF	ZT	458.9[d]	1.85[c]	4.03[f]
	CT	463.2[d]	2.02[bc]	4.36[d]
I_1	ZT	491.9[c]	2.07[bc]	4.21[e]
	CT	489.1[c]	2.41[ab]	4.93[a]
I_2	ZT	534.8[b]	2.14[ab]	4.00[gf]
	CT	533.5[b]	2.47[abc]	4.62[b]
I_3	ZT	581.1[a]	2.32[ab]	3.99[g]
	CT	578.7[a]	2.61[a]	4.51[c]
Analysis of variance $P > F$				
Tillage		0.95	0.06	<0.01
Irrigation		<0.01	0.01	<0.01
Tillage × Irrigation		0.69	0.81	<0.01

RF = Rainfed; I_1 = Irrigation at CRI stage; I_2 = Irrigation at CRI and flowering stage; I_3 = Irrigation at CRI, tillering and flowering stage; WUE = Water use efficiency.

Table 10. Effect of different irrigation regime treatments on total water use, grain yield, and water use efficiency (WUE) for wheat under zero tillage (ZT) and conventional tillage (CT) treatments during the 2004-2005 cropping season.

Irrigation regime Treatment	Tillage Treatment	Total water use* (mm)	Grain yield (Mg·ha^{-1})	WUE (kg·grain·ha^{-1}·mm^{-1})
RF	ZT	820	2.48^b	3.02^e
	CT	820	2.03^f	2.48^c
I$_1$	ZT	820	2.85^a	3.47^d
	CT	820	2.22^d	2.71^a
I$_2$	ZT	820	2.41^c	2.93^f
	CT	820	2.08^e	2.54^b
I$_3$	ZT	820	2.48^b	3.02^e
	CT	820	2.06^{ef}	2.51^c
Analysis of variance $P > F$				
Tillage		-	<0.01	<0.01
Irrigation		-	<0.01	<0.01
Tillage × Irrigation		-	<0.01	<0.01

*No irrigation other than pre-sown was given to wheat crop during 2004-2005 season as enough rainfall received during this season.

different for the 2004-2005 and 2005-2006 cropping seasons. Irrespective of the tillage system, however, grain yield and WUE increased with increased level of irrigation (except for the 2004-2005 cropping season which received frequent rains) and values were higher with the three irrigation treatments for the cropping seasons which had deficit rain. Although grain yield was inconsistent between the tillage systems over years, economic costs were lower for the ZT tillage system which implies it may be the best system for occasional increases in yield and consistently lower input costs. Furthermore, this ZT system conserves soil moisture and reduces soil erosion as residues are left on the plots.

REFERENCES

[1] P. K. Sharma and S. K. De Datta, "Physical Properties and Processes of Puddled Rice Soils," *Advanced Soil Science*, Vol. 5, 1986, pp. 139-178. doi:10.1007/978-1-4613-8660-5_3

[2] P. K. Sharma, J. K. Ladha and L. Bhushan, "Soil Physical Effects of Puddling in Rice-Wheat Cropping Systems. Improving the Productivity and Sustainability of Rice-Wheat Systems: Issues and Impacts," ASA Special publication, Vol. 65, 2003.

[3] P. K. Sharma and R. M. Bhagat, "A Simple Apparatus for Measuring Clod Breaking Strength," *Journal of the Indian Society of Soil Science*, Vol. 41, 1993, pp. 422-425.

[4] R. Lal, "Soils and Sustainable Agriculture: A Review," *Agronomy for Sustainable Development*, Vol. 28, No. 1, 2008, pp. 57-64.

[5] J. He, N. J. Kuhn, X. M. Zhang, X. R. Zhang and H. W. Li, "Effects of 10 Years of Conservation Tillage on Soil Properties and Productivity in the Farming—Pastoral Ecotone of Inner Mongolia, China," *Soil Use Manage*,

Vol. 25, No. 2, 2009, pp. 201-209. doi:10.1111/j.1475-2743.2009.00210.x

[6] P. R. Hobbs, L. W. Harrington, C. Adhkari, G. S. Giri, S. R. Upadhya and B. Adhikari, "Wheat and Rice in Nepal Tarai: Farm Resources and Production Practices in Rupandhi District," Nepal Agricultural Research Council and CIMMYT, Mexico, 1996.

[7] A. P. Regmi, J. K. Ladha, Pathak, H. Pasuquin, C. Bueno, D. Dawe, P. R. Hobbs, D. Joshy, S. L. Maskey and S. P. Pandey, "Yield and Soil Fertility Trends in a 20-Year Rice-Rice-Wheat Experiment in Nepal," *Soil Science Society of America Journal*, Vol. 66, 2002, pp. 857-867. doi:10.2136/sssaj2002.0857

[8] V. H. D. Zuazo and C. R. R. Pleguezuelo, "Soil-Erosion and Runoff Prevention by Plant Covers, A Review," *Agronomy for Sustainable Development*, Vol. 28, No. 1, 2008, pp. 65-86.

[9] M. Aslam, A. Majid, N. I. Hashmi and P. R. Hobbs, "Improving Wheat Yield in the Rice-Wheat Cropping System of the Punjab through Zero Tillage," *Pakistan Journal of Agricultural Research*, Vol. 14, No. 1, 1993, pp. 8-11.

[10] B. T. Barnes and F. B. Ellis, "Effects of Different Methods of Cultivation and Direct Drilling, and Dispersal of Straw Residues on Populations of Earthworms," *Journal of Soil Science*, Vol. 30, 1979, pp. 669-679. doi:10.1111/j.1365-2389.1979.tb01016.x

[11] R. S. Mehla, J. K. Verma, R. K. Gupta and P. R. Hobbs, "Stagnation in Productivity of Wheat in Indo-Gangetic Plains: Zero-Till-Seed-Cum-Fertilizer Drill as an Integrated Solution," Rice-Wheat Consortium Paper Series 8, Rice-Wheat Consortium for the Indo-Gangetic Plains, New Delhi, 2000.

[12] P. Smith, D. S. Powlson, M. J. Glendining and J. U. Smith, "Preliminary Estimates of the Potential for Carbon Mitigating in European Soils through No-Till Farming," *Global Change Biology*, Vol. 4, No. 6, 1998, pp. 679-685.

[13] S. D. Verma, "Characterization and Genesis of Soils of Himachal Pradesh," Ph.D. Thesis, Department of Soil Science, Himachal Pradesh Krishi Vishvavidyalaya, Palampur, H.P., India, 1979.

[14] G. W. Gee and D. Or, "Particle-Size Analysis," In: J. H. Dane and G. C. Topp, Eds., *Methods of Soil Analysis*, Part 4, Physical Methods, SSSA Book Series 5, SSSA, Madison, 2002, pp. 255-294.

[15] A. Flint and L. E. Flint, "Particle Density," In: J. H. Dane and G. C. Topp, Eds., *Laboratory Methods of Soil Analysis*, Part 4: Physical Methods, SSSA, Madison, 2002, pp. 229-240.

[16] R. A. Singh, "Soil Physical Analysis," Kalyani Publishers, New Delhi, 1980, p. 165.

[17] W. D. Reynolds, D. E. Elrick, E. G. Youngs and A. Amoozegar, "Field Methods (Vadose and Saturated Zone Techniques)," In: J. H. Dane and G. C. Topp, Eds., *Methods of Soil Analysis*, Part 4, Soil Science Society of America Book Series No. 5, Soil Science Society of America Madison, 2002, pp. 817-826.

[18] W. H. Green and G. A. Ampt, "Studies on Soil Physics. Part I. The Flow of Air and Water through Soils," *Journal of Agricultural Research*, Vol. 4, No. 1, 1911, pp. 1-24.

[19] J. R. Philip, "The Theory of Infiltration: 4. Sorptivity and Algebraic Infiltration Equations," *Soil Science*, Vol. 84, No. 3, 1957, pp. 257-264. doi:10.1097/00010694-195709000-00010

[20] B. Clothier, D. Scotter and J. P. Vandervaere, "Unsaturated Water Transmission Parameters Obtained from Infiltration," In: J. H. Dane and G. C. Topp, Eds., Part 4, *Methods of Soil Analysis*, Soil Science Society of America, Madison, 2002, pp. 879-898.

[21] SAS Institute, "SAS User's Guide," Statistics, SAS Inst., Cary, 1999.

[22] Y. Singh, A. K. Bhardwaj, P. K. Singh, A. Saxena, V. Singh, S. P. Singh and A. Kumar, "Effect of Rice Estab-lishment Methods, Tillage Practices in Wheat and Fertilization on Soil Physical Properties and Rice-Wheat System Productivity on a Silty Clay Mollisol of Utranchal," *Indian Journal of Agricultural Science*, Vol. 72, No. 4, 2002, pp. 200-205.

[23] A. R. Barzegar, M. A. Asoodar, A. R. Eftekhar and S. J. Herbert, "Tillage Effects on Soil Physical Properties and Performance of Irrigated Wheat and Clover in Semi Arid Region," *Journal of Agronomy*, Vol. 3, No. 4, 2004, pp. 237-242. doi:10.3923/ja.2004.237.242

[24] B. T. Barnes and F. B. Ellis, "Effects of Different Methods of Cultivation and Direct Drilling, and Disposal of Straw Residues on Populations of Earthworms," *Journal of Soil Science*, Vol. 30, 1979, pp. 669-679.

[25] W. Ehlers, "Observations on Earthworm Channels and Infiltration on Tilled and Untilled Loess Soil," *Soil Science*, Vol. 119, 1975, pp. 242-249. doi:10.1097/00010694-197503000-00010

[26] I. J. Packer, G. J. Hamilton and I. White, "Tillage Practices to Conserve Soil and Improve Soil Conditions," *Soil Conservation Journal NSW (Aust)*, Vol. 40, 1984, pp. 78-87.

[27] P. W. Unger and P. K. Cassel, "Tillage Implement Disturbance Effects on Soil Properties Related to Soil and Water Conservation: A Literature Review," *Soil Tillage Research*, Vol. 19, 1991, pp. 363-382. doi:10.1016/0167-1987(91)90113-C

[28] R. J. Loch and K. J. Coughlan, "Effects of Zero Till and Stubble Retention or Some Properties of Cracking Clay," *Australian Journal of Soil Research*, Vol. 22, 1984, pp. 91-98. doi:10.1071/SR9840091

[29] P. K. Sharma and S. K. De Datta, "Rainwater Utilization Efficiency in Rainfed Lowland Rice," *Advance in Agronomy*, Vol. 52, 1994, p. 101. doi:10.1016/S0065-2113(08)60622-7

12

Study of Heavy Metal Accumulation in Sewage Irrigated Vegetables in Different Regions of Agra District, India

Preeti Parashar[*]**, Fazal Masih Prasad**

School of Chemical Sciences, Department of Chemistry, St. John's College, Agra, India.
Email: [*]preetiavn@gmail.com

ABSTRACT

Heavy metal contamination of soil resulting from sewage irrigation is a cause of serious concern due to the potential health impacts of consuming contaminated products. In this study an assessment made of the impact of sewage irrigation on heavy metal contamination of Spinach, Cabbage, Beetroot, Reddish, Okra, Tomato, and Cucumber is widely cultivated and consumed in urban India, particularly by the poor. A field study was conducted at seven major sites that were irrigated by either treated, (Dhandupura) or untreated wastewater in the suburban areas of Agra, India. Samples of irrigation water, soil, and the edible portion of all the vegetables were collected monthly during the winter seasons and were analyzed for Fe, Cd, Cu, Zn, and Pb. Heavy metals in irrigation water were below the internationally recommended (WHO) maximum permissible limits set for agricultural use for all heavy metals except Cd at all the sites. Similarly, the mean heavy metal concentrations in soil were below the Indian standards for all heavy metals, but the maximum value of Cd recorded during January was higher than the standard. However, in the edible portion of spinach, the Cd concentration was higher than the permissible limits of the Indian standard during summer, whereas Pb concentrations were higher in winter seasons. Results of correlation analysis were computed to assess the relationship between individual heavy metal concentration in the vegetable samples. The study concludes that the use of treated and untreated wastewater for irrigation has increased the contamination of Cd, Pb in edible portion of vegetables causing potential health risk in the long term from this practice. The study also points to the fact that adherence to standards for heavy metal contamination of soil and irrigation water does not ensure safe food. Fe was measured abundant in soil whereas Pb and Cd were found more in untreated sites as compared to treated site. Correlation, paired T-test and ANOVA were also carried out for pre post harvested soil and vegetables.

Keywords: Heavy Metals; AAS; Sewage; Pearson Correlation; Paired T-Test; ANOVA

1. Introduction

Agra is situated in the extreme southwest corner of Uttar Pradesh, India. It stretches across 26"44'N to 27"25'N and 77"26'E to 78"32'E. It is situated at the bank of Yamuna. It has limited forest area having sporting trees of Babul, Ber, Neem, Peepal. Agra suffers from extremities of climate with scorching hot summers and chilly winters. Like most of the cities of North India the weather and climate of Agra is extreme and tropical. In Agra and surrounding areas summers are extremely hot and the maximum temperature goes to 47°C while during winters it remains cold and foggy. During monsoon it becomes hot and humid. The soil, water and vegetable samples were collected during winter season in the month of December. The vegetable crops were irrigated in sewage water.

Wastewater irrigation is known to contribute significantly to the heavy metal contents of soils [1]. Plant species have a variety of capacities in removing and accumulating heavy metals, so there are reports indicating that some species may accumulate specific heavy metals, causing a serious health risk to human health when plants based food stuff are consumed [2]. Disposal of sewage water and industrial wastes is a great problem. Often it is drained to the agricultural lands where it is used for growing crops including vegetables. These sewage effluents are considered not only a rich source of organic matter and other nutrients but also they elevate the level of heavy metals like Fe, Mn, Cu, Zn, Pb, Cr, Ni, Cd and Co in receiving soils [3].

Heavy metals are one of a range of important types of contaminants that can be found on the surface and in the tissue of fresh vegetables. Prolonged human consumption of unsafe concentrations of heavy metals in foodstuffs may lead to the disruption of numerous biological and biochemical processes in the human body. Heavy metal

---

[*]Corresponding author.

accumulation gives rise to toxic concentrations in the body, while some elements (e.g. arsenic, cadmium, chromium) act as carcinogens and others (e.g. mercury and lead) are associated with developmental abnormalities in children.

Vegetables are an important part of human's diet. In addition to a potential source of important nutrients, vegetables constitute important functional food components by contributing protein, vitamins, iron and calcium which have marked health effects [4]. Vegetables, especially those of leafy vegetables grown in heavy metals contaminated soils, accumulate higher amounts of metals than those grown in uncontaminated soils because of the fact that they absorb these metals through their leaves [5]. [6] investigated the concentrations of heavy metals such as Cd, lead (Pb), Zinc (Zn), Cu and Ni in different vegetables, grown in various parts of Turkey. The levels of heavy metals (lead, cadmium, copper and zinc) were examined in selected fruits and vegetables sold in the local markets of Egypt [7,8] studied the contents of heavy metals in vegetables grown in an industrial area of North Greece.

According to [9], Zn deficiency in plants increased iron concentration. [10] have reported generally much more pronounced toxic effect of cadmium on plant growth parameters under iron deficiency conditions than under normal iron supply.

Plant studies have shown that although zinc is an essential element for higher plants, it is considered phytotoxic in elevated concentrations, directly affecting crop yields and soil fertility. Soil concentrations range from 70 - 400 mg/Kg. Total zinc is classified as critical, above which toxicity is considered likely [11]. Zinc is an essential element needed by our body in small amounts. We are exposed to zinc compounds in food. The average daily zinc intake through the diet ranges from 5.2 to 16.2 milligrams (milligram = 0.001 gm). Food may contain levels of zinc ranging from approximately 2 parts of zinc per million (2 ppm) parts food (e.g. leafy vegetables) to 29 ppm (meats, fish, poultry).

Cadmium and its compounds may travel through soil, but its mobility depends on several factors such as pH and amount of organic matter which will vary depending on the local environment. Generally cadmium binds strongly to organic matter where it will be immobile in soil and be taken up by plant life, eventually entering the food supply. Leafy vegetables like cauliflower, cabbage, spinach, etc., grow quite well in the presence of sewage water [12] whereas vegetables such as radish are sensitive to sewage water [13]. Vegetables grown by the use of sewage water contain many heavy metals, which cause serious health hazards to the community and animals as well [14,15]. This concern is of special importance, where un-treated sewage is applied for longer periods to grow vegetables in urban lands. Heavy metal bioaccumulation in the food chain can be especially highly dangerous to human health. These metals enter the human body mainly through two routes namely: inhalation and ingestion, and with ingestion being the main route of exposure to these elements in human population.

The aim of research is to assess the effect of heavy metal accumulation in sewage irrigated soil and vegetable crop.

2. Method and Material

The analysis of various parameters for the soil samples collected from the varying plots of different sites e.g. (Balkeshwar, Bichpuri, Dhandupura, Etmadpur, Gopalpura, Nunhai, Sikandra) of Agra region was conducted during the winter season before and after harvesting the various vegetable crops. The soil in Agra region is basically alluvial. The climate is neither too wet nor too dry. For the detection of heavy metals in sewage water AAS technique was used.

2.1. Sample Preparation

Reagents used:
1) Air: Cleaned and dried through a filter air;
2) Acetylene: Standard, commercial graded;
3) Metal free water: All the reagents and dilutions were made in metal free water;
4) Methyl isobutyl ketone (MIBK): Reagent grade MIBK is purified by redistillation before use;
5) Ammonium pyrrolidine dithiocarbonate (APDC) solution: 4 gm APDC is dissolved in 100 ml water;
6) Conc. HNO_3;
7) Standard metal solutions: Five standard solutions of 0.01, 0.1, 1, 10 and 100 mg/l concentrations of metals for instrument calibration and sorption study are prepared by diluting their stock solution of 1 gm/l i.e. 1 ml = 1 mg metal.

2.1.1. Water Samples

The sewage water samples were collected from different sites. The samples were collected in 1-litre precleaned (with 50% HNO_3 and than thrice with deionized water) plastic bottles and acidified with 5 ml conc. HNO_3 per litre of wastewater for the analysis of heavy metals.

2.1.2. Soil Samples

About 0.5 Kg composite samples from 10 - 20 sub samples were taken in plastic bags of 1 Kg capacity by quartering technique taken in zig zag along different sections of the area until the whole area was covered. A representative composite soil sample from agricultural is made up of 10 - 20 sub samples from a uniform field. Sub samples are taken with a soil auger after cleaning ground tightly with wooden pestle in a wooden mortar. The composite sample is spread to be air dried and extraneous material such as leaves, twigs, rocks etc. were removed. The whole of composite sample is spread in a uniform way and go on quartering and taking diagonal quarters rejecting the other too till approximately 500 gm soil sample is obtained. The sample is filtered through a 2 mm plastic sieve. About 0.5 Kg filtered soil sample was stored in a clean polythene bag with proper labelling for analysis.

2.1.3. Vegetable Samples

Plant samples were collected in paper bags. Each sample was given a particular identification number. The collected plant samples were washed with distilled water, wiped and dried in oven at 60°C and ground in a micro grinding mill and stored.

The heavy metal analysis is done to detect heavy metals in water, soil and vegetable samples.

2.2 Statistical Analysis

Descriptive statistical analysis including 2-way ANOVA, pearson correlation, significance (0.01 and 0.05) was done by SPSS software.

3. Results and Discussion

Sewage contained on an average 1.20 mg·Kg^{-1} Cd, over 8 times the permissible level by the EU standards (0.2 mg·Kg^{-1}); Cu concentrations were 29.07 mg·Kg^{-1}, which is little higher than EU Standard (20 mg·Kg^{-1}); concentrations of Pb were 6.77 mg·Kg^{-1}, over 22 times the permissible levels allowed by both EU standards and UK guidelines (0.3 mg·Kg^{-1}); Zn concentrations were 221 mg·Kg^{-1}, over 4 times the guideline value (50 mg·Kg^{-1}). All the plants contained concentrations of heavy metals above the permissible levels. Furthermore the concentrations observed in this study were higher than those reported by other workers who have examined vegetation from other contaminated sites. The plants grown on the soil polluted with sewage-effluents were found to record higher uptake of heavy metals when compared to plants grown on normal soils [16]. The Fe concentration in sewage is high as soil in Agra is alluvial and has capacity to absorb iron. Cd and Pb content is also high in sewage. Zn, Cu and Cd were highly significant (p < 0.01) (**Table 1**). The treated site (Dhandupura) has comparatively lower amount of Cd as compared to other sites. In pre and post harvested soil Fe, Zn, Pb, Cu and Cd are highly significant (p < 0.01) (**Tables 2** and **3**). Cu is negatively correlated to Cd concentration in all the vegetables and vice versa (**Table 4**).

The Fe content in sewage, pre and post harvested soil post < pre < sewage, Zn content post < pre < sewage, Pb content sewage < post < pre, Cu content sewage < pre < post, Cd content sewage < post < pre.

The Fe content in vegetables tomato < okra < brinjal

Table 1. Showing ANOVA for comparison of heavy metals of different sites of sewage.

	Between sites	0.915	6	0.152	2.104	0.078	*
Fe	Error	2.537	35	0.072			
	Total	3.451	41				
	Between sites	2347.682	6	391.280	15.640	0.000	**
Zn	Error	875.643	35	25.018			
	Total	3223.326	41				
	Between sites	355.420	6	59.237	4.178	0.003	**
Pb	Error	496.231	35	14.178			
	Total	851.651	41				
	Between sites	924.134	6	154.022	9.106	0.000	**
Cu	Error	591.994	35	16.914			
	Total	1516.128	41				
	Between sites	1.099	6	0.183	1.576	0.183	NS
Cd	Within error	4.067	35	0.116			
	Total	5.165	41				

Table 2. Showing ANOVA for comparison of heavy metals of different site in pre harvested soil.

Fe	Between sites	2.381	6	0.397	5.016	0.001	**
	Error	2.769	35	0.079			
	Total	5.150	41				
Zn	Between sites	1665.300	6	277.550	7.471	0.000	**
	Error	1300.320	35	37.152			
	Total	2965.621	41				
Pb	Between sites	2010.395	6	335.066	20.570	0.000	**
	Error	570.124	35	16.289			
	Total	2580.519	41				
Cu	Between sites	3279.855	6	546.642	26.801	0.000	**
	Error	713.883	35	20.397			
	Total	3993.737	41				
Cd	Between sites	7.305	6	1.217	5.185	0.001	**
	Error	8.219	35	0.235			
	Total	15.524	41				

Table 3. Showing ANOVA for comparison of heavy metals of different site in post harvested soil.

Fe	Between groups	1.949	6	0.325	4.674	0.001**
	Within groups	2.432	35	0.069		
	Total	4.381	41			
Zn	Between groups	860.047	6	143.341	7.283	0.000**
	Within groups	688.826	35	19.681		
	Total	1548.873	41			
Pb	Between groups	1336.214	6	222.702	11.949	0.000**
	Within groups	652.336	35	18.638		
	Total	1988.550	41			
Cu	Between groups	4572.126	6	762.021	69.481	0.000**
	Within groups	383.858	35	10.967		
	Total	4955.983	41			
Cd	Between groups	5.851	6	0.975	5.695	0.000**
	Within groups	5.993	35	0.171		
	Total	11.843	41			

and cucumber < reddish < spinach and beetroot, Zn content as brinjal and copper < okra < tomato < reddish < spinach < beetroot < Pb content as tomato < okra < brinjal and cucumber < reddish < beetroot < spinach. Cu as okra < tomato < cucumber < brinjal < reddish < spinach < beetroot Cd as okra < brinjal and cucumber < tomato < reddish < beetroot < spinach. The mean values and range of heavy metals in sewage pre and post harvested soil and vegetables are shown in **Tables 5-7**. In sewage Zn, Pb and Cu were highly significant W.R.T different sites. Shown in **Table 1**. All the heavy metals in pre and post harvested soil were highly significant W.R.T different sites. Shown in **Tables 2** and **3**. The correlation between vegetables and sites was calculated through pearson correlation. Fe is highly significant with Zn, Cu and Cd of different sites. Cu is negatively correlated with Cd (r = −0.009) shown in **Table 4**.

The conc. of heavy metals in sewage and soil is shown in **Figure 1**. It is concluded from the figure that Zn concentration is highest in sewage. It is also clear that pre harvested soil accumulates more heavy metals as compared to post harvested soil. Cd is accumulated in traces in sewage as well as soil. (**Figure 1**). Spinach, reddish and beetroot are more prone to absorb more heavy metals from soil (**Figure 2**).

The heavy metal content in Akaki water (Ni—12.1, Cu—38.4, Pb—35.5 and Cd—2.5 $\mu g \cdot L^{-1}$) [17] was higher than the natural elemental levels in freshwater [18].

4. Conclusion and Recommendation

From the study it is revealed that, untreated sewage and industrial effluents are the main source of pollution to soil and irrigation with contaminated sewage water containing variable amounts of heavy metals leads to increase in concentration of metals in soil and vegetables, which is grown using the polluted water. Concentration of metals in vegetables will provide baseline data and there is a need for intensive sampling for quantification of results throughout the country. Since cucumber is the least accumulator of metals and metalloids, it may be less risky to eat cucumber than eating beetroot and spinach, from health standpoint. To avoid entrance of metals into the food chain, municipal or industrial waste should not

Table 4. Correlation between different components of vegetable in different sites.

Component	(r)	Fe	Zn	Pb	Cu	Cd
Fe	Pearson corr.	1	0.414**	0.207	0.396**	0.308*
	Sig.		0.003	0.154	0.005	0.032
Zn	Pearson corr.	0.414**	1	0.192	0.092	0.363*
	Sig.	0.003		0.187	0.531	0.010
Pb	Pearson corr.	0.207	0.192	1	0.078	0.368**
	Sig.	0.154	0.187		0.597	0.009
Cu	Pearson corr.	0.396**	0.092	0.078	1	−0.009
	Sig.	0.005	0.531	0.597		0.954
Cd	Pearson corr.	0.308*	0.363*	0.368**	−0.009	1

Table 5. Showing heavy metal conc. ($mg \cdot Kg^{-1}$) in sewage water of different sites.

Sewage water (Heavy metals)	1	2	3	4	5	6	7	Overall average
Iron (%)	0.98 - 1.9 (1.21)	0.97 - 1.17 (1.05)	0.87 - 1.43 (1.11)	0.93 - 1.12 (1.03)	1.23 - 1.76 (1.43)	1.09 - 1.89 (1.4)	0.9 - 1.99 (1.2)	0.87 - 1.99 (1.20)
Zinc ($mg \cdot Kg^{-1}$)	45.34 - 60.11 (53.49)	56.56 - 71.78 (63.27)	49.9 - 57.76 (53.82)	65.87 - 78.98 (71.82)	56.78 - 76.76 (70.4)	54.76 - 61.67 (57.93)	65.71 - 73.56 (70.64)	45.34 - 78.98 (63.05)
Lead ($mg \cdot Kg^{-1}$)	21.0 - 32.66 (29.12)	30.43 - 40.78 (35.06)	30.13 - 37.87 (32.95)	32.65 - 40.43 (36.78)	23.76 - 39.76 (30.91)	28.54 - 36.01 (32.27)	23.73 - 29.32 (27.92)	21.0 - 40.78 (32.14)
Copper ($mg \cdot Kg^{-1}$)	29.09 - 32.87 (30.98)	25.9 - 35.76 (30.15)	17.34 - 29.02 (24.45)	28.76 - 40.64 (35.72)	19.09 - 29.54 (22.07)	30.13 - 39.23 (34.1)	21.54 - 30.1 (26.02)	17.34 - 40.64 (29.07)
Cadmium ($mg \cdot Kg^{-1}$)	0.84 - 1.78 (1.13)	0.67 - 1.75 (1.02)	0.54 - 1.2 (0.94)	0.98 - 1.9 (1.34)	1.1 - 2.01 (1.49)	0.98 - 1.52 (1.2)	0.67 - 1.92 (1.15)	0.54 - 2.01 (1.18)

Table 6. Showing heavy metal conc. (mg·Kg^{-1}) in pre harvested soil of different sites.

Heavy metals in soil	1	2	3	4	5	6	7	Overall range
Iron (%)	0.9 - 2.01 (1.41)	0.99 - 1.9 (1.42)	0.54 - 0.73 (0.68)	0.87 - 1.61 (1.23)	0.98 - 1.61 (1.24)	0.9 - 1.19 (1.07)	0.89 - 1.26 (1.05)	0.54 - 2.01 (1.15)
Zinc (mg·Kg^{-1})	45.76 - 55.09 (49.69)	48.03 - 58.98 (53.71)	47.61 - 54.9 (50.32)	47.69 - 76.43 (62.29)	42.6 - 63.6 (55.23)	47.9 - 54.16 (50.11)	62.19 - 70.69 (67.27)	42.6 - 76.43 (55.51)
Lead (mg·Kg^{-1})	34.9 - 46.98 (41.19)	35.76 - 43.98 (39.41)	21.66 - 36.94 (28.7)	24.36 - 31.93 (28.45)	21.65 - 36.0 (27.1)	17.21 - 24.64 (20.93)	20.17 - 29.79 (24.85)	17.21 - 46.98 (30.09)
Copper (mg·Kg^{-1})	29.04 - 40.32 (35.42)	29.76 - 48.2 (40.62)	16.69 - 18.96 (17.78)	16.72 - 21.93 (19.44)	12.34 - 18.98 (15.32)	22.96 - 36.19 (30.19)	16.94 - 32.8 (24.55)	12.34 - 48.2 (26.18)
Cadmium (mg·Kg^{-1})	0.92 - 2.98 (1.33)	0.8 - 1.65 (1.09)	0.07 - 0.32 (0.22)	0.29 - 1.61 (0.79)	0.36 - 2.41 (1.59)	0.22 - 1.01 (0.7)	0.56 - 1.07 (0.76)	0.07 - 2.98 (0.92)

Table 7. Showing heavy metal conc. (mg·Kg^{-1}) in post harvested soil of different sites.

Heavy metals in soil	1	2	3	4	5	6	7	Overall range
Iron (%)	0.89 - 1.09 (0.97)	0.91 - 1.82 (1.11)	0.49 - 0.64 (0.58)	0.47 - 1.72 (1.11)	0.8 - 1.02 (0.94)	0.89 - 1.17 (1.0)	0.19 - 0.94 (0.56)	0.19 - 1.82
Zinc (mg·Kg^{-1})	48.1 - 58.91 (52.48)	40.81 - 52.02 (47.37)	45.15 - 48.71 (47.53)	49.61 - 66.17 (58.64)	42.68 - 59.61 (50.74)	49.69 - 54.61 (51.95)	57.16 - 61.92 (59.36)	40.81 - 66.17
Lead (mg·Kg^{-1})	30.12 - 42.05 (37.18)	29.01 - 34.04 (31.68)	21.61 - 36.11 (28.57)	25.61 - 30.99 (28.22)	12.16 - 32.19 (24.78)	18.96 - 22.9 (20.59)	18.1 - 27.1 (20.09)	12.16 - 42.05
Copper (mg·Kg^{-1})	29.45 - 38.11 (33.28)	32.96 - 41.72 (36.77)	17.27 - 16.69 (18.22)	18.86 - 22.41 (20.36)	10.86 - 17.31 (13.02)	31.69 - 49.62 (43.98)	19.79 - 26.59 (22.28)	10.86 - 49.62
Cadmium (mg·Kg^{-1})	0.92 - 1.67 (1.12)	0.87 - 1.18 (0.97)	0.89 - 1.27 (1.05)	0.21 - 1.66 (0.65)	0.28 - 2.9 (1.6)	0.98 - 1.12 (1.06)	0.27 - 0.37 (0.31)	0.21 - 2.9

Table 8. Showing heavy metal conc. (mg·Kg^s) in vegetables of different sites.

Heavy metals in veg.	Fe (%) (mg·Kg^{-1})	Zn (mg·Kg^{-1})	Pb (mg·Kg^{-1})	Cu (mg·Kg^{-1})	Cd (mg·Kg^{-1})
Spinach	0.92 - 1.98 (1.38)	54.12 - 71.89 (63.91)	25.76 - 36.98 (31.42)	18.35 - 39.56 (24.72)	0.14 - 1.98 (1.09)
Brinjal	(0.61 - 1.09) (0.91)	49.09 - 60.12 (53.92)	20.12 - 32.56 (27.76)	15.23 - 31.35 (20.89)	0.12 - 0.97 (0.47)
Cucumber	0.61 - 1.09 (0.87)	50.12 - 60.12 (54.84)	20.12 - 32.56 (26.27)	15.23 - 29.56 (20.38)	0.12 - 0.97 (0.5)
Reddish	0.9 - 1.9 (1.12)	53.12 - 70.12 (60.84)	23.06 - 34.26 (28.99)	18.45 - 38.67 (24.35)	0.19 - 1.52 (0.87)
Beetroot	0.99 - 1.98 (1.4)	54.89 - 71.98 (63.03)	23.98 - 36.23 (30.93)	19.01 - 40.13 (24.95)	0.17 - 1.9 (1.3)
Tomato	0.69 - 0.97 (0.86)	49.78 - 69.03 (59.25)	19.23 - 30.12 (24.51)	16.23 - 29.12 (21.31)	0.12 - 0.98 (0.61)
Okra	0.26 - 0.98 (0.72)	48.67 - 62.87 (48.67)	18.34 - 31.23 (23.34)	16.45 - 25.3 (19.64)	0.19 - 0.89 (0.68)

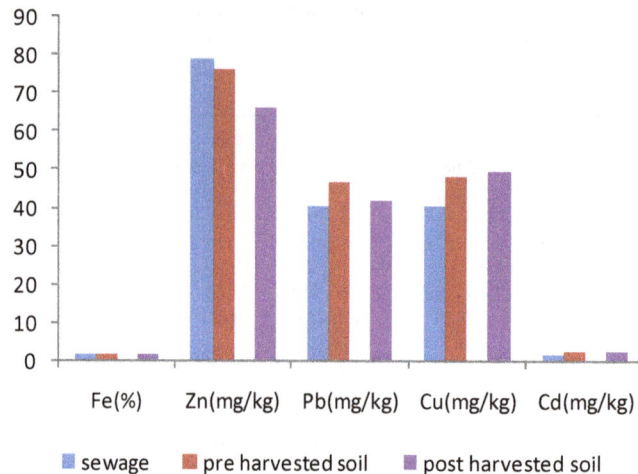

Figure 1. Showing comparison of conc. of heavy metals in sewage, pre and post harvested soil.

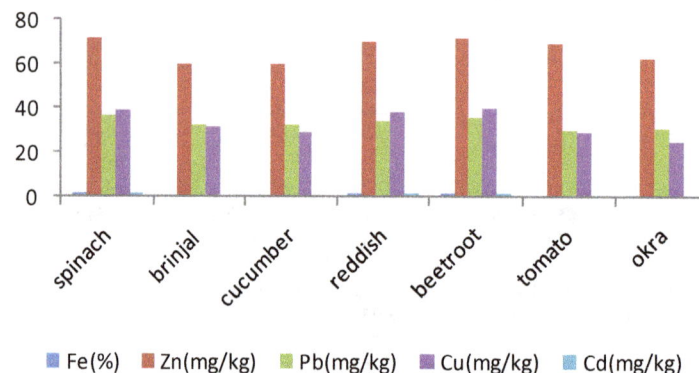

Figure 2. Showing comparison of conc. of heavy metals in various vegetables.

be drained into rivers and farmlands without prior treatment. Apart from treating the discharge that enters into the farms, it is also imperative to utilize alternative measures of cleaning up the already contaminated substrates. Continuous monitoring of soil, plant and water quality together with prevention of metals entering vegetables is a prerequisite in order to prevent potential health hazards to human beings.

REFERENCES

[1]   F. Mapanda, E. N. Mangwayana, J. Nyamangara and K. E. Giller, "The Effect of Long-Term Irrigation Using Wastewater on Heavy Metal Contents of Soils under Vegetables in Harare, Zimbabwe," *Agriculture, Ecosystems & Environment*, Vol. 107, 2005, pp. 151-165. doi:10.1016/j.agee.2004.11.005

[2]   W. Wenzel, and F. Jackwer, "Accumulation of Heavy Metals in Plants Grown on Mineralized Solids of the Austrian Alps," *Environmental Pollution*, Vol. 104, No. 1, 1999, pp. 145-155. doi:10.1016/S0269-7491(98)00139-0

[3]   K. P. Singh, D. Mohon, S. Sinha and R. Dalwani, "Impact Assessment of Treated/Untreated Waste Water Toxicants Discharge by Sewage Treatment Plants on Health, Agri-

cultural and Environmental Quality in Waste Water Disposal Area," *Chemosphere*, Vol. 55, No. 2, 2004, pp. 227-255. doi:10.1016/j.chemosphere.2003.10.050

[4]   S. Arai, "Global View on Functional Foods: Asian Perspectives," *British Journal of Nutrition*, Vol. 88, Suppl. 2, 2002, pp. S139-S143.

[5]   M. S. Al Jassir, A. Shaker and M. A. Khaliq, "Deposition of Heavy Metals on Green Leafy Vegetables Sold on Roadsides of Riyadh City, Saudi Arabea," *Bulletin of Environmental Contamination and Toxicology*, Vol. 75, No. 5, 2005, pp. 1020-1027. doi:10.1007/s00128-005-0851-4

[6]   D. Demirezen and A. Ahmet, "Heavy Metal Levels in Vegetables in Turkey Are within Safe Limits for Cu, Zn, Ni and Exceeded for Cd and Pb," *Journal of Food Quality*, Vol. 29, No. 3, 2006, pp. 252-265. doi:10.1111/j.1745-4557.2006.00072.x

[7]   M. A. Radwan and K. A. Salama, "Market Basket Survey for Some Heavy Metals in Egyptian Fruits and Vegetables," *Food and Chemical Toxicology*, Vol. 44, No. 8, 2006, pp. 1273-1278. doi:10.1016/j.fct.2006.02.004

[8]   K. Fytianos, G. Katsianis, P. Triantafyllou and G. Zachariadis, "Accumulation of Heavy Metals in Vegetables Grown in an Industrial Area in Relation to Soil," *Bulletin of Environmental Contamination and Toxicology*, Vol. 67,

No. 3, 2001, pp. 423-430. doi:10.1007/s001280141

[9]   I. Cakmak, "Possible Roles of Zinc in Protecting Plant Cells from Damage by Reactive Oxygen Species," *New Phytologist*, Vol. 146, No. 2, 2000, pp. 185-205. doi:10.1046/j.1469-8137.2000.00630.x

[10] A. Siedlecka and Z. Krupa, "Interaction between Cadmium and Iron. Accumulation and Distributionof Metals and Changes in Growth Parameters of Phaseolus Vulgaris L. Seedlings," *Acta Societatis Botanicorum Poloniae*, Vol. 65, No. 3-4, 1996, pp. 277-282.

[11] B. J. Alloway, A. P. Jackson and H. Morgan, "The Accumulation of Cadmium by Vegetables Grown on Soils Contaminated from a Variety of Sources," *Science of the Total Environment*, Vol. 91, 1990, pp. 223-236. doi:10.1016/0048-9697(90)90300-J

[12] B. Nrgholi, "Investigation of the Firoza Bad Waste Water Quality-Quantity Variation for Agricultural Use," Final Report, Iranian Agriculture Engineering Research Institute, Karaj, 2007.

[13] G. Ellen, J. W. Loon and K. Tolsma, "Heavy Metals in Vegetables Grown in the Netherlands and in Domestic and Imported Fruits," *Z Lebensm Unters Forsch*, Vol. 190, No. 1, 1990, pp. 34-39. doi:10.1007/BF01188261

[14] M. Qadir, A. Ghafoor, S. I. Hssain, G. Murtaza and T. Mahmood, "Copper Concentration in City Effluents Irrigated Soils and Vegetables," *Pakistan Journal of Soil Science*, 1999, pp. 91-102.

[15] M. Mostachari, "Investigation of Qazvin Soils and Plants Pollution with Heavy Metals during Irrigation with Waste Water," *Proceeding of 75 Water and Soil Conference*, Persian, 2002.

[16] T. Adhikari, M. C. Manna, M. V. Singh and R. H. Wanjari, "Bioremediation Measure to Minimize Heavy Metals Accumulation in Soils and Crops Irrigated with City Effluent Food," *Agriculture & Environment*, Vol. 2, No. 1, 2004, pp. 266-270.

[17] P. C. Prabu, "Impact of Heavy Metal Contamination of Akaki River of Ethiopia on Soil and Metal Toxicity on Cultivated Vegetable Crops," *Electronic Journal of Environmental Agricultural and Food Chemistry*, Vol. 8, No. 9, 2009, pp. 818-827.

[18] H. Lokeshwari and G. T. Chandrappa, "Imapct of Heavy Metal Contamination of Bellandur Lake on Soil and Cultivated Vegetation," *Current Science*, Vol. 91, No. 5, 2006, pp. 622-627.

13

Using Automation Controller System and Simulation Program for Testing Closed Circuits of Mini-Sprinkler Irrigation System

Hani A. Mansour[1*], **Hany M. Mehanna**[1], **Mohamed E. El-Hagarey**[2], **Ahmehd S. Hassan**[3]

[1]Water Relations Field Irrigation Department,
Agricultural and Biological Division, National Research Center, Cairo, Egypt
[2]Soil and Water Resources Conservation Department,
Desert Research Center, Cairo, Egypt
[3]Irrigation Department, Agriculture Engineering Research Institute, Agricultural Research Center, Cairo, Egypt
Email: *mansourhani2011@gmail.com

ABSTRACT

The field experiments were conducted at the experimental farm of Faculty of agricultural, southern Illinois University SIUC, USA. The project makes the irrigation automated. With the use of low cost sensors and the simple circuitry makes currently project a low cost product, which can be bought even by a poor farmer. This research work is best suited for places where water is scares and has to be used in limited quantity and this proposal is a model to modernize the agriculture industries at a mass scale with optimum expenditure. In the field of agricultural engineering, use of sensor method of irrigation operation is important and it is well known that closed circuits of Mini-sprinkler irrigation system are very economical and efficient. Closed circuits are considered one of the modifications of Mini-sprinkler irrigation system, and added advantages to Mini-sprinkler irrigation system because it can relieve low operating pressures problem at the end of the lateral lines. In the conventional closed circuits of Mini-sprinkler irrigation system, the farmer has to keep watch on irrigation timetable, which is different for different crops. Using this system, one can save manpower, water to improve production and ultimately profit. The data could be summarized in following: Irrigation methods under study when using lateral length 60 m could be ranked in the following ascending order according the values of the predicted and measured head losses CM1M-SIS < CM2M-SIS.The correlation (Corr.) coefficients were used to compare the predicted and measured head losses along the lateral lines of all the closed circuits designs. Generally, the values of correlation analysis were (>0.90) were obtained with 0% field slope 60 m length (experimental conditions) for all closed circuits.The interaction between irrigation methods: at the start there are significant differences between CM2M-SIS and CM1M-SIS.

Keywords: Automation; Controller; Simulation; Program; Mini-Sprinkler; Irrigation; Lateral; Closed; Circuits

1. Introduction

The continuous increasing demand of the food requires the rapid improvement in food production technology. In a country like India, where the economy is mainly based on agriculture and the climatic conditions are isotropic, still we are not able to make full use of agricultural resources. The main reason is the lack of rains & scarcity of land reservoir water. The continuous extraction of water from earth is reducing the water level due to which lot of land is coming slowly in the zones of un-irrigated land. Another very important reason of this is due to un-planned use of water due to which a significant amount of water goes waste. In the modern closed circuits of Mini-sprinkler irrigation system systems, the most significant advantage is that water is supplied near the root zone of the plants Mini-sprinkler by Mini-sprinkler due to which a large quantity of water is saved. At the present era, the farmers have been using irrigation technique in India through the manual control in which the farmers irrigate the land at the regular intervals. This process sometimes consumes more water or sometimes the water reaches late due to which the crops get dried. Water deficiency can be detrimental to plants before

---

*Corresponding author.

visible wilting occurs. Slowed growth rate, lighter weight fruit follows slight water deficiency. This problem can be perfectly rectified if we use automatic micro controller based closed circuits of Mini-sprinkler irrigation systemin which the irrigation will take place only when there will be intense requirement of water.

Irrigation system uses valves to turn irrigation ON and OFF. These valves may be easily automated by using controllers and solenoids. Automating farm or nursery irrigation allows farmers to apply the right amount of water at the right time, regardless of the availability of labor to turn valves on and off. In addition, farmers using automation equipment are able to reduce runoff from over watering saturated soils, avoid irrigating at the wrong time of day, which will improve crop performance by ensuring adequate water and nutrients when needed. Automatic Closed circuits of Mini-sprinkler irrigation system is a valuable tool for accurate soil moisture control in highly specialized greenhouse vegetable production and it is a simple, precise method for irrigation. It also helps in time saving, removal of human error in adjusting available soil moisture levels and to maximize their net profits.

The entire automation work can be divided in two sections, first is to study the basic components of irrigation system thoroughly and then to design and implement the control circuitry. So we will first see some of the basic platform of closed circuits of Mini-sprinkler irrigation system (**Figure 1**).

1.1. Definition of Irrigation

Irrigation is the artificial application of water to the soil usually for assisting in growing crops. In crop production it is mainly used in dry areas and in periods of rainfall shortfalls, but also to protect plants against frost.

1.2. Types of Irrigation

Surface irrigation, Localized irrigation (Drip, mini sprinkler, bubbler, etc.), Closed circuits of Mini-sprinkler irrigation system and Sprinkler irrigation.

Closed circuits of Mini-sprinkler irrigation system also known as Mini-sprinkler irrigation system or micro-irrigation is an sprinkler irrigation method which minimizes the use of water and fertilizer by allowing water to Mini-sprinkler slowly to the roots of plants, either onto the soil surfaceor directly onto the root zone, through a network of valves, pipes, tubing, and emitters., [1].

1.3. Concept of Modern Irrigation System

The conventional irrigation methods like overhead sprinklers, flood type feeding systems usually wet the lower leaves and stem of the plants. The entire soil surface is saturated and often stays wet long after irrigation is completed. Such condition promotes infections by leaf mold fungi. The flood type methods consume large amount of water and the area between crop rows remains dry and receives moisture only from incidental rainfall. On the contrary the Mini-sprinkler or closed circuits of Mini-sprinkler irrigation system is a type of modern irrigation technique that slowly applies small amounts of water to part of plant root zone. Closed circuits of drip irrigation system are invented by the Egyptian thesis of researcher Hani Mansour, 2010s [2]. Water is supplied frequently, often daily to maintain favorable soil moisture condition and prevent moisture stress in the plant with proper use of water resources.

Closed circuits of Mini-sprinkler irrigation system save water because only the plant's root zone receives moisture. Little water is lost to deep percolation if the proper amount is applied. Closed circuits of Mini-sprin-

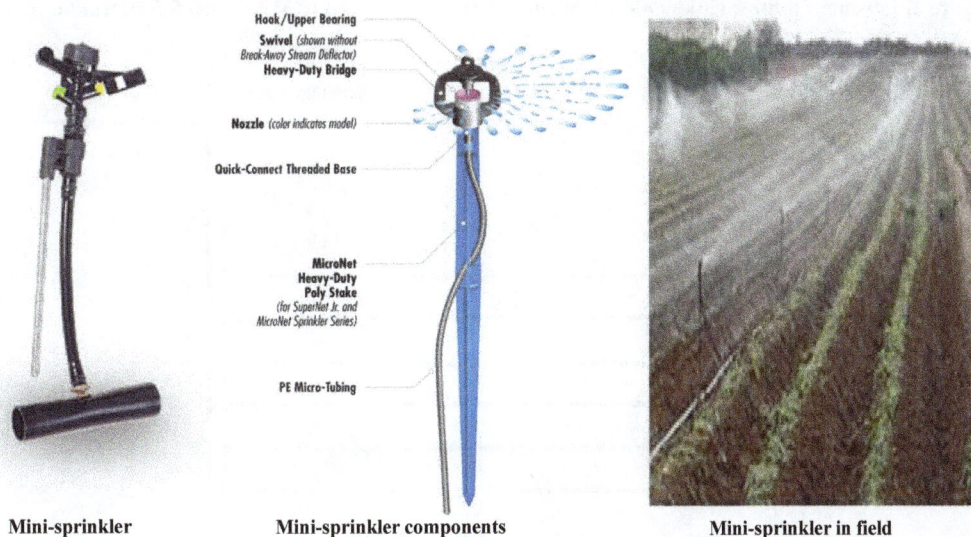

Hook/Upper Bearing
Swivel (shown without Break-Away Stream Deflector)
Heavy-Duty Bridge

Nozzle (color indicates model)

Quick-Connect Threaded Base

MicroNet Heavy-Duty Poly Stake (for SuperNet Jr. and MicroNet Sprinkler Series)

PE Micro-Tubing

Mini-sprinkler   **Mini-sprinkler components**   **Mini-sprinkler in field**

Figure 1. Min-sprinkler irrigation system.

kler irrigation system are popular because it can increase yields and decrease both water requirements and labor. Closed circuits of drip irrigation system require about half of the water needed by sprinkler or surface irrigation. Lower operating pressures and flow rates result in reduced energy costs. A higher degree of water control is attainable. Plants can be supplied with more precise amounts of water. Disease and insect damage is reduced because plant foliage stays dry [2]. Operating cost is usually reduced. Federations may continue during the irrigation process because rows between plants remain dry. Fertilizers can be applied through this type of system. This can result in a reduction of fertilizer and fertilizer costs. When compared with overhead sprinkler systems, closed circuits of Mini-sprinkler irrigation system lead to less soil and wind erosion. Closed circuits of Mini-prinkler irrigation system can be applied under a wide range of field conditions. A typical closed circuits of drip irrigation system assembly is shown in **Figures 2** and **3** be-

low [1]. The goal of this study is to design and simulate a new solution for modernizing the irrigation operation of agriculture industries.

2. Material and Methods

The field experiments were conducted at the experimental farm of Faculty of agricultural, southern Illinois University SIUC, USA. The design of field experiments was split in randomized complete block design with three replicates. The field tests carried out using line length 60m and the following tow Mini-sprinkler irrigation circuits (DIC): 1) one manifold for lateral lines or closed circuits with one manifold of Mini-sprinkler irrigation system (CM1M-SIS) and 2) closed circuits with two manifolds for lateral lines (CM2M-SIS) **Figures 2** and **3**.

Irrigation networks include the following components are:

1) Control head: It was located at the water source

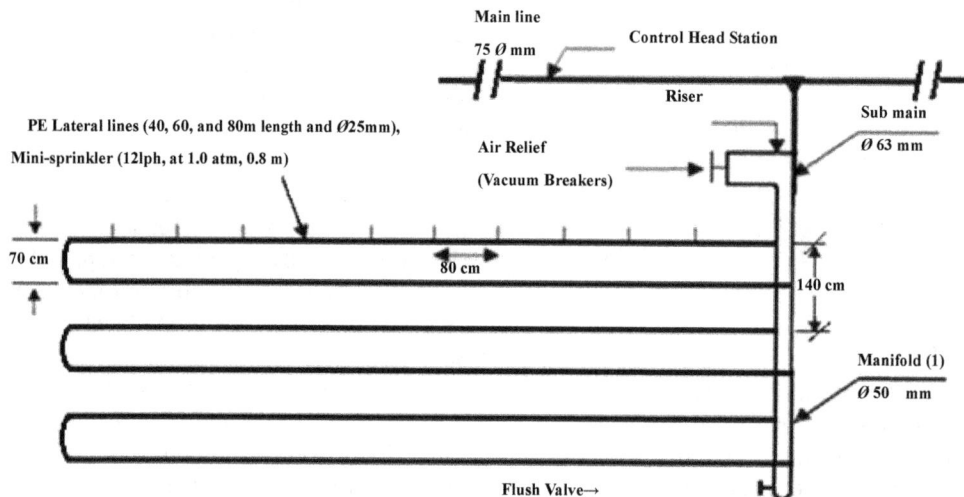

Figure 2. Layout of mini-sprinkler closed circuits with one manifold (CM1M-SIS) for lateral lines.

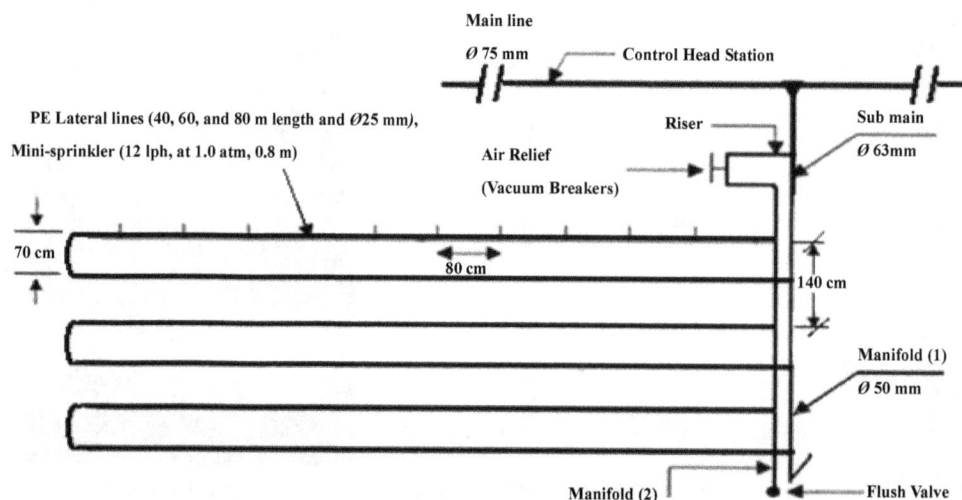

Figure 3. Layout of mini-sprinkler closed circuit with two manifolds (CM2M-SIS) for lateral lines.

supply. It consists of centrifugal pump 3"/3", driven by electric engine (pump discharge of 80 m^3/h and 40 m lift), sand media filter 48" (two tanks), screen filter 2" (120 mesh), back flow prevention device, pressure regulator, pressure gauges, flow-meter, control valves and chemical injection; 2) Main line: PVC pipes of 75 mm in (ID) Ø to convey the water from the source to the main control points in the field; 3) Sub-main lines: PVC pipes of 75mm in (ID) Ø were connected to with the main line through a control unit consists of a 2" ball valve and pressure gauges; 4) Manifold lines: PVC pipes of 50 mm in (ID) Ø were connected to the sub main line through control valves 1.5"; 5) Lateral lines: PE tubes of 25 mm in (ID) Ø were connected to the manifolds through beginnings stalled on manifolds lines; 6) Mini-sprinklers: These online PE tubes 25mm in (ID) Ø, mini sprinkler discharge of 12 lh^{-1} at 1 atm, Operating pressure and 50 cm spacing in-between [1].

The components of closed circuits of the Mini-sprinkler system include, supply lines, control valves, supply and return manifolds, Mini-sprinkler lateral lines, emitters, check valves and air relief valves/vacuum breakers.

2.1. Irrigation Scheduling

Intervals of irrigation (*I*) in day were calculated using the following equations:

$$I = d/Etc \qquad (1)$$

Where: d = net water depth applied per each irrigation (mm), and *Etc* = crop evapotranspiration (mm/day).

$$d = AMD \cdot ASW \cdot Rd \cdot P \qquad (2)$$

Where: AMD = allowable soil moisture depletion (%), ASW = available soil water, (mm water/m depth), Rd = effective root zone depth (m), or irrigation depth (m), and P = percentage of soil area wetted (%).

$$AW \ (v/v \ \%) = ASW \ (w/w \ \%) \cdot B.D. \qquad (3)$$

Where: B.D. = Soil bulk density (gm·cm^{-3}).

Irrigation Intervals used was 4 days under both closed circuits of Mini-sprinkler irrigation systems.

Design of automation controller based closed circuits of Mini-sprinkler irrigation system: (presented for Drip irrigation by [3]:

The key elements that should be considered while designing a mechanical model.

Pressure (The force pushing the flow): Most products operate best between 1.0 and 1.5 bars of pressure. Normal household pressure is 10 m (1.0 bar).

Water Supply & Quality: City and well water are easy to filter for closed circuits of Mini-sprinkler irrigation system systems. Pond, ditch and some well water have special filtering needs. The quality and source of

water will dictate the type of filter necessary for your system.

Flow: We can measure the output of your water supply with a one or five gallon bucket and a stopwatch. Time how long it takes to fill the bucket and use that number to calculate how much water is available per hour. Gallons per minute × 60 = number of gallons per hour.

Elevation: Variations in elevation can cause a change in water pressure within the system. Pressure changes by one pound for every 2.3 foot change in elevation. Pressure-compensating emitters are designed to work in areas with large changes in elevation.

Soil Type and Root Structure: The soil type will dictate how a regular Mini-sprinkler of water on one spot will spread. Sandy soil requires closer emitter spacing as water percolates vertically at a fast rate and slower horizontally. With a clay soil water tends to spread horizontally, giving a wide distribution pattern. Emitters can be spaced further apart with clay type soil. A loamy type soil will produce a more even percolation dispersion of water. Deep-rooted plants can handle a wider spacing of emitters, while shallow rooted plants are most efficiently watered slowly (low gap emitters) with emitters spaced close together. On clay soil or on a hillside, short cycles repeated frequently work best. On sandy soil, applying water with higher gap emitters lets the water spread out horizontally better than a low gap emitter.

Timing: Watering in a regular scheduled cycle is essential. On clay soil or hillsides, short cycles repeated frequently work best to prevent runoff, erosion and wasted water. In sandy soils, slow watering using low output emitters is recommended. Timers help prevent the too-dry/too-wet cycles that stress plants and retard their growth. They also allow for watering at optimum times such as early morning or late evening.

Watering Needs: Plants with different water needs may require their own watering circuits. For example, orchards that get watered weekly need a different circuit than a garden that gets watered daily. Plants that are drought tolerant will need to be watered differently than plants requiring a lot of water.

The components of an automation controller unit based closed circuits of Mini-sprinkler irrigation system are as following: A) Control Head Station, B) Flow Meter, C) Control, flushing Valves, D) Chemical Injection Unit (Fertigation unit), E) Manifolds and Mini-sprinkler lines with Emitters, F) Moisture and Temperature Sensors and G) An automation controller (The brain of the system).

The signal send by the sensor is boosted unto the required level by corresponding amplifier stages. Then the amplified signal is fed to A/D converters of desired resolution to obtain digital form of sensed input for an automation controller use.

Sensor: LCD module can be used in the system to

monitor current readings of all the sensors and the current status of respective valves. The solenoid valves are controlled by an automation controller though relays. A Chemical injection unit is used to mix required amount of fertilizers, pesticides, and nutrients with water, whenever required. Varying speed of pump motor can control pressure of water. It can be obtained with the help of PWM output of an automation controller unit. A flow meter is attached for analysis of total water consumed. The required readings can be transferred to the Centralized Computer for further analytical studies, through the serial port present on an automation controller unit. While applying the automation on large fields more than one such an automation controller units can be interfaced to the Centralized Computer the an automation controller unit has in-built timer in it, which operates parallel to sensor system. In case of sensor failure the timer turns off the valves after a threshold level of time, which may prevent the further disaster. Automation controller unit may warn the pump failure or insufficient amount of water input with the help of flow meter.

2.2. Automation Controller Unit Is Now Explained in Flowing Details

The automated control system consists of moisture sensors, temperature sensors, Signal conditioning circuit, Digital to analog converter, LCD Module, Relay driver, solenoid control valves, etc. The unit is expressed in **Figure 4** above. The important parameters to be measured for automation of irrigation system are soil moisture and temperature. The entire field is first divided in to small sections such that each section should contain one moisture sensor and a temperature sensor. RTD like PT100 can be used as a temperature sensor while Densitometer can be used as the moisture sensor to detect moisture contents of soil showed in **Figure 5**. These sensors are buried in the ground at required depth. Once the soil has reached desired moisture level the sensors send a signal to the micro controller to turn off the relays, which control the valves.

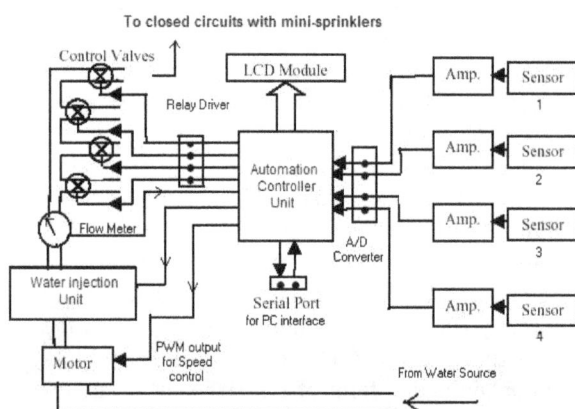

Figure 4. Controller unit.

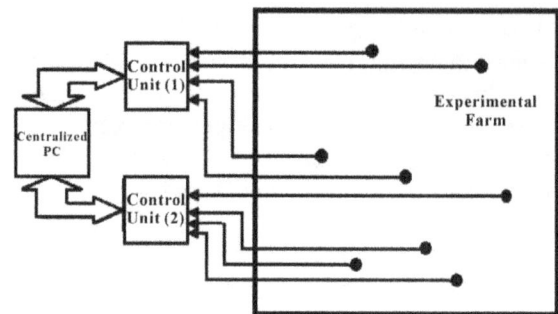

Figure 5. Application to field.

2.3. Using Simulation Program for Hydraulic Calculations

Hydro-Calc irrigation system planning software is designed to help the user to define the parameters of an irrigation system, **Figure 6** Showing Hydro-Calc program. The user will be able to run the program with any suitable parameters, review the output, and change input data in order to match it to the appropriate irrigation system set up. Some parameters may be selected from a system list; whereas other are entered by the user according to their own needs so they do not conflict with the program's limitations. The software package includes an opening main window, five calculation programs, one language setting window and a database that can be modified and updated by the user.

2.4. Hydro-Calc Includes Several Sub-Programs as Following

Figure 7 indicate that the Emitters program calculates the cumulative pressure loss, the average flow rate, the water flow velocity etc. in the selected emitter. It can be changed to suit the desired irrigation system parameters. The Sub-Main program calculates the cumulative pressure loss and the water flow velocity in the sub-main distributing water pipe (single or telescopic). It changes to suit the required irrigation system parameters. The Main Pipe program calculates the cumulative pressure loss and the water flow velocity in the main conducting water pipe (single or telescopic). It changes to suit the required irrigation system parameters. The Shape Wizard program helps transfer the required system parameters (Inlet Lateral Flow Rate, Minimum Head Pressure) from the Emitters program to the Sub-Main program. The Valves program calculates the valve friction loss according to the given parameters. The Shifts program calculates the irrigation rate and number of shifts needed according to the given parameters.

The Emitters program is the first application which can be used in the frame of Hydro-Calc software program. There are 4 basic type of emitters which can be used: Mini-sprinkler Line, On line, Sprinklers and Micro-

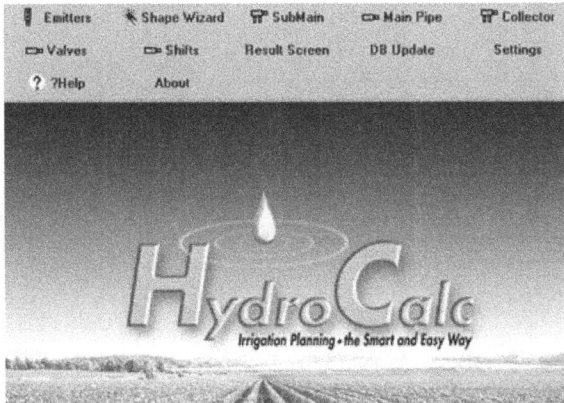

Figure 6. Hydro-calc irrigation planning.

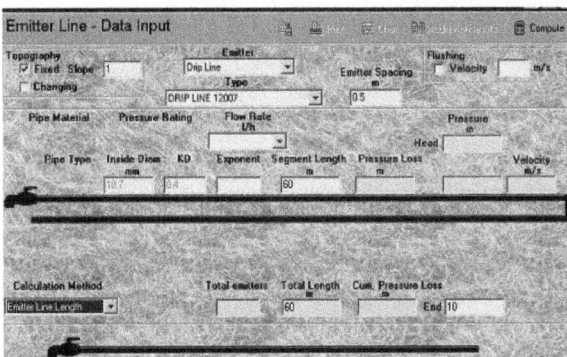

Figure 7. Hydro-calc working sheet before computation procedure.

Sprinklers. According to the previous selection the user can opt for a specific emitter which can be a pressure compensated or a non-pressure compensated. Each emitter has its own set of nominal flow rate values available. After the previous mentioned fields were completed, the program automatically fills the following fields: "Inside Diameter", "KD" and "Exponent", values which cannot be changes unless the change will be made in the database.

Figure 8 Showed that the steps of Flow chart for Hydro-Calc simulation program are components, planning, design, and calculating the hydraulic analysis of Mini-sprinkler irrigation system at different slopes or levels. The segment length is next field in which the user must introduce a value. The end pressure represents the actual value for calculation of pressure at the furthest emitter. There are some common values for this field: around 10 m for Mini-sprinkler, around 20 m for mini-sprinklers, between 20 - 30 m for sprinklers and around 2 m when using the flushing system. There are 2 more options which can be filled before starting the computation, options which can also be used with their default values. The Flushing field can be used if the user intends to calculate a system that includes and lateral flushing. Flushing option will work only in subsequently will be used the "Emitter Line Length" calculation method. The second option is about topography. Default value is 0%. Topography field has 2

sub-fields: fixed slope and changing slope. Usually the slopes values are not exceeding 10%. In many cases the slope is not uniform.

2.5. Validation of Measured Data with Calculated Data by Hdro-Calc

The emission rate for 10 emitters tested for each Lateral line for lengths (40, 60 and 80 m) at three stages. First, middle and end on the line were calculated theoretically using the following procedure (**Figure 8**). The head loss due to friction and insertion of emitters was calculated and then the pressure head at every emitter was determined. The emission from every emitter was calculated using the characteristic equation developed for pressure head vs. discharge for each product.

Costat program was used to carry out statistical analysis. Treatments mean were compared using the technique of analysis of variance (ANOVA) and the least significant difference (L.S.D) between systems at 1% had been done. The randomized complete block design according to [4].

3. Results and Discussions

3.1. Validation of Lateral Lines Hydraulic Analysis by Hydro-Calc Simulation Program When Lateral Lines Did Not Slope (0%) and Sloped down by (7%)

Validation of the Hydro-Calc Simulation Program

The discharge rates and pressures at the Mini-sprinkler head were measured under field conditions at three sites along the lateral lines (start, middle and end) for CM2M-SIS, CM1M-SIS, and TDIS with lateral line length 60 m and for two different slopes of the Mini-sprinkler line (0 and 7%). Empirical measurements were used to validate the Mini-sprinkler simulation program (Hydro-Calc Simulation program copyright 2009 developed by NETA-FIM, USA). Hydro-Calc, was a computer simulation Program used for planning and design of Mini-sprinkler or sprinkler irrigation systems. Modification of closed circuit Mini-sprinkler lateral lines irrigation depended on hydraulic equations such as, Hazen-William's equations, Pernolli's equations, etc. The data inputs provided to Hydro-Calc were shown in **Table 1**. The empirical data depended on the laboratory measurements of pressures and discharge, as well as the field uniformity.

The predicted outputs of Hydro-calc simulation program (exponent (X), pressure head loss (m), velocity (m/s), and pressure along the lateral line Mini-sprinkler pers) were shown in **Table 2** and **Figures 9-12**.

3.2. Predicted and Measured Head Loss Analysis along the Lateral Mini-Sprinkler Line of Closed Circuit Designs with No Slope 0%

The predicted head loss analysis along the lateral Mini-

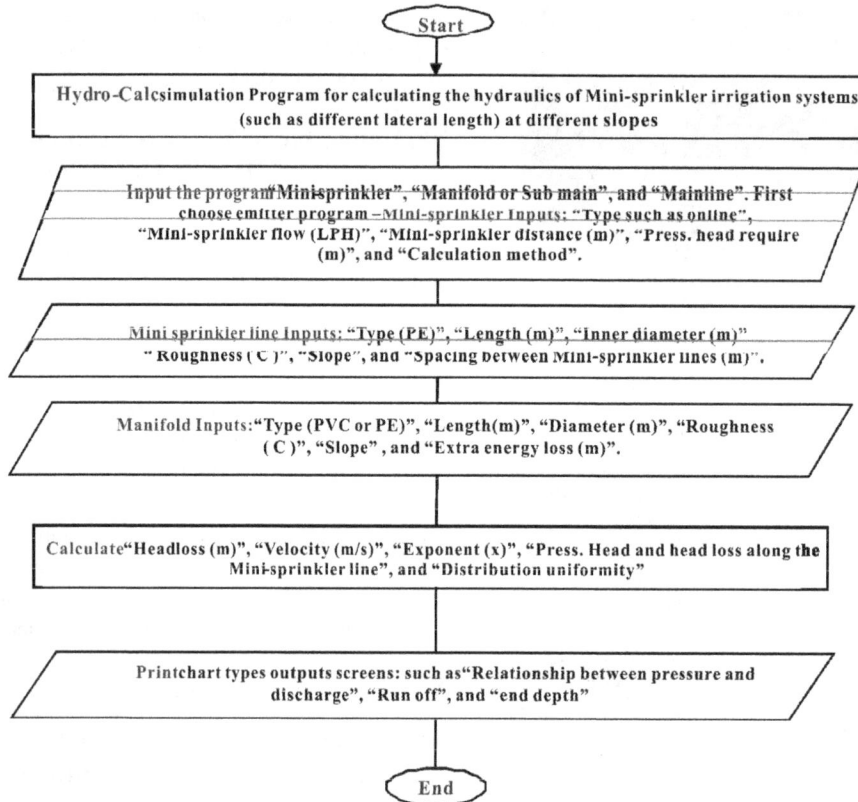

Figure 8. Flow chart components of hydro-calc simulation program for planning, design, and calculating the hydraulic analysis of mini-sprinkler irrigation system at different slopes or levels.

Table 1. Inputs for the hydro-calc simulation program for closed circuit designs Mini-sprinkler irrigation systems.

Manifold		Mini-sprinkler line		Mini-sprinkler	
Name	Value	Name	Value	Name	Value
Pipe type:	PVC	Tubes type	PE	Mini-sprinkler type	Online
Pipe length:	-	Tubes lengths:	60	Mini-sprinkler flow (Lph)	12.0
Pipe diameter:	0.05 m	Inner diameter	0.0235 m	Mini-sprinkler distance	0.50 m
(C) Pipe roughness:	150	(C) Pipe roughness	150	Press head require (m)	10.0 m
Slope:	0 m/m	Slope	0 or 0.03 m/m	Calculation method	Flow rate variation
Extra energy losses:	0.064	Spacing	0.7 m	-	-

Table 2. Predicted exponent (x), head loss (m) and velocity (m/s) by the hydro-calc simulation program for closed circuits mini-sprinkler irrigation design with different slopes (0% and 7%).

Field slope, (%)	Irrigation connection design					
	CM2M-SIS			CM1M-SIS		
	Exponent (x)	Head loss (m)	Velocity (m/s)	Exponent (x)	Head loss (m)	Velocity (m/s)
	0.72	0.64	1.58	0.69	0.73	1.55
0	0.65	1.48	1.63	0.61	1.55	1.57
	0.58	3.00	1.92	0.52	3.11	1.88
	0.76	0.45	1.51	0.71	0.76	1.51
7	0.68	1.34	1.57	0.64	1.55	1.55
	0.61	2.92	1.89	0.58	3.00	1.74

sprinkler's line had been calculated by Hydro-calc simulation program for closed circuits Mini-sprinkler irrigation systems CM2M-SIS and CM1M-SIS with no slope (0%) but with different lateral line length 60 m. Those predictions were tabulated in **Table 3**.

Table 3, **Figures 9** and **10** showed the relationships among the predicted and measured head losses as well as the correlations Under CM2M-SIS and CM1M-SIS methods with no slope 0%. The irrigation methods could be ranked in the following ascending order CM1M-SIS < CM2M-SIS.

The correlation (Corr.) coefficients were obtained to compare the significant of the predicted and measured head loss along the lateral lines of the two closed circuits designs. Generally, the values of correlation analysis were (>0.90) when 0% field slope.

Data in **Table 3** show the head pressures (bar) along the lateral lines of the different closed circuit designs CM2M-SIS and CM1M-SIS. Clearly the irrigation closed circuits designs under study could be ranked in the following ascending order CM1M-SIS<CM2M-SIS according to the values of the pressure head. Possibly this was due to increased friction losses for the - TDIS method.LSD$_{0.01}$ values in **Table 3** under CM2M-SIS and CM1M-SIS showed there was no significant difference between both start and end values. LSD$_{0.01}$ values in **Table 3** shown that Under CM2M-SIS there is no significant difference between both start and end values of pressure head but there are significant differences between middle value and both start and end pressure head values. While under CM1M-SIS there are significant

Table 3. Pressure head analysis along the lateral lines in mini-sprinkler irrigation closed circuits CM2M-SIS and CM1M-SIS methodswhen slope 0% level.

Distance along laterals (m)	Mini-sprinkler irrigation circuits			
	CM2M-SIS		CM1M-SIS	
	Predicted	Measured	Predicted	Measured
1	0.93	0.93	0.94	0.90
6	0.88	0.91	0.93	0.88
12	0.83	0.90	0.90	0.86
18	0.83	0.85	0.87	0.84
24	0.82	0.82	0.85	0.82
30	0.81	0.79	0.82	0.80
36	0.83	0.82	0.81	0.81
42	0.83	0.86	0.80	0.83
48	0.85	0.88	0.82	0.85
54	0.89	0.90	0.84	0.87
60	0.94	0.92	0.87	0.89
Average	0.85	0.87	0.86	0.86
LSD$_{0.01}$	**0.03**	**0.07**	**0.05**	**0.02**

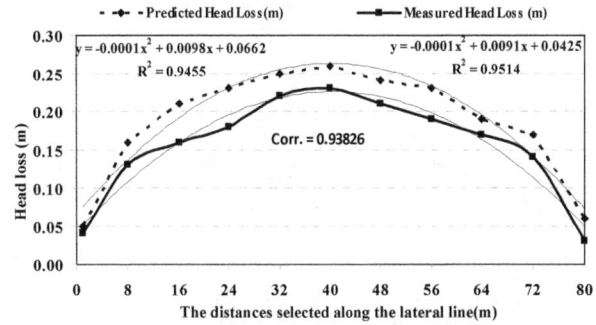

Figure 9. The relationship between different lateral line length 60 m and both the predicted and measured head losses when the slope was 0% with the CM2M-SIS design.

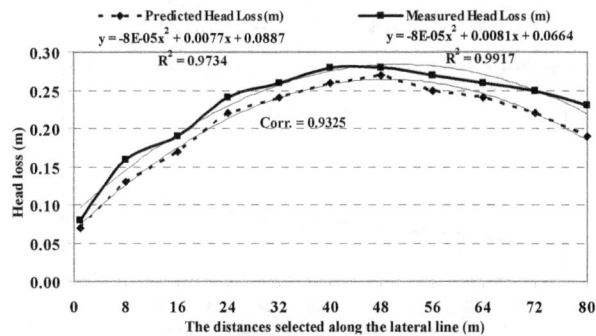

Figure 10. The relationship between different lateral line lengths (40, 60, 80 m) and both the predicted and measured head losses when the slope was 0% with the CM1M-SIS design.

differences between the all pressure head values of start, data are supported by [5-11].

3.3. Predicted and Measured Head Loss Analysis along the Lateral Mini-Sprinkler Line of Closed Circuits with a 7% Downward Slope from the Manifold to the Terminus

The predicted head loss analysiswhen lines sloped by 7% along the lateral Mini-sprinklers line had been calculated by Hydro-calc simulation program for closed circuits Mini-sprinkler irrigation systems CM2M-SIS and CM1M-SIS. **Figures 11** and **12** and **Table 4** showed the relationship between predicted and measured head losses as well as the correlations. Irrigation methods under study when using lateral length 60 m could be ranked in the following ascending order according the values of the predicted and measured head losses CM1M-SIS < CM2M-SIS.

While by using Lateral length 80m the values of the predicted and measured head losses under irrigation methods could be ranked in the following ascending orders;CM2M-SIS < CM1M-SIS. This may be attributed to the different numbers of mini-sprinklers or how many mini-sprinklers were built-in with every lateral line

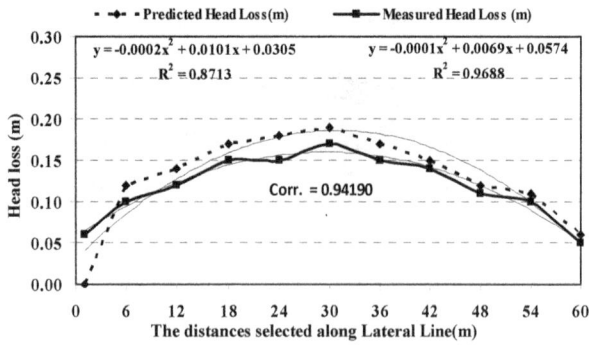

Figure 11. The relationship between different lateral line length 60 m and both the predicted and measured head loss when the line sloped 7% down with the CM2M-SIS design.

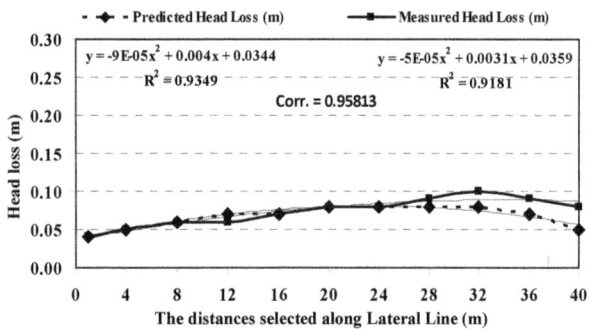

Figure 12. The relationship between different lateral line length 60 m and both the predicted and measured head loss when sloped 7% down with the CM1M-SIS design.

Table 4. Pressure head analysis along the lateral lines in Mini-sprinkler irrigation closed circuits CM2M-SIS and CM1M-SIS method when slope 7% level.

Distance along laterals (m)	Mini-sprinkler irrigation circuits			
	CM2M-SIS		CM1M-SIS	
	Predicted	Measured	Predicted	Measured
1	0.94	0.95	0.94	0.91
6	0.90	0.91	0.93	0.89
12	0.86	0.87	0.90	0.88
18	0.85	0.85	0.87	0.84
24	0.84	0.85	0.85	0.81
30	0.83	0.83	0.82	0.78
36	0.84	0.85	0.81	0.79
42	0.86	0.87	0.80	0.80
48	9.89	0.88	0.82	0.81
54	0.90	0.90	0.84	0.82
60	0.93	0.96	0.87	0.83
Average	**0.88**	**0.88**	0.86	0.83
LSD$_{0.01}$	**0.09**	**0.04**	**0.02**	**0.04**

length.

The correlation (Corr.) coefficients were used to compare the predicted and measured head losses along the-

lateral lines of all the closed circuits designs. Generally, the values of correlation analysis were (>0.90) were obtained with 0% field slope 60 m length (experimental conditions) for all closed circuits.

Data in **Table 4** and **Figures 11** and **12** shows the head pressures(bar) along the lateral lines of the different closed circuit designs CM2M-SISand CM1M-SIS.When using lateral length 60 m under CM2M-SIS and CM1M-SIS methods. According to Lateral length 60m the values of the pressure head under irrigation methods could be arranged in the following ascending orders CM1M-SIS < CM2M-SIS. This may be attributed to the decreased the head loss lateral line length by using the modified method CM2M-SIS and CM1M-SIS. LSD$_{0.01}$ values in **Table 4** show that Under CM2M-SIS and CM1M-SIS there is no significant difference between both start and end values of pressure head (bar) but there are significant differences between middle value and both start and end pressure head values. The interaction between irrigation methods: at the start there are significant differences between CM2M-SIS and CM1M-SIS. While at both of end and middle there are significant differences between all irrigation methods. These data are agreed well with the following references [5-12].

4. Conclusion

Automation controllerforclosed circuits of Mini- sprinkler irrigation system proves to be a real time feedback control system which monitors and controls all the activities of closed circuits of Mini-sprinkler irrigation system efficiently. The present proposal is a model to modernize the agriculture industries at a mass scale with optimum expenditure. Using this system, one can save manpower, water to improve production and ultimately profit. The data could be summarized in following: Irrigation methods under study when using lateral length 60 m could be ranked in the following ascending order according the values of the predicted and measured head losses CM1M-SIS < CM2M-SIS.The correlation (Corr.) coefficients were used to compare the predicted and measured head losses along the lateral lines of all the closed circuits designs. Generally, the values of correlation analysis were (>0.90) were obtained with 0% field slope 60m length (experimental conditions) for all closed circuits. The interaction between irrigation methods: at the start there are significant differences between CM2M-SIS and CM1M-SIS.

5. Acknowledgements

This paper is part of the mission of Dr. Hani Abdel-Ghani Abdel-Ghani Mansour to USA. So authors would to thanks WRFI Dept., Agricultural Division, National Research Center, Prof. Dr. David A. Lightfoot, PSAS Dept., SIUC, USA., and Ministries of Higher Education and

Scientific Research in Egypt.

REFERENCES

[1] H. A. Mansour, "Design Considerations for Closed Circuits of Drip Irrigation System," Ph.D. Thesis, Faculty of Agriculture, Ain Shams University, Cairo, 2012.

[2] H. A. Mansour, M. Y. Tayel, D. A. Lightfoot and A. M. El-Gindy, "Energy and Water Saving by Using Closed Circuits of Mini-Sprinkler Irrigation Systems," *Agriculture Science Journal*, Vol. 1, No. 3, 2010, pp. 1-9. http://www.scirp.org/journal/as/

[3] P. Ashok and K. Ashok, "Microcontroller Based Drip Irrigation System," 1st BTech, ECE, 2010. http://www.yuvaengineers.com/?p=583

[4] R. G. D. Steel and J. H. Torrie, "Principles and Procedures of Statistics, a Biometrical Approach," 2nd Edition, McGraw Hill Inter. Book Co., Tokyo, 1980.

[5] H. A. Mansour and A. S. Aljughaiman, "Water and Fertilizers Use Efficiency of Corn Crop under Closed Circuits of Drip Irrigation System," *Journal of Applied Sciences Research*, Vol. 8, No. 11, 2012, pp. 5485-5493.

[6] M. Y. Tayel, A. M. El-Gindy and H. A. Mansour, "Effect of Drip Irrigation Circuit Design and Lateral Line Lengths III—On Dripper and Lateral Discharge," *Journal of Applied Sciences Research*, Vol. 8, No. 5, 2012, pp. 2725-2731.

[7] M. Y. Tayel, A. M. El-Gindy and H. A. Mansour, "Effect of Drip Irrigation Circuit Design and Lateral Line Lengths IV—On Uniformity Coefficient and Coefficient of Variation," *Journal of Applied Sciences Research*, Vol. 8, No. 5, 2012, pp. 2741-2748.

[8] M. Y. Tayel, H. A. Mansour and D. A. Lightfoot, "Effect of Drip Irrigation Circuit Design and Lateral Line Lengths I—On Pressure Head and Friction Loss," *Agriculture Science*, Vol. 3, No. 3, 2012, pp. 392-399. http://www.scirp.org/journal/as

[9] M. Y. Tayel, H. A. Mansour and D. A. Lightfoot, "Effect of Drip Irrigation Circuit Design and Lateral Line Lengths II—On Flow Velocity and Velocity Head," *Agriculture Science*, Vol. 3, No. 4, 2012, pp. 531-537. http://www.scirp.org/journal/as

[10] C. M. Burt, A. J. Clemens, T. S. Strelkoff, K. H. Solomon, R. D. Blesner, L. A. Hardy and T. A. Howell, "Irrigation Performance Measures: Efficiency and Uniformity," *Journal of Irrigation and Drainage Engineering*, Vol. 123, No. 6, 1997, pp. 423-442. doi:10.1061/(ASCE)0733-9437(1997)123:6(423)

[11] N. Mizyed and E. G. Kruse, "Emitter Discharge Evaluation of Subsurface Trickle Irrigation Systems," *Transactions of the ASAE*, Vol. 32, No. 4, 1989, pp. 1223-1228.

[12] A. G. Smajstrla and G. A. Clark, "Hydraulic Performance of Micro-Irrigation Mini-Sprinkler Tape Emitters," ASAE Paper No. 92-2057, St Joseph, 1992. http://www.toromicro-irrigation.com

14

Effect of Saline Water Application through Different Irrigation Intervals on Tomato Yield and Soil Properties

Mahmoud Rahil[1*], Hajaj Hajjeh[1], Alia Qanadillo[2]

[1]Faculty of Agricultural Science and Technology, Palestine Technical University-Kadoorie, Tulkarm, Palestine; [2]Palestinian National Agriculture Research Center, Jenin, Palestine.
Email: *mrahail@yahoo.com

ABSTRACT

A field study was conducted on the experimental farm of ministry of agriculture, located at Palestine Technical University-Kadoorie, to investigate the effects of saline water irrigation through three irrigation intervals on yield of tomato crop and soil properties. The land was prepared and divided into 12 treatments, each of 48 square meters on the first of April. Tomato seedlings were planted on 25 April 2010; the seedlings were irrigated with fresh water for a period of 10 days after planting. Three levels of saline water irrigation (3, 5, 7 dS/m) plus fresh water as control were applied during the growing season. The four irrigation water treatments were applied through three irrigation intervals (every day, every second day and every three days). Gravimetric soil moisture content and soil electrical conductivity were monitored every two weeks during the growing period. Yield measurements were taken for total fruit yield, marketable yield as a percent of total yield, and average fruit weight of each treatment. Results of this study indicated that, plant treatments irrigated with saline water gave the highest yield for treatments irrigated every day compared to the treatments irrigated every second day and every three days. Statistical analysis showed significant differences in yield reduction between every second day and every three days irrigation intervals under 5 and 7 dS/m saline irrigation levels, while there was no significant difference between irrigation intervals under 3 dS/m salinity level.

Keywords: Saline Water; Irrigation Interval; Soil; Tomato; Electrical Conductivity

1. Introduction

The declining availability of fresh water has become a worldwide problem, which endorses the development of alternative, secondary quality water resources for agricultural use. In Palestine, besides water scarcity, water quality is deteriorating and water salinity is increasing due to uncontrolled discharges of untreated or poorly treated wastewater, over-abstraction of the aquifers, and the excessive use of fertilizers in agriculture. Field drainage water, urban wastewater, domestic gray water and saline water are reused and recycled for irrigation in many parts of the world. When saline water is used, several factors have to be considered: plant tolerance, irrigation system, water management strategies, irrigation intervals and soil properties.

Salinity can negatively affect plants through three major components: osmotic, nutritious, and toxic stresses [1,2]. When exposed to salinity, growth, development, and yield of most cultivated crops tend to decline, with consequent reduction in their economic value [3]. However, the response pattern of many crop species may substantially change due to environmental conditions (e.g., soil properties and weather) as well as by agricultural practices [4] (e.g., irrigation methods). Many studies have reported substantial increases in crop yield as a result of suitable irrigation management under saline conditions [3,5-9].

Several studies have indicated that when saline water is used for irrigation, more attention should be given to minimize root-zone salinity [10-12]. Others have indicated the need to select appropriate irrigation systems and practices that will supply a sufficient quantity of water to the root zone to meet the evaporative demand and to minimize salt accumulation inside the root zone [13]. Other approach is to select crops and varieties that can tolerate a degree of water and salinity stress [14,15].

The objective of this work is to study the effects of irrigation with three levels of saline water (3, 5, 7 dS/m)

---

*Corresponding author.

and with fresh water as control, through three irrigation intervals (every day, every second day, and every three days) using drip irrigation system on growth and yield of tomato crop and on soil salinity.

2. Materials and Methods

A field study was conducted in the experimental farm of the ministry of agriculture, located at Palestine Technical University-Kadoorie, to investigate the effect of irrigation with three saline water levels through three irrigation intervals on growth and yield of tomato crop and soil salinity. To conduct this study, the experimental field was prepared and divided into 12 treatments, each of 48 square meters, at the beginning of April 2010. Basic fertilizers at rate of 200 kg/du superphosphate and 150 kg/du ammoniac were applied before planting. A drip irrigation system was installed and characterized by emitter discharge of 4 l/hr and spacing between emitters and laterals at 40 and 100 cm, respectively.

Tomato seedlings were cultivated in an open field, on 25 April 2010. An amount of 101 kg/du compound fertilizer (20-8-11), and 28 kg/du compound fertilizer (14-14-14) were applied during the growing period.

Plants were irrigated with three saline water levels (3, 5, 7 dS/m) and with fresh water as control. Irrigation water treatments were applied through three irrigation intervals (every day, every second day and every three days) with a total of 12 treatments. Saline water was prepared by mixing fresh water with sodium chloride salt in three tanks. Same amount of irrigation water were applied for each treatment during the growing period (Figure 1). At the beginning of the growing period, all treatments were irrigated with fresh water for a period of 10 days.

Soil samples were taken at initial condition at depths of 0 - 20, 20 - 40 & 40 - 60 cm, and every two weeks during the experimental period at depth of 0 - 30 cm. Soil samples were analyzed at a central laboratory in Nablus to study the (soil texture, soil moisture content, soil electrical conductivity, soil pH, Ca, Mg, and Na). Soil electrical conductivity was analyzed using saturation past.

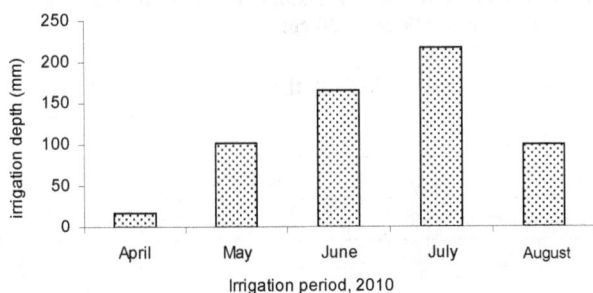

Figure 1. The monthly amount of irrigation water that applied for each treatment during the growing period.

(Table 1) shows the physical and chemical properties of the soil at initial condition. Plant measurements were taken for total yield, marketable yield, and average fruit weight of each treatment.

3. Results and Discussion

3.1. Yield Production of Tomato

The results of this study showed that, the treatments irrigated with fresh water (control) gave the highest yield compared to the treatments irrigated with different saline water levels, particularly for plants irrigated every second day. Under saline irrigation, the highest yield observed for treatments irrigated every day compared to the treatments irrigated every second day and every three days (Figure 2). Reduced total and marketable fruit yield with increasing salinity level was a consequence of reduction in fruit fresh weight (Table 2). [16-18] reported

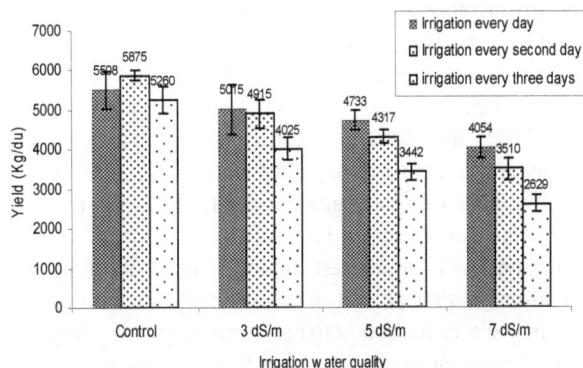

Figure 2. Impact of three levels of saline water irrigation through three irrigation intervals on tomato yield.

Table 1. The physio-chemical properties of the soil.

Depth (cm)	Texture	EC$_e$ (dS/m)	pH	TDS ppm	Ca^{2+} ppm	Mg^{2+} ppm	Na^{1+} ppm	K^{1+} ppm	Cl^{1-} ppm
0 - 20	Sandy clay	0.28	7.3	181.3	3.7	4.7	22.5	3.3	132.9
20 - 40	Sandy clay	0.27	7.3	171.5	3.6	5.1	23.4	3.1	138.4
40 - 60	Sandy clay	0.25	7.3	161.6	3.4	4.7	22.5	2.9	141.2

Table 2. The average fruit weight of tomato under saline water irrigation through three irrigation intervals.

Irrigation salinity level	Irrigation intervals		
	Every day	Every second day	Every three days
	Fruit weight (g)	Fruit weight (g)	Fruit weight (g)
Control	133	134	130
3 dS/m	130	122	120
5 dS/m	126	115	113
7 dS/m	102	97	92

that, the numbers of fruits was not affected by moderate salinity, and the yield reduction was entirely due to smaller fruit. Others found that, the number of harvested fruit per plant reduced with salinity [19-21].

3.2. Relative Yield Reduction

Total and marketable fruit yield decreased with increasing salinity level, and clearly followed Maas and Hoffman model [22], particularly in treatments irrigated every three days. Moreover, there were differences in yield reduction through the three irrigation intervals. Tomato yield reduced 9% under 3 dS/m saline irrigation level when plants irrigated every day, compared to the yield reduction of 16% and 23%, through every second day and every three days irrigation intervals, respectively, under the same salinity level. Under 5 dS/m salinity level the yield reduced 14%, when plants irrigated every day, while the yield reduced 27% and 35%, for treatments irrigated every second day and every three days, respectively, under the same salinity level. In addition to that, total yield reduced only 26%, under 7 dS/m salinity level when plants irrigated every day, compared to yield reduction of 40% and 50% for treatments irrigated every second day and every three days, respectively (**Table 3**). Statistically, there were no significant differences in yield reduction between irrigation intervals under 3 dS/m saline irrigation level. In contrast, under 5 and 7 dS/m saline irrigation, statistical analysis showed significant differences in yield reduction between every day and every three days irrigation intervals (**Figure 3**). These results indicated that, when using highly saline water for irrigation purposes, short irrigation interval (every day) is recommended to decrease the yield reduction caused by salinity stress.

3.3. Soil Moisture Content

The soil moisture content status depended on the irrigation intervals and salinity levels of irrigation water. Irrespective of the irrigation intervals, the gravimetric soil moisture content (θ_w) of the treatments irrigated with fresh water (control) was lower than that under different levels of saline water irrigation (**Figure 4**). This explains the potential of plant to uptake much water under fresh

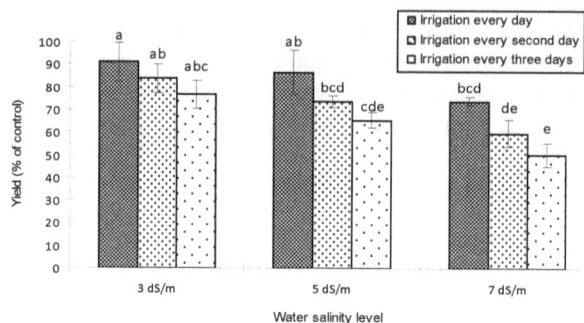

Figure 3. Impact of saline water irrigation levels through three irrigation intervals on tomato yield reduction. Treatments with the same letters do not differ significantly (P < 0.05).

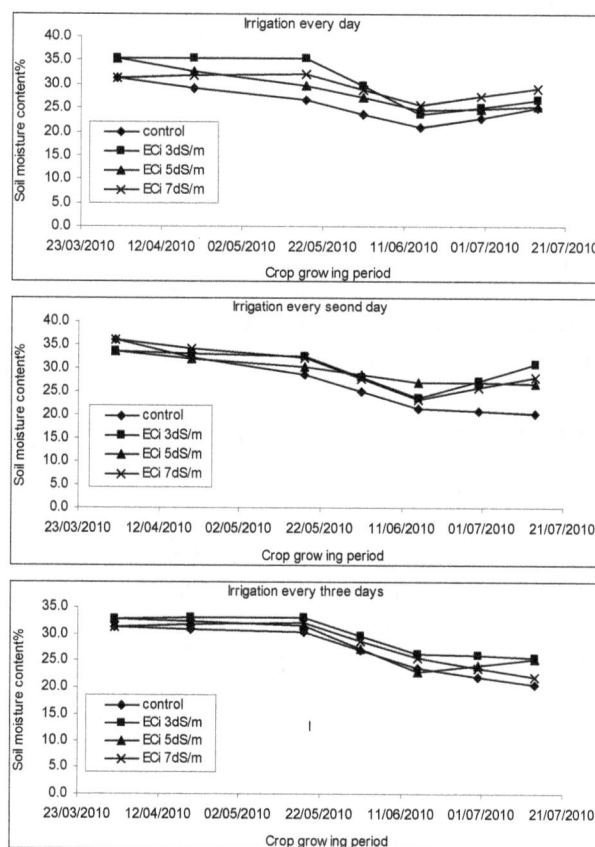

Figure 4. Impact of saline water irrigation through three irrigation intervals on soil moisture content during the growing period at depth of 0 - 30 cm.

Table 3. Total yield of tomato and percent of yield reduction under saline water irrigation through three irrigation intervals.

Irrigation salinity level	Irrigation intervals					
	Every day		Every second day		Every three days	
	Total yield (Kg/du)	Yield reduction (%)	Total yield (Kg/du)	Yield reduction (%)	Total yield (Kg/du)	Yield reduction (%)
3 dS/m	5015 ± 605	9	4915 ± 356	16	4025 ± 278	23
5 dS/m	4733 ± 247	14	4317 ± 168	27	3442 ± 207	35
7 dS/m	4054 ± 260	26	3510 ± 286	40	2629 ± 203	50

water irrigation without water stress. [23,24] found that, tomato plants irrigated with saline water transpire less water than when fresh water is used. [25] indicated that, irrespective of irrigation interval, the volumetric soil moisture under saline water treatment was higher than that under good quality water treatments.

3.4. Soil Electrical Conductivity and pH

The results of soil analysis before planting showed that the initial electrical conductivity of the soil was 0.28 dS/m, while the electrical conductivity reached up to 4 dS/m at the end the experiment for the three irrigation intervals, with the highest salinity level at the treatments irrigated every day and every second days mainly under 7 dS/m saline irrigation level (**Figure 5**). The low irrigation intervals imposed a more rapid salt accumulation in the root zone, which was ascribed to restriction of the volume of drainage solution.

[26-28] reported that one consequence of reducing irrigation water use by deficit irrigation is the greater risk of increased soil salinity due to reduced leaching. The soil pH was in the normal range and was not affected by the saline irrigation during the growing period (**Figure 6**).

4. Conclusions

Evidence from this study indicated that, under moderate

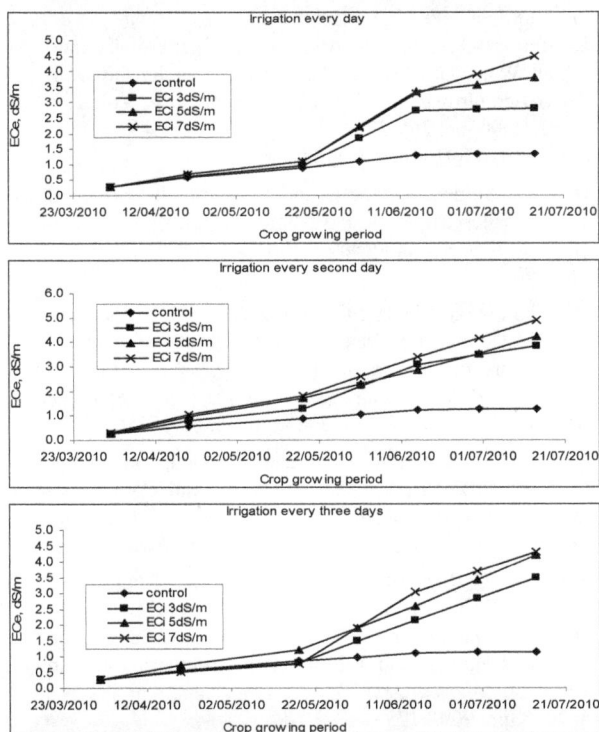

Figure 5. Impact of saline water irrigation through three irrigation intervals on soil electrical conductivity during the growing period at depth of 0 - 30 cm.

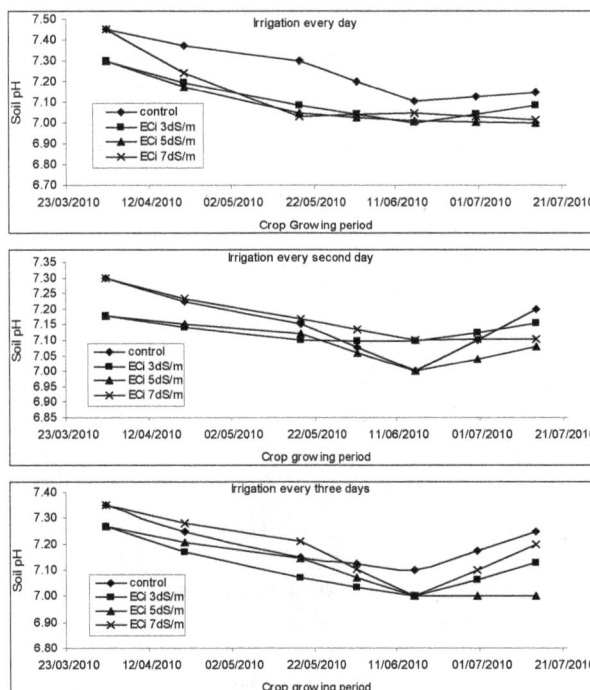

Figure 6. Impact of saline water irrigation through three irrigation intervals on soil pH during the growing period at depth of 0 - 30 cm.

saline irrigation (3 dS/m) the tomato yield production was not significantly affected by irrigation intervals, while under highly saline irrigation (5, 7 dS/m), the yield production significantly affected by irrigation intervals. Accordingly, when highly saline water is used for irrigation, it is recommended to use short irrigation interval (one day interval) instead of applying irrigation every three or four day as it is practiced by the farmers. However, the short irrigation interval practice normally reduces the plant stress under saline irrigation.

Regarding the salt accumulation in the root zone, the results of this study showed an increase in soil salinity that reached up to 4 dS/m at the end of the growing period, particularly under highly saline irrigation. This emphasizes the need for conservation of soil properties besides the yield production of tomato under long-term saline irrigation.

REFERENCES

[1]   A. Lauchli and E. Epstein, "Plant Responses to Saline and Sodic Conditions," In: K. K. Tanji, Ed., *Agricultural Salinity Assessment and Management. Manuals and Reports on Engineering Practice*, ASCE, New York, 1990, pp. 113-137.

[2]   R. Munns, "Physiological Processes Limiting Plant Growth in Saline Soil: Some Dogmas and Hypotheses," *Plant, Cell & Environment*, Vol. 16, No. 1, 1993, pp. 15-24. doi:10.1111/j.1365-3040.1993.tb00840.x

[3]   D. Pasternak and Y. De Malach, "Crop Irrigation with

Saline Water," In: M. Pessarakli, Ed., *Handbook of Plant and Crop Stress*, Marcel Dekker, Inc., New York, 1995, pp. 599-622.

[4] M. Shannon and C. Grieve, "Tolerance of Vegetable Crops to Salinity," *Horticulture Science*, Vol. 78, No. 1-4, 1999, pp. 5-38.
doi:10.1016/S0304-4238(98)00189-7

[5] A. Bustan, M. Sagi, Y. De Malach and D. Pasternak, "Effects of Saline Irrigation Water and Heat Waves on Potato Production in an Arid Environment," *Field Crops Research*, Vol. 90, No. 2-3, 2004, pp. 275-285.
doi:10.1016/j.fcr.2004.03.007

[6] N. Malsh, T. Flowers and R. Ragab, "Effect of Irrigation Systems and Water Management Practices Using Saline and Non-Saline Water on Tomato Production," *Agricultural Water Management*, Vol. 78, No. 1-2, 2005, pp. 25-38.

[7] S. Jalota, A. Sood, G. Chahal and B. Choudhury, "Crop Water Productivity of Cotton," *Agricultural Water Management*, Vol. 84, No. 1-2, 2006, pp. 137-146.
doi:10.1016/j.agwat.2006.02.003

[8] M. Ali, M. Hoque, A. Hassan and M. Khair, "Effect of Deficit Irrigation on Yield, Water Productivity and Economic Returns of Wheat," *Agricultural Water Management*, Vol. 92, No. 3, 2007, pp. 151-161.

[9] K. Nagaz, I. Toumi, M. Masmoudi and N. Ben Mechlia, "Soil Salinity and Barley Production under Full and Deficit with Saline Water in Arid Conditions of Southern Tunisia," *Research Journal of Agronomy*, Vol. 2, No. 3, 2008, pp. 90-95.

[10] U. Shani and L. Dudley, "Field Studies of Crop Response to Water and Salt Stress," *Soil Science Journal*, Vol. 65, No. 5, 2001, pp. 1522-1528.
doi:10.2136/sssaj2001.6551522x

[11] O. Gideon, Y. DeMalach, L. Gillerman, I. David and S. Lurie, "Effect of Water Salinity and Irrigation Technology on Yield and Quality of Pears," *Biosystem Engineering*, Vol. 81, No. 2, 2002, pp. 237-247.

[12] N. Katerji, J. van Hoorn, A. Hamdy and M. Mastrorilli, "Comparison of Corn Yield Response to Plant Water Stress Caused by Salinity and by Drought," *Agricultural Water Management*, Vol. 65, No. 2, 2004, pp. 95-101.

[13] R. Munns, "Comparative Physiology of Salt and Water Stress," *Plant, Cell & Environment*, Vol. 25, No. 2, 2002, pp. 239-250.

[14] G. Hammer and I. Broad, "Genotype and Environment Effects on Dynamics of Harvest Index during Grain Filling Sorghum," *Agronomy Journal*, Vol. 95, No. 1, 2003, pp. 199-206. doi:10.2134/agronj2003.0199

[15] L. Feitosa, J. Cambraia, M. Oliva and H. Ruiz, "Changes in Growth and in Solute Concentration in Sorghum Leaves and Roots during Salt Stress Recovery," *Environmental and Experimental Botany*, Vol. 54, No. 1, 2005, pp. 69-76.

[16] C. Sonneveld and G. Welles, "Yield and Quality of

[17] Y. Li, C. Stanghellini and H. Challa, "Effect of Electrical Conductivity and Transpiration on Production of Greenhouse Tomato (*Lycopersicon esculentum* L.)," *Horticulture Science*, Vol. 88, No. 1, 2001, pp. 11-29.
doi:10.1016/S0304-4238(00)00190-4

[18] R. Eltez, Y. Tuzel, A. Gul, I. Tuzel and H. Duyar, "Effect of Different EC Levels of Nutrient Solution on Greenhouse Tomato Growing," *Acta Horticulture*, Vol. 573, 2002, pp. 443-448.

[19] P. Adams and L. Ho, "Effect of Constant and Fluctuating Salinity on the Yield, Quality and Calcium Status of Tomatoes," *Journal of Horticulture Science*, Vol. 64, No. 6, 1989, pp. 725-732.

[20] W. Van Ieperen, "Effect of Different Day and Night Salinity Levels on Vegetative Growth, Yield and Quality of Tomato," *Journal of Horticulture Science*, Vol. 71, 1996, pp. 99-111.

[21] J. Magan, M. Gallardo, R. Thompson and P. Lorenzo, "Effect of Salinity on Fruit Yield and Quality of Tomato Grown in Soil-Less Culture in Greenhouse in Mediterranean Climatic Conditions," *Agricultural Water Management*, Vol. 95, No. 9, 2008, pp. 1041-1055.

[22] E. Maas and G. Hoffman, "Crop Salt Tolerance: Current Assessment," *Journal of Irrigation*, Vol. 103, No. 2, 1977, pp. 115-134.

[23] T. Soria and J. Cuartero, "Tomato Fruit Yield and Water Consumption with Saline Water Irrigation," *Acta Horticulture*, Vol. 458, 1997, pp. 215-219.

[24] R. Romero-Aranda, T. Soria and J. Cuartero, Tomato "Plant Water-Uptake and Plant-Water Relationships under Saline Growth Conditions," *Plant Science*, Vol. 160, No. 2, 2000, pp. 265-272.
doi:10.1016/S0168-9452(00)00388-5

[25] B. Ould Ahmed, M. Inoue and S. Moritani, "Effect of Saline Water Irrigation and Manure Application," *Agricultural Water Management*, Vol. 97, No. 1, 2010, pp. 165-170.

[26] G. Schoups, J. Hopman, C. Young, J. Vrugt, W. Wallender, K. Tanji and S. Pandy, "Sustainability of Irrigated Agriculture in the San Joaquin Valley," *Proceedings of the National Academy of Sciences of the United States of America*, Vol. 102, No. 43, 2005, pp. 15352-15356.

[27] H. Kaman, C. Kirda, M. Cetin and S. Topcu, "Salt Accu-Mulation in the Root Zone of Tomato and Cotton Irrigated with Partial Root-Drying Technique," *Irrigation and Drainage*, Vol. 55, No. 5, 2006, pp. 533-544.

[28] S. Geerts, D. Raes, M. Garcia, O. Condori, J. Mamani, R. Miranda, J. Cusicanqui, C. Taboada and J. Vacher, "Could Deficit Irrigation Be a Sustainable Practice for Quinoa (*Chenopodium quinoa* Willd.) in the Southern Bolivian Altiplano," *Agricultural Water Management*, Vol. 95, No. 8, 2008, pp. 909-917.

15

Economic Potential of Compost Amendment as an Alternative to Irrigation in Maine Potato Production Systems[*]

John M. Halloran[1], Robert P. Larkin[1], Sherri L. DeFauw[2], O. Modesto Olanya[1], Zhongqi He[1]

[1]USDA Agricultural Research Service, New England Plant, Soil, and Water Laboratory, Orono, ME, USA; [2]Department of Agricultural Economics and Rural Sociology, The Pennsylvania State University, University Park, PA, USA
Email: john.halloran@ars.usda.gov

ABSTRACT

Potato productivity in the northeastern US has been relatively constant for over 50 years, raising questions about what factors are limiting productivity. Research was initiated in 2004 to identify key constraints to potato productivity by evaluating Status Quo (SQ), Soil Conserving (SC), and Soil Improving (SI) cropping systems under both rainfed and irrigated management, and it was found that addition of compost or irrigation substantially increased yield. In this study, we employed partial budgeting to determine cost differences and their impact on net revenue for these cropping systems. Differences in systems were primarily associated with rotation length, tillage operations, compost and application expenses, and water management practices. When compost (as composted dairy manure) was annually applied at 19 $Mg \cdot ha^{-1}$ and evaluated over the entire 3-year crop rotation cycle, the compost-amended rainfed SI system was more expensive to maintain than the irrigated SC system if compost cost exceeded $3.63 Mg^{-1}. Average marketable yields were used to calculate gross and net revenue for each system. Because average potato yield for the irrigated SQ system (28.4 $Mg \cdot ha^{-1}$) equaled that in the rainfed SI system (28.3 $Mg \cdot ha^{-1}$), we were able to compare cost of irrigation versus compost for achieving comparable yield. The compost-amended SI system under rainfed management generated more net revenue from the potato crop than the irrigated SQ system when compost costs were less than $7.42 Mg^{-1}. When compared to the commonly used rainfed SQ system, rainfed SI achieved higher net revenue as long as compost cost was less than $22.95 Mg^{-1}. The rainfed SI system achieved higher net revenue than the irrigated SC system when compost cost was $9.43 Mg^{-1}or less, but generated greater net revenue than the rainfed SC system regardless of compost costs, due to substantially higher yields associated with compost amendment. This investigation demonstrates that compost is a potentially viable substitute to irrigation for potato in the northeastern US; however, such potential is highly dependent on suitable compost sources and application costs.

Keywords: Compost; Cropping Systems; Economic Potential; Irrigation; Partial Budgeting; Potato Production; Water Stress; Yield

1. Introduction

The Maine potato industry is a major contributor to the State's economy, annually generating $540 million in direct, indirect and induced impacts, and supporting employment for more than 6000 people [1]. However, the industry has contracted to about one-third of the land base used in the 1950s; during the late1940s Maine was the largest producer of potatoes in the country [2]. Much of the reduction can be attributed to increased irrigated production promoting substantially elevated yields in other regions of the United States such as the Pacific Northwest.

Other factors have also contributed to the drop in Maine's competitive position. As shown in **Figure 1** [2], potato productivity in Maine has remained relatively constant over the past five decades, while productivity in other Fall potato producing states has increased. This raises the question, "what factors are limiting productivity in Maine potato systems?"

One possible limitation may be related to the soil itself. The potential benefits of incorporating biological amendments (manure, compost, green manures, etc.) to soil are well recognized. In potato systems, these amendments increase soil carbon and nitrogen, improve soil structural

[*]Mention of trade names or commercial products in this article is solely for the purpose of providing specific information and does not imply recommendation or endorsement by the US Department of Agriculture.

characteristics, improve moisture-holding capacity, and may confer disease-suppressive advantages [3-9]. These attributes may be especially appealing to producers with limited water and/or degraded soil resources. Successfully implemented, soil amendments can have a positive impact on profitability and risk.

Farmers look not only at the biological potential of a crop, but also the variability in yields from season to season [10]. Yield fluctuations are caused by variable weather conditions, pest pressures, nutrient levels and other factors [11]. Due to its shallow root system, Potato (*Solanum tubersum* L.) is very sensitive to weather-related variations [12,13]. Wide fluctuations in seasonal yield that may be compounded by constraints in the plant-available water supply from degraded potato production soils make management more difficult, and also increase the level of economic risk.

The standard industry practice in Maine potato cropping systems that has prevailed for the past several decades has been a 2-year rotation with barley (or another small grain). During the potato phase of the rotation the soil is intensively tilled and undergoes other mechanical operations (hilling, spraying, fertilization, vine-kill, and harvesting). This intensity can lead to reductions in soil organic carbon concentration and structural stability and adversely affect yield potential [14-16].

Porter *et al.* [3] conducted long-term evaluations of soil management practices and supplemental irrigation and their impacts on soil properties, tuber yield, and quality in Maine. They hypothesized that water stress can reduce potato yield and quality. Treatments consisted of soil management (a green manure at the start of the rotation, and annual applications of manure, and compost) under both rainfed and irrigated conditions. After the first growing season, amended plots showed improved soil properties (e.g. increased soil organic matter, increased cation exchange capacity). After two years of amend-

ment contributions, bulk density decreased. Amendment and irrigation both resulted in significant yield increases. Opena and Porter [13] evaluated the use of soil amendments rich in organic carbon in combination with supplemental irrigation to promote potato root growth. The amended systems had increased root length density, which was correlated with tuber yield. Supplemental irrigation showed positive impacts in one out of two growing seasons.

Under intensively tilled short rotations, managing soil carbon and nitrogen can be difficult. Griffin and Porter [16] examined the results of a long-term trial to assess the effect of cover crops, green manure, and amendment application frequency on soil carbon and nitrogen pools. They concluded cover crop and green manure had little impact on total carbon and nitrogen content. However, a single application of paper mill sludge, animal manure and/or compost increased carbon and nitrogen by 25% - 53%. It was found that these latter management practices led to large increases in total particulate organic matter and microbial biomass carbon and nitrogen.

As previously mentioned, one criterion producers evaluate for crop selection is stability in yields. Mallory and Porter [17] reported on potato yield variability from a long-term (13 years) potato cropping systems study conducted under contrasting soil management strategies. Of particular interest were yield differences between amended and non-amended potato systems. They found the amended systems increased total yield by up to 54% with US no. 1 yield gains up to 36%. In addition, the amended soil management system promoted greater yield stability, particularly in seasons with low rainfall.

A recurring theme of these studies is that increasing soil organic matter can decrease yield risk. Porter *et al.* [3] suggested that incorporating biological amendments may be an alternative to supplemental irrigation. Although adequate precipitation is generally received in Maine, the timing of that rainfall often does not occur at critical stages of crop growth. Benoit and Grant [18] computed a plant water deficit index and compared this to the actual evapotranspiration over the growing season. Based on 30 years of weather data, they found that even in the wettest years, potato still faced periods of five days or more with insufficient water. In subsequent research, Benoit and Grant [19] found that potato yields were also adversely affected by too much water and recommended the adoption of both supplemental irrigation and improved drainage. The Maine Irrigation Guide [20] projected that in 9 out of 10 years potatoes will benefit from supplemental irrigation. Authors of the guide also reported that irrigation will increase yield of US no.1 tubers by an estimated 38%.

Given the interest in both supplemental irrigation and organic amendments in Maine, it is surprising that little

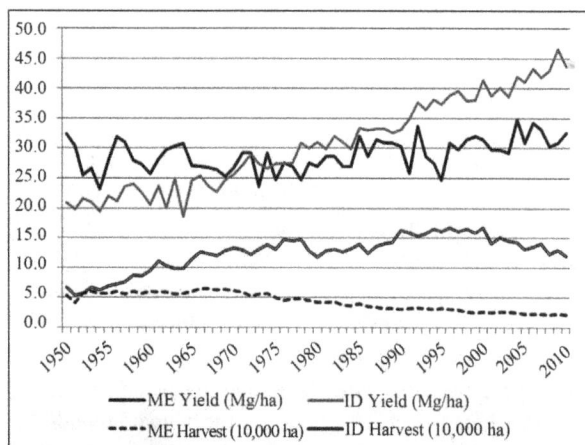

Figure 1. Historical fall potato yields and total harvested areas for Maine and Idaho from 1950 to 2010.

research has been conducted on the economics of these cultural practices. An exception is the research conducted by Dalton *et al.* [21] they compared the risk management benefits of supplemental irrigation versus federal crop insurance programs and uninsured rainfed crop production. An expected utility framework was used to differentiate the programs. They found for large-scale contiguous field operations (>81 ha), there are risk management benefits for water development. The impact on smaller operations with respect to managing risk was less certain. Dalton *et al.* [21] concluded that supplemental irrigation is preferred by the risk-averse producer over buying crop insurance. They also found that the risk management benefits are technology dependent. Thus, while supplemental irrigation can reduce production risk, the ability of producers to deploy this practice is dependent on scale factors (small vs. large) and technology (fixed vs. portable system).

The analysis by Dalton *et al.* [21] demonstrates the complexity of the irrigation decision. Unless conditions are favorable (*i.e.*, appropriate scale and technology), irrigation may not be economically beneficial due to uncertainty regarding weather events, future costs and revenues generated by this practice. The objectives of our investigation were to determine if improving soil properties by adding compost can achieve the same risk reduction benefits and if compost amended potato systems were comparable in cost to supplemental irrigation.

2. Material and Methods

2.1. Field Study

Field experiments were established in 2004 that encompass different categories of cropping systems designed to identify limitations to sustainability. Systems were designed and managed as 1) Status Quo (SQ); 2) Soil Conserving (SC); and 3) Soil Improving (SI) (**Table 1**). The SQ system consisted of a barley-potato 2-year rotation (typical for potato growers). The SC system was a 3-year rotation of no-till barley underseeded with timothy (forage grass) in Year 1, timothy alone in Year 2, and potato followed by straw mulch in Year 3. The SI system built upon the SC system by adding composted dairy manure at a dry weight equivalent of 19 Mg·ha^{-1} in each phase of the rotation. All rotation entry points were grown each year under both irrigated and rainfed management (thus potato crops were harvested each year); 5 replications were arranged in a split-block design, with water management as the main block and cropping system as the sub-plot. Each cropping system by water management sub-plot combination was 4 m wide by 16 m long, with a 16 m buffer between sub-plots to ensure the separation of water management treatments which were applied with a lateral, overhead sprinkler irrigation system. Irrigation

water (1.25 cm) was applied to all irrigated treatments, and application was triggered when 25% of the tensiometers placed in irrigated plots registered 0.5 MPa. Insects and weeds were controlled in all plots using commercially available pesticides (and tillage) following University of Maine recommendations. Each system was evaluated by our interdisciplinary team for plant growth and productivity, soil chemical-physical-biological properties, tuber diseases, soil-borne diseases, foliar diseases, economics, and their interactions. Additional details of the set-up and methodologies of the field experiment, as well as system effects on soilborne disease and soil microbial communities are available in Larkin *et al.* [7]. Rotations and treatments were maintained each year and data were collected from potato crops for this study from 2004 to 2008.

2.2. Economic Analysis

A partial budgeting approach was employed to determine cost differences and their impact on net revenue [22]. Partial budgeting includes only those costs that vary from one enterprise (system) to another. For example, since all systems received the same fertilization regime these costs are not included. Costs that varied were associated with tillage operations, compost and its application, and water management practices (*i.e.*, rainfed or irrigated).

To evaluate the differences in costs related to potato cropping systems utilizing compost versus irrigated and non-irrigated systems, associated costs were determined on an annual basis as well as for the full 3-year rotation period. Costs for the 3-year rotation includes all costs for each crop in the rotation (costs of seed, planting, tillage, crop maintenance), including crops that derive no revenues (such as timothy forage crop). We evaluated the cropping systems with compost costs ranging from $0 Mg^{-1} to $27.21 Mg^{-1}. Irrigation costs were based on values from Dalton *et al.* [21] adjusted by the spring 2008 producer price index for agriculture from the Bureau of Labor Statistics. The cost for spreading compost was based on the average NASS custom rate from Pennsylvania.

Table 1. Characteristics of the cropping systems evaluated in this study comparing a typical 2-year rotation (Status Quo, SQ) with 3-year Soil Conserving (SC) and Soil Improving (SI) alternative systems.

Cropping system	Year one	Year two	Year three
Status Quo (SQ) (2-yr)	Barley/ red clover	Potato	N/A
Soil Conserving (SC)	Barley/timothy	Timothy sod	Potato
Soil Improving (SI)	Barely/timothy + Compost	Timothy + Compost	Potato + Compost

Additional economic data collected over a 5-year cropping period (2004-2008) included marketable yields and revenue. Marketable yield for potatoes was determined as the weight of all tubers that were 114 g or greater, which corresponds to commercial standards. Revenue was determined from the marketable yield based on average market prices from NASS figures for Maine for each harvest year. Average marketable yields were used to calculate gross and net revenue for each system over the course of the study.

3. Results and Discussion

3.1. Costs and Revenues

Differences in costs related to potato cropping systems utilizing compost versus irrigated and non-irrigated systems were determined on an annual basis as well as for the 3-year rotation period. The relative annual cost of irrigation was estimated at 9.94% of the total potato production costs, whereas costs associated with compost included both the cost of spreading it on the fields (estimated at 2.6% of production costs) and the cost of acquiring the compost (which ranged from 0% to 27.1% of total potato production costs as compost cost increased from 0 to $27.21 Mg^{-1}). Cost of the rainfed SI cropping system ranged from 90% to 165% of the cost of the irrigated SQ barley-potato cropping system (**Figure 2(a)**). The rainfed SI system became more costly than the irrigated SQ system once compost costs exceeded $3.63 Mg^{-1}. Cost of the rainfed SI system ranged from 115% to 210% of the cost of rainfed SQ system over a 3-year cropping cycle (**Figure 2(a)**). As shown in **Figure 2(b)**, the SI system was always higher in cost compared to the SC system over the 3-year rotation interval, except when the compost cost dropped below $1.81 Mg^{-1} and the SC system was irrigated.

Gross and net revenue for each system were calculated from average marketable yields. Interestingly, average potato yield for the irrigated SQ system (28.4 Mg·ha^{-1}) equaled that obtained in the rainfed SI system (28.3 Mg·ha^{-1}). Consequently, this parity in yield outcomes provided a rather unique opportunity to compare costs of irrigation vs. compost to achieve the same overall yield. In this comparison, the SI system under rainfed management generated more net revenue from the potato crop than the irrigated SQ system if compost costs were under $7.42 Mg^{-1} (**Table 2**).

The rainfed SI system was also compared to the rainfed SQ system. With compost cost ranging from 0 to $27.21 Mg^{-1}, net revenue from potato in the SI system ranged from $391.94 to $208.06 (**Table 2**). Net revenue from rainfed potato in the SI system exceeded that in the SQ system when compost cost less than $22.95 Mg^{-1}. Comparing only the potato years, this suggests that

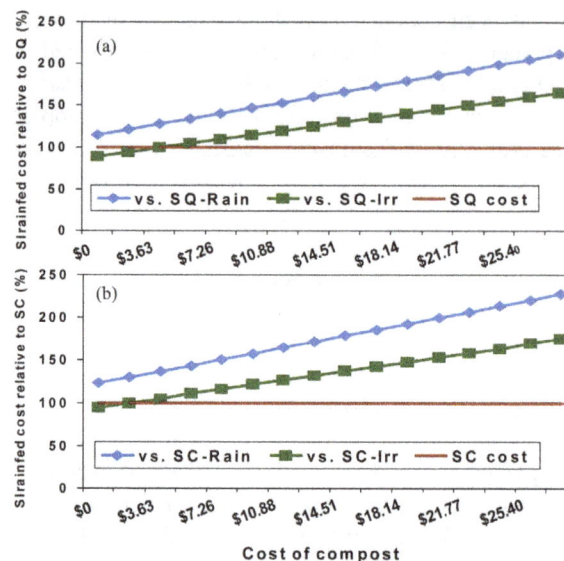

Figure 2. Comparison of production costs for Soil Improving (SI) rainfed cropping system (with compost) in relation to A) Status Quo (SQ) (standard 2-year rotation) and B) Soil Conserving (SC) (3-year system, without compost) cropping systems under both rainfed and irrigated conditions for the entire 3-year cropping cycle, and expressed as a percentage of costs for each system. Horizontal (red) line represents same cost as comparison system (100%).

growers using the traditional barley-potato rotation under rainfed management could possibly increase their net revenue by applying compost when the cost of this amendment was below $22.95 Mg^{-1}.

For making compost versus irrigation comparisons over an entire cropping system, the two most appropriate management strategies to consider were the rainfed SI system and the irrigated SC system. This is because the same 3-year barley-timothy-potato rotation was used in both systems, but compost was added only to the SI system. The SQ system (a 2-year rotation) restricted systems level comparisons with the 3-year rotation systems due to the unequal rotation length and associated cost disparities. When considering just the potato crop alone, the rainfed SI system returned greater net revenue than the irrigated SC system at all compost prices evaluated (**Table 2**). If all three crops (barley-timothy-potato) were included in the analysis, net revenue was less negative in the rainfed SI system compared to the irrigated SC system when compost cost was below $9.43 Mg^{-1} (**Table 2**). In this case, although production costs were higher for the compost-amended SI system than the SC system even at relatively low compost cost (**Figure 2(b)**), overall revenues were higher for the SI system due to substantially higher yields associated with compost amendment. For example, average marketable potato yield for the rainfed SI system yield was 28.3 Mg·ha^{-1} compared to 21.7 Mg·ha^{-1} in the irrigated SC system, which represents an

Table 2. Net revenue generated from three cropping systems under irrigated or rainfed conditions in relation to variable costs of compost for a single potato crop and over the entire cropping cycle for each cropping system.

Cropping system[a]	Cost of Compost ($/Mg)	Net Revenue from One Potato Crop ($/ha)	Net Revenue from 3-yr cropping system ($/ha)
SI rainfed	0.00	391.94	89.05
	1.81	351.94	(30.95)
	3.63	311.94	(150.95)
	5.44	271.94	(270.95)
	7.26	231.94	(390.95)
	9.07	191.94	(510.95)
	10.88	151.94	(630.95)
	12.70	111.94	(750.95)
	14.51	71.94	(870.95)
	16.33	31.94	(990.95)
	18.14	(8.06)	(1110.95)
	19.95	(48.06)	(1230.95)
	21.77	(88.06)	(1350.95)
	23.58	(128.06)	(1470.95)
	25.40	(168.06)	(1590.95)
	27.21	(208.06)	(1710.95)
SQ irrigated		228.51	259.86[b]
SQ rainfed		(114.17)	(83.65)[b]
SC irrigated		(237.47)	(534.80)
SC rainfed		(173.87)	(112.67)

[a]Cropping systems: SQ = Status Quo system, 2-year (barley/red clover-potato) standard rotation; SC=Soil Conserving system, 3-year (barley/timothy-timothy-potato), reduced tillage; SI = Soil Improving system, same as SC (3-year, barley/timothy-timothy-potato), but with yearly compost amend- ment (composted dairy manure at 19 Mg·ha^{-1}) added. [b]Since SQ system is a 2-year rather than a 3-year rotation, these values reflect net revenues over the 2-year cropping system instead of 3 years.

average yield increase of 30%. It also should be noted that in these 3-year cropping systems, the middle year of the rotation consisted of timothy, a forage grass crop, which yielded no revenue. Thus the net revenue over the 3-year rotation is substantially lower than for the potato crop year alone due to this lack of revenue for the timothy rotation crop.

3.2. Risk and Stability

In 2004, each entry point of each crop rotation was initiated so that yields were available from five full years of potato crops, although the effect of the full rotation was not observed until the 2006-2008 growing seasons. When the response to compost was evaluated by comparing yield in the rainfed SI system to yield in the irrigated and rainfed SQ systems, in only one year was the response

negative when compared to the irrigated SQ system; it was never negative when compared to the rainfed SQ system. Using these same comparisons, the response to compost resulted in at least a 20% yield increase in five out of ten potato crops (both irrigated and non-irrigated), and over 10% increase in yield in eight out of ten potato crops (**Figure 3**). This represents an average yearly yield increase due to compost amendment of 11% and 33% compared to the irrigated and non-irrigated SQ system, respectively, over the full five years of the study. The yield benefit from compost amendment was even greater when compared to the SC system, in which the yield from the composted SI rainfed system was at least 40% greater than that from the SC system in four out of ten potato crops, and at least 15% greater in every potato crop from 2004 to 2008 (**Figure 3**), representing an average yield increase due to compost amendment of 28% and 47% compared to the irrigated and non-irrigated SC system, respectively.

3.3. Additional Considerations

Our economic analysis indicates that the SI cropping system, as presently configured, does not offer the most economically advantageous system in the short-term (less than or equal to 4 years from implementation of amendment additions). There does appear to be a greater likelihood of conferring an advantage for the potato crop when compared to the conventional SQ rotation in rainfed circumstances, and an advantage over the conventional SQ rotation in the irrigated system depending on compost costs. The strength of the compost system comes from utilizing the compost as a moisture regulator in the soil, to buffer the crops through high moisture levels and conserve moisture in the soil through moisture

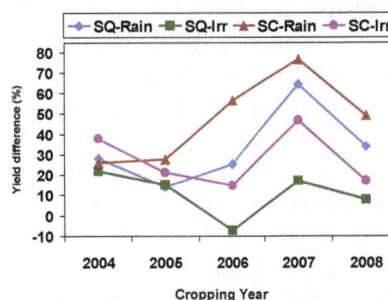

Figure 3. Relative change in marketable yield (%) comparing tuber production from SI rainfed cropping system to other cropping systems (SQ and SC, rainfed and irrigated) in potato crops from 2004 to 2008. SQ = Status Quo system, 2-year (barley/red clover-potato) standard rotation; SC = Soil Conserving system, 3-year (barley/timothy-timothy-potato), reduced tillage; SI = Soil Improving system, same as SC (3-year, barley/timothy-timothy-potato), but with yearly compost amendment (composted dairy manure at 19 Mg·ha^{-1}) added.

deficient periods.

Previous researchers have quantified that in three out of every four years, there is at least one 5-day period where moisture levels are lower than optimal for potato production [18]. This indicates that supplemental applications of water may be needed, but generally only over very short periods of time during the growing season. Thus, if compost can improve soil quality so that surplus water drains more effectively [23,24], and plant-available water holding capacity is enhanced [25], then compost amendments may serve as an important alternative to supplemental irrigation for improving yield and reducing risk. And use of compost amendments as an alternative to irrigation does not have the high initial development costs inherent in establishing an irrigation system.

Other studies have indicated that compost-enriched soil can also reduce erosion, alleviate soil compaction, increase soil microbial abundance and diversity as well as help control diseases and pest infestations in plants [7-8,26-30]. Potato is in the top tier of crops with the highest erosion risk [31,32]. Additionally, in some production settings harvest erosion rates may be of the same order of magnitude (approximately 10 $Mg \cdot ha^{-1} \cdot yr^{-1}$) as water and tillage erosion on sloping land. Tiessen *et al.* [33] reported soil losses of 20 - 100 $Mg \ ha^{-1} \cdot yr^{-1}$ from convex landscape positions in commercial potato production systems of Atlantic Canada; these observations were linked to crop yield reductions of up to 40%. Recent geospatial assessments of agri-environmental indicators (combining farmland and erodibility classifiers) based on a 3-year production footprint showed that close to 85% of Maine potato systems soils were either "potentially highly erodible" or "highly erodible" [34,35]. These soils require the highest standards in soil conservation practices. Investigators in Atlantic Canada found that the use of a bulk yield monitoring device with GPS helped resolve the yield benefits of an alternative residue management strategy compared to the conventionally managed portion of the field [36].

High-resolution investigations that detail crop yield in topographic/hydrodynamic contexts and assess fine-scale soil resource risks (preferably farmscape-assemblage to watershed- or subregional-level) for key "food security" cropping systems such as potato using geographic information systems (GIS) based approaches are needed [37]. Geospatial frameworks help resolve patterns and trends in production environments (at multiple scales). Implementation of these technologies will help farmers build detailed archives on site-specific production constraints that may enable improvements in adaptive management strategies which enhance yield and increase whole-farm profitability. Longer-term beneficial outcomes from compost use tailored to potato systems have the potential to improve soil quality and health and increase plant productivity while possibly reducing alternative management costs such as deep tillage to alleviate compaction and pest/pathogen applications.

Another value that can be attributed to compost is its plant nutrition content. This could significantly lower fertilization costs. The value of the P and K fertilizer used for these cropping trials varied from year to year due to market conditions. For example, application of 2.5 cm of mature compost containing 1% P_2O_5 will add 840 kg of this nutrient per hectare [38]. In addition, since K is not incorporated into soil organic matter, K furnished by finished compost is much more available for plant uptake compared to N and P [38]. The potential offset of costs by using compost to meet P and K crop requirements ranged from approximately $185 ha^{-1} to over $370 ha^{-1}. Considering this fertilizer cost offset makes the rainfed SI system more profitable than the irrigated SC system at any compost cost when evaluated for a single potato crop (**Table 3**), and at compost costs less than approximately $15 Mg^{-1} for the entire three-year rotation system (**Table 3**).

Table 3. Net revenue generated by the Soil Improving (SI) cropping system under rainfed conditions and with variable compost costs (as updated to reflect lower fertilizer costs) compared with the Soil Conserving (SC) cropping system under both rainfed and irrigated conditions for one potato crop as well as the full 3-year cropping system.

Cropping system	Cost of Compost ($/Mg)	Net Revenue from One Potato Crop ($/ha)	Net Revenue from 3-yr cropping system ($/ha)
SI rainfed[a]	0.00	597.94	463.05
	1.81	557.94	343.05
	3.63	517.94	223.05
	5.44	477.94	103.05
	7.26	437.94	(16.95)
	9.07	397.94	(136.95)
	10.88	357.94	(256.95)
	12.70	317.94	(376.95)
	14.51	277.94	(496.95)
	16.33	237.94	(616.95)
	18.14	197.94	(736.95)
	19.95	157.94	(856.95)
	21.77	117.94	(976.95)
	23.58	77.94	(1096.95)
	25.40	37.94	(1216.95)
	27.21	(2.06)	(1336.95)
SC irrigated		(237.47)	(534.80)
SC rainfed		(173.87)	(112.67)

[a]Cropping systems: SC = Soil Conserving system, 3-year (barley/timothy-timothy-potato), reduced tillage; SI=Soil Improving system, same as SC (3-year, barley/timothy-timothy-potato), but with yearly compost amendment (composted dairy manure at 19 $Mg \cdot ha^{-1}$) added.

4. Conclusions

Our analyses indicate that compost amendments may be a viable alternative to supplemental irrigation. One major advantage is that it requires no large capital outlay. For example, as Dalton *et al.* [21] found, a field slightly over 80 ha could cost as much as $200,000 when water development costs were included. This represents a large opportunity cost. In addition, the majority of potato production areas in the Northeast U.S. present significant challenges for irrigation (e.g. lack of surface or ground water source, undulating topography, irregularly-shaped fields, and a high number of non-adjacent fields).

Consideration of weather-related impacts and the probability that in three out of every four years, there is at least one 5-day period where moisture levels are lower than optimal for potato production [18], compost amendments may serve as an important alternative to supplemental irrigation for improving yield and reducing risk.. However, a noteworthy constraint to the SI system is the cost of compost and its application in the short-term (≤ 4 years). Current market rates for purchasing compost in Maine can run as much as $30 to $40 Mg^{-1} (Mark Hutchinson, University of Maine, personal communication), which would make compost application much less economically feasible for potato growers. However, with on-farm composting, local availability, and cooperative associations with livestock farms and compost producers, compost costs may be substantially reduced to more favorable levels, making the use of compost far more attractive. Further research is also needed to determine if the same or similar yield benefits from adding compost can be attained at reduced application rates or application frequencies, perhaps also coupled with the feasibility of site-specific applications. Additional considerations involve customization of the compost mixture for a particular application and soil type based on parameters that include maturity, stability, pH level, density, particle size, moisture, salinity, and organic content. Other system modifications such as inclusion of more marketable rotation crops may also serve to improve overall system profitability.

REFERENCES

[1]   Planning Decisions, Inc., "A Study of the Maine Potato Industry: Its Economic Impact," Portland, 2003.

[2]   US Department of Agriculture, Economic Research Service, "US Potato Statistics," 2011. http://www.ers.usda.gov/Briefing/Potatoes

[3]   G. Porter, G. Opena, W. Bradbury, J. McBurnie and J. Sisson, "Soil Management and Supplemental Irrigation Effects on Potato: I. Soil Properties, Tuber Yield, and Quality," *Agronomy Journal*, Vol. 91, No. 3, 1999, pp. 416-425. doi:10.2134/agronj1999.00021962009100030010x

[4]   R. D. Peters, A. V. Sturz, M. R. Carter and J. B. Sanderson, "Influence of Crop Rotation and Conservation Tillage Practices on the Severity of Soil-Borne Potato Diseases in Temperate Humid Agriculture," *Canadian Journal of Soil Science*, Vol. 84, No. 4, 2004, pp. 397-402. doi:10.4141/S03-060

[5]   M. R. Carter, J. B. Sanderson, D. A. Holstrom, J. A. Ivany and K. R. DeHaan, "Influence of Conservation Tillage and Glyphosate on Soil Structure and Organic Carbon Fractions Through the Cycle of a 3-Year Potato Rotation in Atlantic Canada," *Soil and Tillage Research*, Vol. 93, No. 1, 2007, pp. 206-221. doi:10.1016/j.still.2006.04.004

[6]   T. S. Griffin, C. W. Honeycutt and R. L. Larkin, "Delayed Tillage and Cover Crop Effects in Potato Systems," *American Journal of Potato Research*, Vol. 86, No. 2, 2009, pp. 79-87. doi:10.1007/s12230-008-9050-2

[7]   R. P. Larkin, C. W. Honeycutt, T. S. Griffin, O. M. Olanya, J. M. Halloran and Z. He, "Effects of Different Potato Cropping System Approaches and Water Management on Soilborne Diseases and Soil Microbial Communities," *Phytopathology*, Vol. 101, No. 1, 2011, pp. 58-67. doi:10.1094/PHYTO-04-10-0100

[8]   R. P. Larkin, C. W. Honeycutt, O. M. Olanya, J. M. Halloran and Z. He, "Impacts of Crop Rotation and Irrigation on Soilborne Diseases and Soil Microbial Communities," In: Z. He, R. P. Larkin and C. W. Honeycutt, Eds., *Sustainable Potato Production: Global Case Studies*, Springer, Amsterdam, 2012, pp. 23-41. doi:10.1007/978-94-007-4104-1_2

[9]   B. Eghball, G. D. Binford, J. F. Power, D. D. Baltensperger and F.N. Anderson, "Maize Temporal Yield Variability under Long-Term Manure and Fertilizer Application: Fractal Analysis," *Soil Science Society of America Journal*, Vol. 59, No. 5, 1995, pp. 1360-1364. doi:10.2136/sssaj1995.03615995005900050023x

[10] G. E. Varvel, "Crop Rotation and Nitrogen Effects on Normalized Grain Yields in a Long-Term Study," *Agronomy Journal*, Vol. 92, No. 5, 2000, pp. 938-941. doi:10.2134/agronj2000.925938x

[11] R. S. Loomis and D. J. Conner, "Crop Ecology: Productivity and Management in Agricultural System," Cambridge University Press, Cambridge, 1992. doi:10.1017/CBO9781139170161

[12] A. J. Haverkort, "Ecology of Potato Cropping Systems in Relation to Latitude and Altitude," *Agricultural Systems*, Vol. 32, No. 3, 1990, pp. 251-272. doi:10.1016/0308-521X(90)90004-A

[13] G. B. Opena and G.A. Porter, "Soil Management and Supplemental Irrigation Effects on Potato: II. Root Growth," *Agronomy Journal*, Vol. 91, No. 3, 1999, pp. 426-431. doi:10.2134/agronj1999.00021962009100030011x

[14] M. R. Carter, J. B. Sanderson and J. A. MacLeod, "Influence of Time of Tillage on Soil Physical Attributes in Potato Rotations in Prince Edward Island," *Soil and Tillage Research*, Vol. 49, No. 1-2, 1998, pp. 127-137. doi:10.1016/S0167-1987(98)00167-6

[15] A. S. Grandy, G. A. Porter and M. S. Erich, "Organic Amendment and Rotation Crop Effects on the Recovery

of Soil Organic Matter and Aggregation in Potato Cropping Systems," *Soil Science Society of America Journal*, Vol. 66, No. 4, 2002, pp. 1311-1319. doi:10.2136/sssaj2002.1311

[16] T. S. Griffin, G. A. Porter and N. G. Winslow, "Altering Soil Carbon and Nitrogen Stocks in Intensively Tilled Two-Year Rotations," *Biology and Fertility of Soils*, Vol. 39, No. 5, 2004, pp. 366-374. doi:10.1007/s00374-004-0725-7

[17] E. B. Mallory and G. A. Porter, "Potato Yield Stability under Contrasting Soil Management Strategies," *Agronomy Journal*, Vol. 99, No. 2, 2007, pp. 501-510. doi:10.2134/agronj2006.0105

[18] G. R. Benoit and W. J. Grant, "Plant Water Deficit Effects on Aroostook County Potato Yields over 30 Years," *American Potato Journal*, Vol. 57, No. 12, 1980, pp. 585-594. doi:10.1007/BF02854128

[19] G. R. Benoit and W. J. Grant, "Excess and Deficient Water Stress Effects on 30 Years of Aroostook County Potato Yields," *American Potato Journal*, Vol. 62, No. 2, 1985, pp. 49-55. doi:10.1007/BF02903462

[20] Central Aroostook Soil and Water Conservation District, "Maine Irrigation Guide 2005," Linda Alverson, Executive Director, Presque Isle.

[21] T. J. Dalton, G. A. Porter and N. G. Winslow, "Risk Management Strategies in Humid Production Regions: A Comparison of Supplemental Irrigation and Crop Insurance," *Agricultural and Resource Economic Review*, Vol. 33, No. 2, 2004, pp. 220-232.

[22] K. D. Olsen, "Farm Management: Principles and Strategies," Wiley & Sons, New York City, 2003.

[23] D. Cox, D. Bezdicek and M. Fauci, "Effects of Compost, Coal Ash, and Straw Amendments on Restoring the Quality of Eroded Palouse Soil," *Biology and Fertility Soils*, Vol. 33, No. 5, 2001, pp. 365-372. doi:10.1007/s003740000335

[24] D. A. Martens and W. T. Frankenberger Jr., "Modification of Infiltration Rates in an Organic Amended Irrigated Soil," *Agronomy Journal*, Vol. 84, No. 4, 1992, pp. 707-717. doi:10.2134/agronj1992.00021962008400040032x

[25] B. J. Foley and L. R. Cooperband, "Paper Mill Residuals and Compost Effects on Soil Carbon and Physical Properties," *Journal of Environmental Quality*, Vol. 31, No. 6, 2002, pp. 2086-2095. doi:10.2134/jeq2002.2086

[26] H. A. J. Hoitink and P. C. Fahy, "Basics for the Control of Soil-Borne Plant Pathogens with Composts," *Annual Review of Phytopathology*, Vol. 24, 1986, pp. 93-114.

[27] US Environmental Protection Agency, "Innovative Uses of Compost: Disease Control for Plants and Animals," EPA530-F-97-043, 1997.
http://www.epa.gov/osw/conserve/rrr/composting/pubs/erosion.pdf

[28] US Environmental Protection Agency, "Innovative Uses of Compost: Disease Control for Plants and Animals," EPA530-F-97-044, 1997.
http://www.epa.gov/osw/conserve/rrr/composting/pubs/disease.pdf

[29] G. S. Abawi and T. L. Widmer, "Impact of Soil Health Management Practices on Soilborne Pathogens, Nematodes and Root Diseases of Vegetable Crops," *Applied Soil Ecology*, Vol. 15, No. 1, 2000, pp. 37-47. doi:10.1016/S0929-1393(00)00070-6

[30] H. A. J. Hoitink, M. S. Krause and D. Y. Han, "Spectrum and Mechanisms of Plant Disease Control with Composts," In: P. J. Stoffella and B. A. Kahn, Eds., *Compost Utilization in Horticultural Cropping Systems*, Lewis Publishers, Boca Raton, 2001

[31] K. Auerswald, G. Gerl and M. Kainz., "Influence of Cropping System on Harvest Erosion under Potato," *Soil and Tillage Research*, Vol. 89, No. 1, 2006, pp. 22-34.

[32] G. Ruysschaert, J. Poesen, G. Verstraeten and G. Govers, "Soil Losses Due to Mechanized Potato Harvesting," *Soil and Tillage Research*, Vol. 86, No. 1, 2006, pp. 52-72. doi:10.1016/j.still.2005.02.016

[33] K. H. D. Tiessen, D. A. Lobb and G. R. Mehuys, "The Canon of Potato Science: 30. Tillage Erosion within Potato Production-Soil Tillage, Earthing Up and Planting," *Potato Research*, Vol. 50, No. 3-4, 2007, pp. 327-330. doi:10.1007/s11540-008-9055-8

[34] S. L. DeFauw, P. J. English, R. P. Larkin, J. M. Halloran and A. K. Hoshide, "Potato Production Systems in Maine: Geospatial Assessments of Agri-Environmental Indicators," *ASA-CSSA-SSSA Conference Proceedings*, 2011.
http://a-c-s.confex.com/crops/2011am/webprogram/Paper64689.html

[35] S. L. DeFauw, R. P. Larkin, P. J. English, J. M. Halloran and A. K. Hoshide, "Geospatial Evaluations of Potato Production Systems in Maine," *American Journal of Potato Research*, Vol. 89, No. 6, 2012, pp. 471-488. doi:10.1007/s12230-012-9271-2

[36] K. R. DeHaan, G. T Vessey, D. A Holmstrom, J. A MacLeod, J. B Sanderson and M. R Carter, "Relating Potato Yield to the Level of Soil Degradation Using a Bulk Yield Monitor and Differential Global Positioning Systems," *Computers and Electronics in Agriculture*, Vol. 23, No. 2, 1999, pp.133-143.
doi:10.1016/S0168-1699(99)00027-7

[37] S. L. DeFauw, Z. He, R.P. Larkin and S. A. Mansour, "Sustainable Potato Production and Global Food Security," In: Z. He, R. P. Larkin and C. W. Honeycutt, Eds., *Sustainable Potato Production: Global Case Studies*, Springer, Dordrecht, 2012, pp. 3-19. doi:10.1007/978-94-007-4104-1_1

[38] J. C. Howell and R. V. Hazzard, "New England Vegetable Management Guide 2012-2013."
http://www.nevegetable.org/index.php/cultural/fertility

16

Effect of Growth Stage-Based Irrigation Schedules on Biomass Accumulation and Resource Use Efficiency of Wheat Cultivars

Muhammad Mubeen[1]*, Ashfaq Ahmad[1], Aftab Wajid[1], Tasneem Khaliq[1], Syeda Refat Sultana[1], Shahid Hussain[1], Amjed Ali[2], Hakoomat Ali[3], Wajid Nasim[4]

[1]Agro-Climatology Laboratory, Department of Agronomy, University of Agriculture, Faisalabad, Pakistan; [2]University College of Agriculture, University of Sargodha, Sargodha, Pakistan; [3]Department of Agronomy, Faculty of Agricultural Sciences and Technology, Bahauddin Zakariya University, Multan, Pakistan; [4]Department of Environmental Sciences, COMSATS Institute of Information Technology, Vehari, Pakistan.
Email: *mubeenagri@gmail.com

ABSTRACT

Climate and weather conditions greatly affect the performance of new wheat cultivars for yield and resource use efficiency. In order to know the effect of irrigation schedules based on growth stage (the most vital criterion in the region) on growth, yield and radiation use efficiency of wheat cultivars in Faisalabad conditions, a study was planned at Agronomic Research Area, University of Agriculture, Faisalabad during 2009-2010. Split plot design with irrigation levels in main plots and cultivars in sub-plots was implied. Irrigation levels were: IT = irrigation at tillering stage, ITS = irrigation at tillering and stem elongation stage, ISB = irrigation at stem elongation and booting stage and ITSBG = irrigation at tillering, stem elongation, booting and grain filling stage. Cultivars selected were: Faisalabad-2008, Lasani-2008, Miraj-2008, Shafaq-2006 and Chakwal-97. Irrigation treatment ITSBG gave higher grain yield (4.23 t·ha^{-1}) followed by ISB (3.60 t·ha^{-1}), however ITSBG was statistically similar to ISB in radiation use efficiency (RUE) for grain yield (RUE$_{GY}$). Similarly the two treatments were statistically at par in maximum leaf area index, total dry matter (TDM) accumulation, cumulative photosynthetically active radiation (PAR), and RUE$_{TDM}$. Hence where less numbers of irrigations are available, irrigation at stem elongation and booting stage is suitable for achieving economic yield. Lasani-2008 produced maximum grain yield (4.37 t·ha^{-1}) compared to other cultivars but it was statistically at par with Shafaq-2006 in plant height, TDM production and RUE$_{TDM}$. Depending on the availability, the two cultivars may be chosen under irrigated conditions of Faisalabad.

Keywords: *Triticum aestivum* L.; Irrigation; Photosynthetically Active Radiation (PAR); Radiation Use Efficiency (RUE); Total Dry Matter (TDM)

1. Introduction

In spite of higher yield potential, average grain yield of wheat in Pakistan is much less than most countries of the world. The yield of wheat depends on many factors, the cultivars and irrigation being the most important ones. Cereal cultivators are encountered with a greater choice of new cultivars both from the public and private sectors, often with little appropriate information available on their performance in the local environment. Wheat cropping areas of Punjab are usually less productive due to improper selection of variety [1]. Climate and weather

conditions greatly affect the performance of new wheat cultivars both for yield and resource use efficiency [2]. Similarly different varieties of wheat respond differently to irrigation treatments. The span of dry period cannot be forecast under arid and semi-arid conditions which mostly prevail in Pakistan and satisfactory grain yields are dependent upon the crop cultivar for its ability to tolerate water stress [3].

Irrigation water is vital for cell turgidity which is associated with photosynthesis, growth of tissues and plant organs [4]. The response of plants to varying degrees of water levels has been a subject of extensive study and evaluation [5,6]. Earlier research showed that irrigation consistently increased wheat yield in Pakistan [7-11]. At tillering, anthesis and grain formation stages availability of water is important for better performance of the crop. At anthesis stress of water reduces pollination and thus less number of grains spike-1 which ultimately result in low grain yield [12] and irrigation at crown root initiation, tillering, jointing, flowering and milking stages influence most the value of growth parameters [13].

The effect of wheat varieties alone and in interaction with deficit irrigation on photosynthetically active radiation utilization (PAR) has been previously tested in a number of studies. Three wheat cultivars (Uqab-2000, AS-2002, and Inqlab-91) differed significantly from one another in leaf area index, leaf area duration, net assimilation rate and radiation use efficiency [14]. Similarly, there were significant differences among cultivars in radiation use efficiency for total dry matter. AS-2002 showed maximum radiation use efficiency ($2.66 \, g \cdot MJ^{-1}$) while minimum radiation use efficiency ($2.16 \, g \cdot MJ^{-1}$) was observed in Iqbal-2000 [15]. The wheat cultivar Uqab-2000 performed very well under Faisalabad conditions. In an experiment conducted by [16], it was found that whether irrigated at jointing stage or not, the difference between Jimai-20 and Lainong-0153 in the amount of intercepted photosynthetically active radiation was non-significant. During the late growing season of winter wheat, irrespective of the irrigation levels, the radiation use efficiency and GY of Jimai-20 were significantly higher than those of Lainong-0153.

Intercepted PAR is the main factor determining both spike and crop growth period, and grain number m^{-2} is linearly related to the accumulated intercepted PAR during this period [17]. Measurements of after-anthesis radiation use efficiency (RUE) not only showed that it was reduced in all cultivars, but also confirmed that during grain filling period the sink size may exert a great effect on post-anthesis RUE through reducing the leaf photosynthetic rates [18].

This paper examines the effect of different irrigation schedules based on growth stage (the most vital criterion in the region) on the growth, yield and radiation use efficiency of different wheat varieties under semiarid conditions of Faisalabad.

2. Materials and Methods

2.1. Experimental Site

The study was conducted in rabi (winter) season 2009-2010 at Agronomic Research Area, University of Agriculture, Faisalabad (Latitude 31.25°N, Longitude 73.06°E and 184.4 m a.s.l.). The soil of the plot is a sandy clay loam. The concentration of organic matter of the plot is low (0.78%) which is peculiar of the area. The analysis of soil before sowing showed that level of rapidly available phosphorous was 6.85 mg/kg, potassium was 189 mg/kg (an appreciable amount which is due to canal irrigation) and nitrogen was 0.071%. The bulk density, field moisture capacity and wilting points were 1.4 $g \cdot cm^{-3}$, 20.2% and 11.6%, respectively. Agriculture in this region is intensified by a double cropping system of winter wheat and autumn maize or rice with high-yielding cultivars and high fertilizer and water inputs.

2.2. Weather Data

All weather data for the whole crop growth period were collected from the meteorological observatory 100 m away from the experimental site. **Table 1** shows the mean monthly data of different climatic factors. The average temperature ranged from 11.1°C to 29.9°C during the crop growth season. April was the warmest month having mean maximum temperature 29.9°C while January was the coldest month with mean minimum temperature of 11.1°C. Total rainfall during the season was 23.5 mm and maximum rainfall occurred in the month of February (11.9 mm).

2.3. Experimental Design

The experiment was conducted in triplicate using a split plot design. Previous studies and recommendations of

Table 1. Summary of meteorological data during the crop growth season November 2009-April 2010.

Month	Temperature Mean (°C)	Rainfall Total (mm)	R. H Mean (%)	Sun shine (hours)	Mean ET (mm)	Wind Speed (km/h)
November	18.2	0.7	64.7	6.3	1.3	3.3
December	14.5	0	64.4	6.6	1.2	2.8
January	11.1	0.8	82.3	4.1	0.8	4.4
February	15.7	11.9	62.7	6.6	2.5	5.0
March	23.5	8.8	57.5	8.7	3.4	3.6
April	29.9	1.3	36.8	9.0	6.0	5.8
Total	-	23.5	-	-	-	-

Department of Agriculture [19-21] recommend that four stages (tillering, stem elongation, booting and grain filling) are critical for wheat at which water application is required the most, so if less number of irrigations are available which irrigation should be skipped for a particular cultivar is a matter of consideration. Hence treatments of water application and skipping irrigations at these growth stages were planned. Split plot design with irrigation levels in main plots and cultivars in sub-plots was implied. More detailed information of the treatments is shown in **Table 2**. The canal water was supplied to the plots using improved earthen watercourse and hence cut throat flume was used to measure the amount of water applied. Between two irrigation plots, there was a 1.5 m wide zone (kept as buffer plot) to minimize the effects of two contiguous plots.

2.3. Crop Husbandry

"Rouni" (soaking) irrigation was applied to the field twelve days before sowing to bring the soil moisture level at field capacity. Deep ploughing was done followed by two cultivations and planking to prepare the seedbed. The crop was sown with the help of single row hand drill. Half of nitrogen and whole of the phosphorous and potash were applied at sowing as a basal dose whereas remaining nitrogen was applied at first irrigation by broadcast method. The rate of the three nutrients is given in **Table 2**.

2.4. Observations

After establishment five plants were tagged in each plot to study growth stages of tillering, stem elongation, anthesis and maturity. A 25 cm long row areawas harvested at 15 days interval. A sub sample of 10 g was used for

measuring leaf area by leaf area meter (CID-202). Appropriate sub samples of different plant fractions (leaf, stem and ear) were dried in an oven at 80°C for 72 h. Leaf area index (LAI) was determined as ratio of leaf area to land area. The fraction of intercepted radiation (Fi) was calculated from measurements of LAI using the exponential equation as suggested by [22].

$$Fi = 1 - \exp(-k \times LAI) \qquad (1)$$

where k is a light transmission co-efficient for total solar radiation and a k value of 0.45 was used for wheat as described by [23]. Photosynthetically active radiation (PAR) was taken equal to half (0.5) of total incident radiation (Si) as suggested by [24]. Multiplying these totals by appropriate estimates of Fi gave the amount of intercepted radiation ($\sum Sa$).

$$Sa = Fi \times Si \qquad (2)$$

Radiation use efficiency for TDM (RUE_{TDM}) and grain yield (RUE_{GY}) was calculated as the ratio of total biomass and grain yield to cumulative intercepted PAR ($\sum Sa$).

At maturity, an area of 6 m^2 was harvested manually from each plot to determine the total biomass and grain yield which were then converted to tons ha^{-1}. Analysis of variance technique was employed to analyze the data. Differences among the treatment means were compared using least significant difference (LSD) at 5% probability level [25].

3. Results and Discussion

3.1. Plant Height

Table 3 showed that application of irrigation had significant effect on plant height. Maximum plant height of

Table 2. Design specifications and agronomic practices.

Treatments				
Main plot = Irrigation			Sub plot = cultivars	
Level	Description		Level	Description
IT	Irrigation at tillering stage		V_1	Faisalabad-2008
ITS	Irrigation at tillering and stem elongation stage		V_2	Lasani-2008
ISB	Irrigation at stem elongation and booting stage		V_3	Miraj-2008
ITSBG	Irrigation at tillering, stem elongation, booting and grain filling stages		V_4	Shafaq-2006
			V_5	Chakwal-97

Planting information								
Sowing date	Design	Replication	Net plot size	R × R	Seed rate (kg·ha⁻¹)	N (kg·ha⁻¹)	P (kg·ha⁻¹)	K (kg·ha⁻¹)
November 12	Split plot	3	2.4 m × 5.0 m	30 cm	100	110	85	65

Table 3. Effect of cultivars and irrigation on yield and radiation-associated traits of wheat.

	Treatments	Plant Height (cm)	Grain Yield (t·ha^{-1})	HI (%)	PAR (MJ·m^{-2})	RUE$_{TDM}$ (g·MJ^{-1})	RUE$_{GY}$ (g·MJ^{-1})
Irrigation	IT[a]	88.07 b	3.01 c	31.36 c	337.66 c	2.48 b	0.82 b
	ITS	95.88 a	3.53 b	34.81 ab	364.14 b	2.76 ab	0.90 b
	ISB	89.91 b	3.61 b	33.29 bc	391.49 a	2.77 ab	0.98 ab
	ITSBG	90.79 b	4.23 a	36.93 a	413.60 a	3.07 a	1.07 a
	S$\bar{x}$	1.08	0.14	0.75	6.86	0.13	0.05
	LSD	3.74	0.48	2.59	23.76	0.32	0.16
Cultivars	V$_1$	93.03 a	3.51 c	33.67 c	376.88 bc	2.77 ab	0.92 bc
	V$_2$	88.53 b	4.37 a	39.03 a	398.63 a	2.95 a	1.10 a
	V$_3$	91.10 ab	3.33 c	31.71 d	365.93 cd	2.83 a	0.90 c
	V$_4$	92.18 a	3.86 b	35.82 b	387.19 ab	2.80 a	0.99 b
	V$_5$	90.95 ab	2.90 d	30.26 d	354.99 d	2.50 b	0.80 d
	S$\bar{x}$	0.92	0.10	0.58	4.11	0.13	0.03
	LSD	2.66	0.30	1.67	11.85	0.27	0.09

[*]Values not sharing common letters differ at 5% level of probability; [a]IT = Irrigation at tillering stage, ITS = Irrigation at tillering and stem elongation stage, ISB = Irrigation at stem elongation and booting stage, ITSBG = Irrigation at tillering, stem elongation, booting and grain filling stage, V1 = Faisalabad-2008, V2 = Lasani-2008, V3 = Mairaj-2008, V4 = Shafaq-2006, V5 = Chakwal-97.

95.88 cm was recorded in ITS (irrigation at tillering and stem elongation stage) and lowest plant height (88.07 cm) was recorded in IT (irrigation at tillering stage) which was statistically at par with ISB (89.91 cm) and ITSBG (90.79 cm). Maximum plant height of 93.03 cm was recorded in cv. Faisalabad-2008 and it was statistically at par with Shafaq-2006. Lowest plant height 88.53 cm was produced by cv. Lasani-2008. These results are in line with those of [26] who reported that plant height was significantly affected by irrigation treatments.

3.2. Leaf Area Index

Leaf area index (LAI) is the main physiological determinant of the crop yield [3]. **Figure 1** presents the effect of irrigation and cultivars on maximum LAI during the season. Application of ITSBG (irrigation at tillering, stem elongation, booting and grain filling stage) significantly produced more LAI during most of the crop season (5.50) over other irrigation levels and it was statistically at par with ISB (irrigation at stem elongation and booting stage). IT (irrigation at tillering stage) produced minimum LAI throughout the crop season (**Figure 1(a)**).

Cultivars had non-significant effect on LAI during most of the crop season (**Figure 1(b)**). Lasani-2008 gave the maximum value 5.23. The maximum LAI continued to increase up to month of February and then gradually declined towards maturity due to leaf senescence. Overall mean value of LAI for all the cultivars remained 5.09.

[4] reported that irrigation treatments significantly affected LAI.

3.3. Total Dry Matter Accumulation

Figure 2(a) indicated that total dry matter (TDM) accumulation was significantly affected by different irrigation levels. An increasing trend in total dry matter accumulation was observed from 40 DAS (22th December) to 120 DAS (12th March) in all treatments. Application of ITSBG (irrigation at tillering, stem elongation, booting and grain filling stage) significantly increased TDM accumulation over ISB (irrigation at stem elongation and booting stage), ITS (irrigation at tillering and stem elongation stage) and IT (irrigation at tillering stage). Final TDM yield was 1146.00 g·m^{-2} in ITSBG, 1086.00 g·m^{-2} in ISB, 1005.00 g·m^{-2} in ITS and 955.50 g·m^{-2} in IT, respectively. [27] reported that total dry matter production increased as they increased the irrigation levels.

The cultivars had non-significant effect on total dry matter accumulation except on 55 DAS, 100 DAS and 120 DAS when cultivars differed significantly (**Figure 2(b)**). Then TDM was leveled off at final harvest. The cv. Lasani-2008 significantly increased TDM accumulation and Shafaq-2006 was statistically at par with Lasani-2008 while Mairaj-2008 and Faisalabad-2008 were statistically at par. Minimum TDM was produced by cv. Chakwal-97. In general, all cultivars showed similar trend in TDM production *i.e.* initially slow accumulation

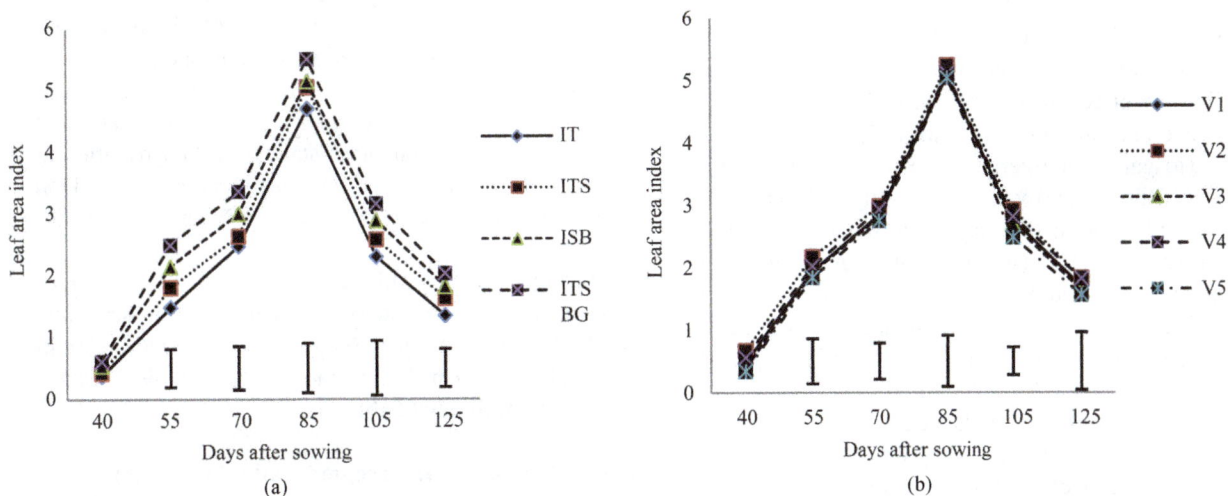

Figure 1. Changes in leaf area index with time as affected by (a) irrigation levels and (b) cultivars; bars indicate LSD at 5%.

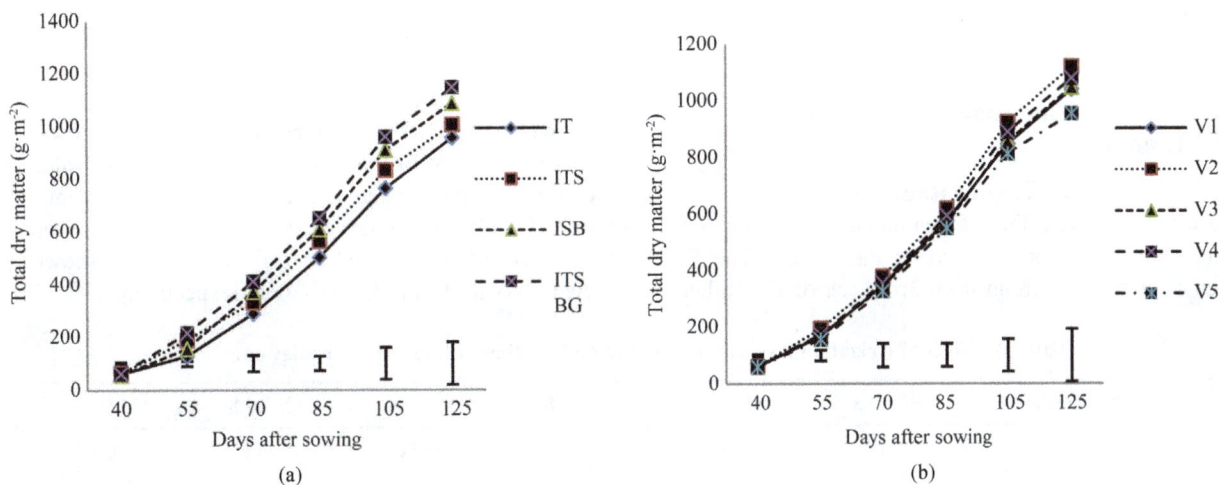

Figure 2. Changes in total dry matter accumulation with time as affected by (a) irrigation levels and (b) cultivars; bars indicate LSD at 5%.

of TDM production was observed from 40 DAS (22 December) to 55 DAS (5 January) and then subsequently fast increase in accumulation of TDM up to 120 DAS (12 March) was recorded. Final TDM was maximum (1120 $g \cdot m^{-2}$) in cv. Lasani-2008 and it was statistically at par with Shafaq-2006 (1079 $g \cdot m^{-2}$). [1] showed that the maximum biomass was obtained in Seher-2006 followed by Uqab-2000; Shafaq-2006 produced relatively less biomass (1.070 $t \cdot ha^{-1}$) but its final TDM was almost similar to that of this study (1.079 $t \cdot ha^{-1}$).

3.4. Grain Yield

Grain yield in wheat is the outcome of number of contributing and inter-related components. These are number of productive tillers per unit area, number of grains ear^{-1} and mean grain weight. Data in **Table 3** showed significant effects of treatments among means of various irriga-

tion levels and cultivars at physiological maturity. Application of ITSBG(irrigation at tillering, stem elongation, booting and grain filling stage) significantly produced more grain yield over other irrigation levels. Both ITS (irrigation at tillering and stem elongation stage) and ISB (irrigation at stem elongation and booting stage) levels were statistically at par in grain yield. The average grain yield remained 3.007 $t \cdot ha^{-1}$ in IT, 3.533 $t \cdot ha^{-1}$ in ITS, 3.609 $t \cdot ha^{-1}$ in ISB and 4.234 $t \cdot ha^{-1}$ in ITSBG. These results corroborate the findings of [4] and [11] who reported that wheat yield increased with increasing irrigation levels.

Maximum mean grain yield was 4.376 $t \cdot ha^{-1}$ (observed in Lasani-2008) and it was statistically different from Miraj-2008, Shafaq-2006, Faisalabad-2008 and Chakwal-97 which produced grain yield of 3.326, 3.860 and 3.514, 2.903 $t \cdot ha^{-1}$, respectively (**Table 3**).

3.5. Harvest Index (%)

Harvest index (HI) shows the efficiency of cultivars to transfer assimilates to economic parts of the crop. Data regarding the effect of different irrigation levels on harvest index are presented in **Table 3**. Application of ITSBG (irrigation at tillering, stem elongation, booting and grain filling stage) was significantly different in HI from IT (irrigation at tillering stage) but it was statistically at par with ITS (irrigation at tillering and stem elongation stage) and ITS was statistically at par with ISB (irrigation at stem elongation and booting stage). ITSBG gave maximum value (36.93%) of HI and IT-produced lowest value (31.36%) of HI. It is eminent from the results that water stress reduced harvest index.

Harvest index among cultivars was also statistically significant. Greater efficiency of conversion of carbohydrates into economic parts was shown by Lasani-2008 which produced 39.03% harvest index value. Lowest efficiency of harvest index was shown by Chakwal-97 (30.26%).

3.6. Intercepted Radiation and Radiation Use Efficiency

3.6.1. Fraction of Intercepted Radiation

Table 4 showed the effect of treatments on the fraction of intercepted radiation (Fi) during the season. The Fi showed an increasing trend from 3rd week of December to 2nd week of February in all treatments; thereafter Fi decreased until maturity. Application of ITSBG (irrigation at tillering, stem elongation, booting and grain filling stage) significantly enhanced Fi over ISB (irrigation at stem elongation and booting stage), ITS (irrigation at tillering and stem elongation stage) and IT (irrigation at tillering stage). ISB was statistically at par with ITS. However, difference in Fi between ITS and IT was significant.

Difference among cultivars was non-significant. Fraction of intercepted radiation reached its maximum value of 0.90 in Lasani-2008, 0.89 in Shafaq-2006, 0.89 in Miraj-2008, 0.89 in Faisalabad-2008 and 0.89 in Chakwal-97, respectively (**Table 4**).

3.6.2. Cumulative Intercepted Radiation (PAR)

Application of irrigation at tillering, stem elongation, booting and grain filling stage (ITSBG) significantly enhanced the cumulative photosynthetically active radiation (PAR) interception (413.60 MJ·m^{-2}) compared with ISB (391.49 MJ·m^{-2}), ITS (364.14 MJ·m^{-2}) and IT (337.66 MJ·m^{-2}) respectively, while ISB was statistically at par with ITSBG. The cultivars showed significant effect on the amount of intercepted PAR. The mean value of cumulative intercepted radiation was 376.72 MJ·m^{-2} (**Table 3**). Overall range of PAR remained between 354.99 MJ·m^{-2} and 398.63 MJ·m^{-2} which were shown by cv. Chakwal-97 and Lasani-2008, respectively.

Table 4. Effect of irrigation levels and cultivars on Fraction of intercepted radiation.

	Treatments	40 DAS	55 DAS	70 DAS	85 DAS	105 DAS	125 DAS
Irrigation	IT	0.15	0.48 c	0.66 c	0.87 c	0.64 c	0.45 c
	ITS	0.17	0.55 b	0.69 c	0.89 b	0.68 bc	0.51 bc
	ISB	0.21	0.61 ab	0.73 b	0.90 b	0.72 ab	0.55 ab
	ITSBG	0.23	0.66 a	0.77 a	0.91 a	0.75 a	0.59 a
	S$\bar{x}$	0.019	0.019	0.012	0.005	0.013	0.02
	LSD	NS	0.069	0.039	0.008	0.048	0.069
Cultivars	V$_1$	0.19 c	0.57	0.71	0.89	0.70	0.53
	V$_2$	0.25 a	0.60	0.73	0.90	0.72	0.55
	V$_3$	0.16 d	0.57	0.71	0.89	0.69	0.51
	V$_4$	0.22 b	0.58	0.72	0.89	0.71	0.55
	V$_5$	0.14 e	0.55	0.70	0.89	0.66	0.49
	S$\bar{x}$	0.006	0.015	0.009	0.002	0.009	0.015
	LSD	0.008	NS	NS	NS	0.026	NS

[*]Values not sharing common letters differ at 5% level of probability; IT = Irrigation at tillering stage, ITS = Irrigation at tillering and stem elongation stage, ISB = Irrigation at stem elongation and booting stage, ITSBG = Irrigation at tillering, stem elongation, booting and grain filling stage, V1 = Faisalabad-2008, V2 = Lasani-2008, V3 = Mairaj-2008, V4 = Shafaq-2006, V5 = Chakwal-97.

3.6.3. Radiation Use Efficiency (Total Dry Matter)

Radiation use efficiency for the final dry matter (RUE_{TDM}) was found to be significant statistically (**Table 3**). ITSBG (irrigation at tillering, stem elongation, booting and grain filling stage) significantly produced more RUE for TDM (3.07 $g \cdot MJ^{-1}$) over other irrigation levels; ISB, ITS and IT gave values of RUETDM as 2.77 $g \cdot MJ^{-1}$, 2.76 $g \cdot MJ^{-1}$ and 2.48 $g \cdot MJ^{-1}$ respectively. [23] reported that water stress caused significant reduction in maximum biomass production by changes in the amount of intercepted PAR. Cultivars showed its potential in terms of RUETDM ranging from 2.50 $g \cdot MJ^{-1}$ to 2.95 $g \cdot MJ^{-1}$. Cultivar Lasani-2008, Shafaq-2006 and Miraj-2008 produced more dry matter (2.95, 2.80 and 2.83 g, respectively) than Faisalabad-2008 (2.77 g) and Chakwal-97 (2.50 g) for each MJ of light [28].

3.6.4. Radiation Use Efficiency (Grain Yield)

Table 3 showed that increasing levels of irrigation had significant effect on RUE for grain yield ITSBG (irrigation at tillering, stem elongation, booting and grain filling stage) gave maximum RUE 1.07 $g \cdot MJ^{-1}$ and it was statistically at par with ISB (irrigation at stem elongation and booting stage) and thought it was statistically different from all treatments of irrigation. Lowest value (0.82 $g \cdot MJ^{-1}$) of RUE for Grain yield was found in IT (irrigation at tillering stage) and it was statistically at par with ITS (irrigation at tillering and stem elongation stage). The mean value of radiation use efficiency for grain remained (0.94 $g \cdot MJ^{-1}$). Cultivars significantly affected the RUE for grain. Radiation used by cv. Lasani-2008 was maximum (1.07 $g \cdot MJ^{-1}$) followed by Shafaq-2006 (0.99 $g \cdot MJ^{-1}$). Lowest value of radiation used by cultivars was found in cv. Chakwal-97 (0.80 $g \cdot MJ^{-1}$) while values of RUE shown by other varieties were cv. Faisalabad-2008 (0.92 $g \cdot MJ^{-1}$) and Miraj-2008 (0.90 $g \cdot MJ^{-1}$) [29].

Results emphasize that different treatments such as Lasani-2008 and ITSBG (irrigation at tillering, stem elongation, booting and grain filling stage) application increased yield by enhancing growth (LAI) than other treatments. This led to higher radiation interception and thus enhanced crop growth rate and increased TDM production in these treatments. [23] found the greatest reduction in RUE was also from early drought with little or no effect from mid to late drought.

In conclusion, results showed that both radiation interception and RUE were the major determinant of crop growth and yield.

REFERENCES

[1] J. Anwar, A. Ahmad, T. Khaliq, M. Mubeen and H. M. Hammad, "Optimization of Sowing Time for Promising Wheat Genotypes in Semiarid Environment of Faisalabad," *Crop Environment*, Vol. 2, 2011, pp. 24-27.

[2] W. Nasim, A. Ahmad, S. A. Wajid, A. Hussain, T. Khaliq, M. Usman, H. M. Hammad, S. R. Sultana, M. Mubeen and S. Ahmad, "Simulation of Different Wheat Cultivars under Agro-Ecological Condition of Faisalabad-Pakistan," *Crop Environment*, Vol. 1, 2010, pp. 44-48.

[3] M. Mubeen, A. Ahmad, A. Wajid, T. Khaliq and A. Bakhsh, "Evaluating CSM-CERES-Maize Model for Irrigation Scheduling in Semi-Arid Conditions of Punjab, Pakistan," *International Journal of Agriculture Biology*, Vol. 15, 2013, pp. 1-10.

[4] A. Wajid, A. Hussain, M. Maqsood, A. Ahmad and M. Awais, "Influence of Sowing Date and Irrigation Levels on Growth and Grain Yield of Wheat," *Pakistan Journal of Agricultural Sciences*, Vol. 39, No. 1, 2002, pp. 22-24.

[5] F. Kajdi, "Effect of Irrigation on the Yield of Winter Wheat Varieties," *Acta Agronomica Ovariensis*, Vol. 35, 1993, pp. 221-231.

[6] E. Khehr, A. F. Matta, S. E. G. Wahba and M. M. El-Koliey, "Effect of Water Regime on Yield of Some Maize Cultivars and Water Relations," Res. Bull 47. Fac. Agric., University of Cairo Crop Res. Inst., 1996, pp. 87-98.

[7] M. A. Bajwa, M. H. Choudhry and A. Sattar, "Influence of Different Irrigation Regimes on Yield and Yield Components of Wheat," *Pakistan Journal of Agricultural Research*, Vol. 14, No. 4, 1993, pp. 361-365.

[8] A. Hussain, M. Maqsood, A. Ahmad, A. Wajid and Z. Ahmad, "Effect of Irrigation during Various Development Stages on Yield, Components of Yield and harvest Index of Different Wheat Cultivars," *Pakistan Journal of Agricultural Sciences*, Vol. 34, 1997, pp. 104-107.

[9] E. A. Waraich, R. Ahmad, A. Ali and Saifullah, "Irrigation and Nitrogen Effects on Grain Development and Yieldin Wheat (*Triticum aestivum* L.),"*Pakistan Journal of Botany*, Vol. 39, No. 5, 2007, pp. 1663-1672.

[10] E. A. Waraich, R. Ahmad, Saifullah, S. Ahmad and A. Ahmad, "Impact of Water and Nutrient Management on the Nutritional Quality of Wheat," *Journal of Plant Nutrition*, Vol. 33, No. 5, 2010, pp. 640-643. doi:10.1080/01904160903575881

[11] N. Sarwar, M. Maqsood, K. Mubeen, M. Shehzad, M. S. Bhullar, R. Qamar and N. Akbar, "Effect of Different Levels of Irrigation on Yield and Yield Components of Wheat Cultivars," *Pakistan Journal Agricultural Sciences*, Vol. 47, No. 4, 2010, pp. 371-374.

[12] A. Nazir, R. H. Qureshi, M. Sarwar and T. Mahmood, "Drought Tolerance of Wheat Genotypes," *Pakistan Journal Agricultural Sciences*, Vol. 24, No. 4, 1987, pp. 231-234.

[13] K. B. Bankar, S. V. Gosavi and V. K. Balsanen, "Effect of Different Irrigation Treatment on Growth and Yield of Wheat Crop Varieties," *International Journal of Agriculture Sciences*, Vol. 4, 2008, pp. 114-118.

[14] H. A. Jahfari, "Modeling the Growth, Radiation Use Efficiency and Yield of New Cultivars under Varying Nitro-

gen Rates," M.Sc. Thesis, University of Agriculture Fais-alabad, 2004.

[15] W. Nasim, "Modeling the Growth, Development, Radiation Use Efficiency and Yield of Different Wheat Cultivars," M.Sc. Thesis, University of Agriculture Faisalabad, 2007.

[16] H. Han, Z. Li, T. Ning, X. Zhang, Y. Shan and M. Bai, "Radiation Use Efficiency and Yield of Winter Wheat under Deficit Irrigation in North China," *Plant Soil Environment*, Vol. 54, No. 7,2008, pp. 313-319.

[17] A. Hussain, M. R. Chaudhry, A. Wajid, A. Ahmad, M. Rafiq, M. Ibrahim and A. R. Goheer, "Influence of Water Stress on Growth, Yield and Radiation Use Efficiency of Various Wheat Cultivars," *International Journal of Agriculture and Biology*, Vol. 6, No. 6, 2004, pp. 1074-1079.

[18] M. M. Acreche and G. A. Slafer, "Grain Weight, Radiation Interception and Use Efficiency as Affected by Sink-Strength in Mediterranean wheat Released from 1940 to 2005," *Field Crop Research*, Vol. 10, No. 2, 2009, pp. 98-105. doi:10.1016/j.fcr.2008.07.006

[19] M. Saeed, "Crop Water Requirements and Irrigation Systems," In: S. Nazir, Ed., *Crop Production*, National Book Foundation, Islamabad, 1994.

[20] M. Maqsood, A. Ali, Z. Aslam, M. Saeed and S. Ahmad, "Effect of Irrigation and Nitrogen Levels on Grain Yield and Quality of Wheat," *International Journal of Agriculture and Biology*, Vol. 4, No. 1, 2002, pp. 164-165.

[21] Government of Punjab, "Paidawarimansubagandum (Production Technology for Wheat) 2009-2010," Directorate of Agricultural Information, Punjab, 2010.

[22] J. L. Monteithand J. F. Elston, "Performance and Productivity of Foliage in the Field in the Growth and Functioning of Leaves," Cambridge University Press, Cambridge, 1983, pp. 499-518.

[23] P. D. Jamieson and M. A. Semenov, "Modeling Nitrogen Uptake and Redistribution in Wheat," *Field Crop Research*, Vol. 68, No. 1, 1998, pp. 21-29. doi:10.1016/S0378-4290(00)00103-9

[24] G. Szeicz, "Solar Radiation for Plant Growth," *Journal of Applied Ecology*, Vol. 11, 1974, pp. 617-636. doi:10.2307/2402214

[25] R. G. D. Steel, J. H. Torrie and D. A. Dickey, "Principles and Procedures of Statistics: A Biometrical Approach," 3rd Edition, McGraw Hill Book.Int, Co, New York, 1997, pp. 400-428.

[26] J. A. Thompson and D. L. Chase, "Effect of Limited Irrigation on Growth and Yield of Semi-Dwarf Wheat in Southern New South Wales," *Australian Journal of Experimental Agriculture*, Vol. 32, No. 6, 1992, pp. 725-730. doi:10.1071/EA9920725

[27] K. G. Mandal, K. M. Hati, A. K. Misra, K. K. Bandyopadhyay and M. Mohanty, "Irrigation and Nutrient Effects on Growth and Water-Yield Relationship of Wheat (*Triticum aestivum* L.) in Central India," *Journal of Agronomy and Crop Science*, Vol. 191, No. 6, 2005, pp. 416-425. doi:10.1111/j.1439-037X.2005.00160.x

[28] W. Nasim, A. Ahmed, M. Tariq and S. A. Wajid, "Studying the Comparative Performance of Wheat Cultivars for Growth and Grains Production," *International Journal of Agronomy and Plant Production*, Vol. 3, No. 9, 2012, pp. 306-312.

[29] W. Nasim, "Modeling the Growth, Development, Radiation Use Efficiency and Yield of Different Wheat Cultivars," M.Sc. Thesis, Department of Agronomy, University of Agriculture, Faisalabad, 2007, p. 130.

17

Effect of Regulated Deficit Irrigation on Productivity, Quality and Water Use in Olive cv "Manzanilla"

Raúl Leonel Grijalva-Contreras[1*], Rubén Macías-Duarte[1], Gerardo Martínez-Díaz[1], Fabián Robles-Contreras[1], Manuel de Jesús Valenzuela-Ruiz[1], Fidel Nuñez-Ramírez[2]

[1]National Research Institute for Forestry, Agricultural and Livestock (INIFAP), Caborca, Sonora, México; [2]Agricultural Science Institute, University Autónoma of Baja California (ICA-UABC), Mexicali, B.C. México.
Email: *rgrijalva59mx@hotmail.com

ABSTRACT

The objective of this experiment was to determine the effect of different regulated deficit irrigation (RDI) strategies on productivity, oil quality and water-use efficiency on olive grown in the Sonoran Desert. The experiments were carried out in 2009 and 2010, and in a ten years old traditional (10 × 5 m) "Manzanilla" olive orchard. The control treatment was irrigated at 100% ETc during the whole season while RDI treatments were applied at 75% ETc or 50% ETc. The two RDI were applied during two phenological stages: at postharvest to evaluate the effect on table olive or from pit hardening to harvest to evaluate the effect on oil olive. Our results indicated that RDI applying 50% ETc during postharvest period reduced significantly fruit set and table olive yield, while applied during pit hardening to harvest period, it decreased oil yield but increased oil content. The RDI applying an ETc of 75% during the postharvest period gave similar table olive yield to the control, and applied form of pit hardening to harvest also gave similar oil yield to the control. The RDI using an ETc of 75% resulted in the highest water-use efficiency for oil or table olive production.

Keywords: *Olea europaea* L.; Water Stress; Yield and Quality; Water Save

1. Introduction

Mexico has about 9000 ha cultivating olive tree, where almost 50% Mexican are located in the arid and semi-arid regions. Caborca and Sonora, the main areas of table olive producer in México, have about 2000 ha planted with this tree and produce approximately 10,000 ton annually, all of which are exported to United State [1].

The climate conditions are characterized by low precipitation (100 to 150 mm per year) occurring mainly during the summer months-high temperature and low atmospheric humidity. These conditions result in an annual ETo of 2200 mm which determines a high water demand by crops [2].

The contribution of precipitation to the water demand of olive trees is insignificant and basically the crop depends exclusively on irrigation. The sustainability of olive production in the area requires improving the crop's water productivity.

Regulated deficit irrigation (RDI) causes a temporary and controlled water deficit in a specific phenological stage. RDI is commonly used in several fruit trees in order to reduce the amount of water applied with minimal or no reduction in fruit production [3]. For olive trees, the second phase of fruit growth, corresponding to the pit hardening period, is the most resistant to water deficit [4]. In the first phase, fruit growth corresponds, when most cell division occurs, and in the third phase, when the olive oil is accumulated, it is sensitive to water stress [5,6].

Some studies indicate that RDI did not affect the yield fruit and fruit weight [7,8], acidity and peroxide value [9]; however it increases oil content, polyphenol concentration and oil stability [7,10]. On the other hand, RDI reduced flowering next year [11] and accelerated ripening [12]. The response to RDI depends on the olive variety [13]. Little information is available about RDI effects on table olive and oil olive quality of the Manzanilla cultivar which is grown in the arid warm climates.

The objective of this work was to determine the effect

*Corresponding author.

of different RDI strategies on productivity, oil quality and water-use efficiency in the cv Manzanilla.

2. Materials and Methods

2.1. Orchard Selection and Management

The experiments were carried out in 2009 and 2010 at the National Research Institute for Forestry, Agricultural and Livestock (INIFAP), located in Caborca, Sonora, Mexico (30°42'55"N, 112°21'28"W and 200 m. a. s. l.), in a ten years old traditional (10 × 5 m) "Manzanilla" olive orchard. The soil was sandy loam with pH 7.96 and electrical conductivity of 1.22 dSm^{-1}. Olive trees were maintained in accordance to commercial recommendations [1]. The trees were drip irrigated by using two emitters for tree (12.0 L·ha^{-1}). Orchard olive was fertilized in both years with 15-15-15 at a rate of 1.5 kg tree^{-1} (468 kg·ha^{-1}) during February and March and with ammonium nitrate (300 kg·ha^{-1}) during the postharvest period. The trees were slightly pruned in November in both years. Table olive was harvested during last week of August and the oil olive was harvested during second week of September. Both harvests were made manually.

2.2. RDI Treatments Applied

Different irrigation treatments, based on RDI strategies were applied in 2009 and 2010. The first experiment consisted in the following treatments: 1) Control (100% ETc), 2) RDI-75 and 3) RDI-50. RDI-2 consisted in applying an ETc of 75% from the second week of August to the third week of November 2009 (postharvest), and RD-3 in applying an ETc of 50% in the same period. The second experiment, conducted in 2010, had the same treatments but the period of water restriction was from third week of May to last week of September, the period from the pit hardening to harvest. In both experiments the cultivar was Manzanilla but in the first experiment was for table olive and in the second experiment for oil olive. ETo was calculated every week based on the meteorological data using an automatic station (Campbell Scientific Ltd., Shepshed, UK) during the whole season.

2.3. Measurement Variable

In the first Experiment the vegetative parameters plant height, shoot length and width canopy were evaluated taking 12 trees per treatment, in mid November of each season. Yield components as floral density floral (flower per cm^{-2}) and fruit set (%), choosing one floral shoot per each side of the tree was also evaluated. Yield (kg tree^{-1}) of table olive and quality fruit were evaluated. Fruit weight, fruit diameter, fruit length and pulp and pit ratio were evaluated taking a random sample of 100 fruits for each tree. Water use efficiency was obtained according the next equation: WUE = Yield kg·ha^{-1}/water applied

(mm). In the second experiment similar variables were evaluated but in addition, fruit oil percentage was measured according to [14,15] and some parameters of oil quality as acidity (% oleic acid), peroxide value (meq. of O_2 per kg) and total polyphenols (ppm of caffeic acid) as described by [15,16]. Also, we recorded weather parameters during both experiments (**Table 1**).

2.4. Statistical Analysis

Experimental design in both years was a randomized complete block and four replications with an experimental unit of 12 trees. Means were separated with the least significant difference test (LSD) at 0.05 probability level.

3. Results and Discussion

3.1. Experiment 1

3.1.1. Vegetative Parameters and Yield Components

RDI strategies during the period of postharvest did not affect significantly plant height, shoot length and canopy size (**Table 2**). Fruit set was statistically affected (p < 0.05) while the density flower was not (**Table 3**). The fruit set percent in Control and RDI-75 was similar (1.43% and 1.38% respectively), while in RDI-50 was statistically reduced to 1.25%. These results do not agree with Tognetti *et al.* [17] who found that RDI at 50% decreased canopy volume and trunk diameter in Koroneiki cultivar. Other study of water stress indicated that a Kc of 75% and 50% decreased 8.6% height plant and 14.2% trunk diameter with respect to control (Kc 100%) in Manzanillo cultivar of six years old [18]. The reduction of

Table 1. Weather parameters recorded during the study period.

Weather parameters	Experiment 1	Experiment 2
Medium temperature (°C)	21.9	30.0
Rainfall (mm)	17.8	74.4
Relative humidity (%)	38.7	38.9
Solar radiation (Kilowatt. m^{-2})	0.153	0.206

Table 2. Vegetative growth parameters in different RDI applied during postharvest on table olive.

Treatments	Plant height (m)	Shoot length (cm)	Canopy width (m)
Control	5.1 ± 0.26	25 ± 1.87	4.2 ± 0.21
RDI-75	5.0 ± 0.19	25 ± 3.53	4.1 ± 0.26
RDI-50	4.7 ± 0.12	23 ± 2.55	3.8 ± 0.14
Significance	ns	ns	ns

Means followed by same letter in a column do not differ significantly (LSD 0.05); ns = non significant.

Table 3. Yield components in different RDI applied on table olive.

Treatments	Floral density (flowers cm^{-2})	Fruit set (%)
Control	3240 ± 530	1.43 ± 0.06 a
RDI-75	2950 ± 337	1.38 ± 0.02 a
RDI-50	3100 ± 420	1.25 ± 0.08 b
Significance	ns	*

Means followed by same letter in a column do not differ significantly (LSD 0.05); ns = non significant; *Significant at 0.05 probability level.

vegetative growth could reduce pruning cost [17]. Floral density was not statistically different among RDI treatments while fruit set percent decreased as the water was reduced. Fruit set for Control and RDI-75 were similar with values of 1.43% and 1.38%, while for RDI-50 was reduced to 1.25% (p < 0.05) with. Other researchers [19] indicated that water stress during whole season decreased drastically fruit set on Koroneiki cultivar when ETc were 50% and 25%. Grattan et al. [20] found the number of fruits per branch, the number of fruits per inflorescence, fruit density and fruit set increased with an increase of ETc up to 71% - 89%.

3.1.2. Table Olive Yield and Fruit Quality

According to **Table 4** there is significant difference among RDI tested on table olive yield (p < 0.01), fruit weigh (p < 0.01) and fruit diameter (p < 0.05). The higher yield was obtained in Control and RDI-75 with 11,040 and 10,280 kg·ha^{-1} without statistical difference between them, while RDI-50 had reduction on the table olive yield of 28.6% and 21.3% in comparison with Control and RDI-75, respectively. The low yield in RDI-50 was due to a lower fruit set in this treatment. Goldhamer [8] found that a reduction of 15% and 25% in water supply during midsummer, did not have a negative impact on canning olive yield, but a reduction by 44%, decreased yield by 10%. Others studies indicated different effects on yield when RDI was imposed and the responses were different among years [7]. Weight and fruit diameter were similar between the control and RDI-75. In both parameters RDI-50 was the better treatment and statistically different to control and RDI-75%. Fruit weight in RDI-50 was 12.2% and 8.1% greater in comparison to the control and RDI-75, respectively, and fruit diameter in RDI-50 was 11.1% greater than the other treatments. The differences in fruit weight and fruit diameters may be due to the different fruit load among treatments. Moriana et al. [5] found that fruit weight decreased drastically in response to the level of water stress and fruit load. During severe deficit irrigation, fruit diameter growth slowed; however, after reintroduction of full irrigation, growth accelerated [21]. RDI did not significantly affect

pulp and pit ratio.

3.2. Experiment 2

3.2.1. Oil Yield and Oil Content

There were statistical difference (p < 0.01) among treatments in oil yield and oil content when RDI strategies were applied during pit hardening to harvest period in the same variety. Oil production per hectare in RDI-75 was 26.5% and 10.4% less in comparison with the Control and RDI-50, respectively. In contrast, the oil content was higher in RDI-50 with 8.7% and 9.6% more that Control and RDI-75, respectively (**Table 5**). The high oil content in RDI-50 is probably as a consequence of lower water content in the olive. Vita et al. [7] found that RDI at 50% ETc applied in the same period decreased 26% oil yield. However, other studies indicated not differences in oil yield when RDI was imposed [12,22]. Goldhamer et al. [23] found that oil content was significantly higher for all RDI strategies applied in comparison to the control, and the most severely stressed had an increase in oil content about 30%.

3.2.2. Oil Quality

No statistical differences among treatments on peroxide value, acidity and total polyphenols were found in the experiment (**Table 6**). Total polyphenols was increased as the amount of supplied water decreased, although without statistical difference. These results are accordance with those previously reported in Arbequina variety [7,9,10].

3.2.3. Water Use Efficiency

The amount of water applied in postharvest period in RDI-75 was 215.8 mm while in the control it was 431.6 mm, in the first experiment. In the second experiment the RDI-75 received 325 mm and the control 650.1 mm of water from pit hardening to harvest. Water use efficiency (WUE) measured as fruit yield or oil yield (kg) per mm of irrigation water, was greater for both periods when RDI-75 was applied, reaching values of 0.83 during pit hardening to harvest period (**Table 7**). These values are less that those reported by Vita et al. [7] for a six years old intensive Arbequina olive orchard. By other side, RDI-75 could save until 12.4% (163 mm) when RDI is applied during pit hardening to harvest period. Goldhamer [8] suggests that the RDI regime that saves about 25% (200 mm) of full ETc may be useful in conserving water while maintaining top yield and high fruit quality.

4. Conclusion

A regulated deficit irrigation applying 75% ETc during the postharvest period allows saving water without affecting table olive yield and quality, while the same wa-

Table 4. Olive table yield and fruit quality in different RDI applied during postharvest.

Treatments	Yield (kg·ha^{-1})	Fruit weight (g)	Fruit diameter (cm)	Pulp-pit ratio
Control	11040 ± 2212 a	4.3 ± 0.26 b	1.6 ± 0.08 b	3.9 ± 0.12
RDI-75	10280 ± 1520 a	4.5 ± 0.22 b	1.6 ± 0.07 b	3.8 ± 0.08
RDI-50	7920 ± 11230 b	4.9 ± 0.18 a	1.8 ± 0.07 a	3.8 ± 0.10
Significance	**	**	*	ns

Means followed by same letter in a column do not differ significantly (LSD 0.05); ns = non significant; *Significant at 0.05 probability level and **Significant at 0.01 probability level.

Table 5. Oil yield and oil content in different RDI applied during pit hardening to harvest period.

Treatments	Oil yield (kg·ha^{-1})	Oil content (%)
Control	1020 ± 71.2 a	10.5 ± 0.22 b
RDI-75	956 ± 62.5 a	10.6 ± 0.24 b
RDI-50	750 ± 55.3 b	11.5 ± 0.21 a
Significance	**	**

Means followed by same letter in a column do not differ significantly (LSD 0.05); **Significant at 0.01 probability level.

Table 6. Oil quality in different RDI applied during pit hardening to harvest period.

Treatments	Peroxide value (meq O$_2$ kg^{-1})	Acidity (% oleic acid)	Polyphenols (ppm of caffeic acid)
Control	15.2 ± 0.90	0.45 ± 0.02	153 ± 12.8
RDI-75	14.8 ± 0.59	0.48 ± 0.01	174 ± 14.3
RDI-50	15.6 ± 0.73	0.43 ± 0.02	180 ± 22.0
Significance	ns	ns	ns

Means followed by same letter in a column do not differ significantly (LSD 0.05); ns = non significant.

Table 7. Water applied and water use efficiency (WUE) during postharvest period and pit hardening to harvest period.

Treatments	Experiment 1		Experiment 2	
	Water applied (mm)	WUE (kg of fruit. mm^{-1})	Water applied (mm)	WUE (kg of oil. mm^{-1})
Control	1416.4	7.79	1305.0	0.78
RDI-75	1308.5	7.85	1142.5	0.83
RDI-50	1200.6	6.60	980.0	0.76

ter application from pit hardening to harvest also saves water without affecting oil yield. Under that water management, the highest water-use efficiency was achieved in this study.

REFERENCES

[1] G. C. R. Leonel, F. F. Adán, N. A. J. A. Cristobal and L. C. Arturo, "El Cultivo del Olivo Bajo Condiciones Desérticas del Norte de Sonora," INIFAP-CIRNO-CECAB, Folleto Técnico No. 41, 2010.

[2] J. A. Ruíz Corral, G. Medina García, J. Grageda Grageda, M. M. Silva Serna and G. Díaz Padilla, "Estadísticas Clima-tológicas Básicas del Estado de Sonora (Periodo 1961-2003)," Libro Técnico No. 1, INIFAP-CIRNO-SAGAR PHA, 2005, pp. 92-93.

[3] M. H. Behboudian, "Deficit Irrigation in Deciduos Orchard," Horticultural Review, Vol. 21, 1997, pp. 105-131.

[4] D. Goldhamer, J. Dunai, L. Ferguson, L. Lavee and I. Klein, "Irrigation Requirement of Olive Trees and Response to Sustained Deficit Irrigation," Acta Horticulturae, Vol. 356, 1994, pp. 172-175.

[5] A. Moriana, D. Pérez-López, A. Gómez-Rico, M. Salvador, N. Olmedilla, F. Ribas and G. Fregapane, "Irrigation Scheduling for Traditional, Low Density Olive Orchard: Water Relation and Influence on Oil Characteristic," Ag-

ricultural Water Management, Vol. 87, No. 2, 2007, pp. 171-179. http://dx.doi.org/10.1016/j.agwat.2006.06.017

[6]   S. Lavee and M. Woodner, "Factor Affecting the Nature of Oil Accumulation in Fruit of Olive (*Olea europaea* L.) Cultivars," *Journal of Horticultural Science*, Vol. 66, No. 5, 1991, pp. 583-591.

[7]   F. Vita Serman, D. Pacheco, A. Olguín Pringles, L. Bueno, A. Carelli and F. Capraro, "Effect of Regulated Deficit Strategies on Productivity, Quality and Water Use Eddiciency in a High-Density 'Arbequina' Olive Orchard Located in an arid Region of Argentina," *Acta Horticulturae*, Vol. 888, 2011, pp. 81-88.

[8]   D. Goldhamer, "Regulated Deficit Irrigation for California Canning Olives," *Acta Horticulturae*, Vol. 474, 1999, Vol. 369-372.

[9]   M. J. Motilva, M. P. Romero, S. Alegre and J. Girona, "Effect of Regulated Deficit Irrigation in Olive Oil Production and Quality," *Acta Horticulturae*, Vol. 474, pp. 377-380.

[10] M. J. Motilva. M. J. Tovar, M. P. Romero, S. Alegre and J. Girona, "Influence of Regulated Irrigation Strategies Applied to Olive Trees (*Arbequina* cultivar) on Oil Yield and Composition during the Fruit Ripening Period," *Journal of the Science of Food and Agriculture*, Vol. 80, No. 14, 2000, pp. 2037-2043. http://dx.doi.org/10.1002/1097-0010(200011)80:14<2037::AID-JSFA733>3.0.CO;2-0

[11] S. Alegre, J. Marsal, M. Mata, A. Arbonés, J. Girona and M. J. Tovar, "Regulated Deficit Irrigation in Olive Trees (*Olea europaea* L. Cv Arbequina) for Oil Production," *Acta Horticulturae*, Vol. 586, 2002, pp. 259-262.

[12] S. Alegre, J. Girona, J. Marsal, A. Arbonés, M. Mata, D. Montagut, F. Teixidó, M. J. Motilva and M. P. Romero, "Regulated Deficit Irrigation in Olive Trees," *Acta Horticulturae*, Vol. 474. 1999, pp. 373-376.

[13] M. Patumi, R. Dandria, G. Fontanazza, G. Morelli, P. Giori and G. Sorentino, "Yield and Oil Quality of Intensive Trained Trees of Three Cultivars of Olive (*Olea europaea* L.) under Different Irrigation Regimens," *Journal of Horticultural Science and Biotechnology*, Vol. 74, No. 6, 1999, pp. 729-737.

[14] Association of Official Agricultural Chemist (AOAC), "Official Methods of Analysis," 14th Edition, Benjamin

Franklin Station, Washington DC, 1985, pp. 490-576.

[15] European Union Commission Regulation EEC 2568/91 on the Characteristics Methods of Olive Oils and Their Analytical Methods. Official Journal Communities, 1995.

[16] A. Carrasco-Pancorbo, L. Cerratini, A. Bendini, A. Segura-Carretero, T. Gallina-Toschi and A. Fernandez, "Analytical Determination of Polyphenols in Olive Oil," *Journal of Separation Science*, Vol. 28, No. 9-10, 2005, pp. 837-858. http://dx.doi.org/10.1002/jssc.200500032

[17] R. Tognetti, R. D'Andia, R. Laurini and G. Morelli, "The Effect of Deficit Irrigation on Crop Yield and Vegetative Development of *Olea europaea* L. cvs. Frantoio and Leccino," *European Journal of Agronomy*, Vol. 25, No. 4, 2006, pp. 356-364. http://dx.doi.org/10.1016/j.eja.2006.07.003

[18] A. Fimbres and M. Castillo, "Comparación y Optimización del Riego por Goteo y Microaspersión en Árboles de Olivo," *Biotecnia*, Vol. 6, No. 2, 2005, pp. 43-50.

[19] J. Nikbakht, M. Thaeri and M. Sakkaki, "Effect of Continuous Deficit Irrigation on Yield and Quality of Koroneiki Olive (*Olea europaea* L.) Cultivar," 21st *International Congress on Irrigation and Drainage*, Teheran, 2011, pp. 356-364.

[20] S. R. Grattan, M. J. Berenguer, J. H. Connel, V. S. Polito and P. M. Vossen, "Olive Oil Production as Influenced by Different Quantities of Applied Water," *Agricultural Water Management*, Vol. 85, No. 1-2, 2006, pp. 133-140. http://dx.doi.org/10.1016/j.agwat.2006.04.001

[21] D. Goldhamer and R. E. Beede, "Effect of Water on Olive Tree Performance," In: G. T. Sibbett and L. Ferguson, Eds., *Olive Production Manual*, 2nd Edition, University of California, Agriculture and Natural Resources, Berkeley, Publication 3353, pp. 71-74.

[22] A. Gómez-Rico, M. D. Salvador, A. Moriana, D. Pérez, N. Almedilla, F. Ribas and G. Fregapane, "Influence of Different Irrigation Strategies in a Traditional Carnicabra cv. Olive Orchard on Virgen Olive Oil Composition and Quality," *Food Chemistry*, Vol. 100, No. 3, 2007, pp. 568-578. http://dx.doi.org/10.1016/j.foodchem.2005.09.075

[23] D. A. Goldhamer, J. Dunai and L. Ferguson, "Irrigation Requirements of Olive Trees and Responses to Sustained Deficit Irrigation," *Acta Horticulturae*, Vol. 356, 1994, pp. 172-175.

18

Microbial Contamination in Vegetables at the Farm Gate Due to Irrigation with Wastewater in the Tamale Metropolis of Northern Ghana

Samuel Jerry Cobbina[1,2]*, Mohammed Clement Kotochi[1], Joseph Kudadam Korese[3], Mark Osa Akrong[4]

[1]Department of Ecotourism & Environmental Management, Faculty of Renewable Natural Resources, University for Development Studies, Nyankpala, Ghana; [2]School of the Environment, Jiangsu University, Zhenjiang, China; [3]Department of Agricultural Mechanization & Irrigation Technology, Faculty of Agriculture, University for Development Studies, Nyankpala, Ghana; [4]Environmental Biology & Health Division, CSIR Water Research Institute, Accra, Ghana.
Email: *cobbinasamuel@yahoo.com

ABSTRACT

The rational for this study was to assess the microbial quality of fresh vegetables at the farm gate of the Water Works road vegetable farm at Gumbihini in the Tamale Metropolis. A total of thirty-six (36) vegetables comprising lettuce, amarantus and cabbages and eight (8) wastewater samples were collected at random and analysed for a period of four months, to assess the microbial contamination level. Samples were analysed for total coliforms, faecal coliforms, *E. coli* and helminthes eggs. All vegetables sampled during the study period recorded high levels of total and faecal coliform bacteria. Mean faecal coliforms for the various vegetables were as follows; lettuce 3.7 ± 0.5 CFU·g^{-1}, amarantus 3.5 ± 0.6 CFU·g^{-1} and cabbage 3.1 ± 0.6 log CFU·g^{-1} fresh weight. FC levels were above the International Commission on Microbiological Specifications for Foods (ICMSF) recommended level of 3 log CFU·g^{-1} fresh weight. *E. coli* were recorded in lettuce (3.3 ± 0.6 log CFU·g^{-1} fresh weight) and amarantus (0.6 ± 0.1 log CFU·g^{-1} fresh weight) but not in cabbages. Lettuce generally recorded high levels of microbial contamination because of the large surface area occupied by its leaves. Two helminth eggs (*Strongiloides stercoralis*) were identified in lettuce while four (*Ascaris lumbricoides*) were identified in wastewater. Microbial loads recorded in wastewater were generally higher than that of vegetables. Since most of these vegetables are eaten fresh or slightly cooked, there is course for concern as public health will be adversely affected. Education of farmers and consumers on food safety has to be intensified to avert a possible outbreak.

Keywords: Helminthes; Faecal Coliform; Irrigation Water; Vegetables; *E. coli*

1. Introduction

The use of wastewater for crop production has been increasing worldwide due to the increasing food demand and the changing climatic conditions that are making food production through rainfed agriculture less reliable [1]. The ever increasing worldwide population, especially in urban and peri-urban areas of the developing economies calls for serious thoughts and approaches in meeting the food demand whiles taking care of the environment for sustainable development. In many countries in Sub-Sahara Africa (SSA), urban wastewater is used to irrigate agricultural lands. This way of disposing of urban

sewage has several advantages. Wastewater contains a lot of nutrients, which increase crop yields without use of fertilizer [2,3]. Furthermore, sewage water is an alternative water source where water is scarce. However, wastewater also contains a variety of chemical substances and microbiological loads from domestic and industrial sources.

Cities in developing countries, including Ghana, are also experiencing this unparalleled population growth. Available reports indicates that, only 4% - 5% of the population in Ghana are linked with—infrequently functional—sewage systems and sewerage treatment plants and as a result, large volumes of wastewater is often released untreated into the environment (streams, drains,

*Corresponding author.

etc.) [4,5]. According to Amoah [6] the use of potable water for vegetable production is constrained because less than 40% of city dwellers are still without good drinking water. The need for year-round production of vegetables in or near urban areas makes irrigation necessary; hence, farmers in search of water for irrigation often rely on the wastewater for irrigation [5]. This practice though beneficial in its contributions to urban food security and livelihoods, raises also public health concerns due to the risks posed from untreated wastewater to farmers and vegetable consumers [7].

Vegetables can become contaminated with pathogenic organism during growth, harvest, postharvest handling or distribution [8,9]. The use of untreated wastewater in irrigation represents an important route for transmission of these pathogenic organisms. The major pathogens associated with the use of highly polluted water are the faecal coliforms, *E. coli* and eggs of some helminthes such as *Ascaris lumbricoides, Trichuris trichura, Hymenolepis diminuta, Fasciola hepatica* and *Strongyloides* whose resistant eggs can be found in the wastewater [2,10,11]. According to Ensink [12], farmers and irrigation workers can acquire helminth infections and parasitic diseases due to direct contact with untreated wastewater and contaminated soils especially if exposed for a long duration. It has also been reported that irrigation of salad crops with untreated wastewater caused excess disease (e.g. shigellosis in England) in those who consumed them [13]. Numerous opportunities exist for attachment and penetration of pathogenic bacteria on vegetables (e.g. lettuce) in the field, as well as during harvesting, processing and marketing, especially when a contaminated product is exposed to water or is damaged [14]. The factors controlling the transmission of disease are agronomic; examples of factors are the type of crop grown, irrigation method used to apply wastewater and the cultural and harvesting practices.

Regarding the situation in the Tamale Metropolis, illegal use of raw wastewater for vegetable production has been observed over the past years. The practice is especially common during the dry seasons when vegetables produced through rainfed agriculture are least available [15]. Previous studies have showed the presence of pathogens in vegetables sold in two (2) major markets and wastewater irrigated sites in Ghana [10]. This study therefore sought to increase knowledge in this area by determining the microbiological loads in vegetables produced at the farm gate in the Tamale Metropolis of Northern Region of Ghana. Specifically, the study aimed at quantifying coliform bacteria, *Escherichia coli* and helminths eggs on vegetables and wastewater used in irrigating the vegetables.

2. Materials and Methods

2.1. Study Area and Sampling

The study was conducted in the Tamale Metropolis which lies between latitude 9°24′00″ N and longitude 0°50′00″ W. A total of thirty-six (36) vegetable samples made up of lettuce, cabbages and amarantus locally known as alefu were collected from the Water Works road vegetable farm in Gumbihini (**Figure 1**). Eight wastewater samples were collected from wastewater used to water vegetables on the farm (**Figure 2**). Sampling was between the months of December 2011 and March 2012. The temperature, pH and electrical conductivity of irrigation water were analyzed insitu using a pH/Conductivity meter 340 i/SET (WTW). Irrigation water was collected in 200 mL sterilized screw-capped glass bottle, kept under ice at a temperature of 4°C and transported to the CSIR Water Research Laboratory for bacteriological analysis.

2.2. Laboratory Analyses

Irrigation water and vegetable samples were analysed quantitatively for the determination of total coliform,

Figure 1. Watering of lettuce at the Waterworks farm.

Figure 2. Wastewater used for watering vegetables.

faecal coliform, *E. coli* and Helminth eggs. 10 g of vegetable sample was weighed into 180 mL of phosphate-buffered saline and rinsed vigorously. The water resulting from the rinsing was used for the determination of total coliform, faecal coliform and *E. coli* analysis (**Figure 3**). The Membrane Filtration (MF) technique was used to analyse these parameters. 10 mL of each of the samples (irrigation water and solution) were separately filtered through a 0.45 μm pore size membrane filter. The filter was then placed on M-endo media, MFC media and Hicrome (Difco) media for the detection of total coliform, faecal coliform and *E. coli* respectively. Incubation was then done at 37°C ± 0.5°C for the determination of total coliform and *E. coli*, and 44°C for faecal coliform determination for 16 - 24 hours. Colonies were counted using a colony counter [16].

Helminth eggs were enumerated from the vegetable samples and irrigation water using the concentration method [17]. This is a modified US-EPA method using $ZnSO_4$ solution (specific gravity, 1.2). About 100 g of vegetable leaves were thoroughly washed in about 1 L of sterile distilled water. The washed water and irrigation water were then poured into a container and allowed to stand overnight to enable the eggs to settle completely. As much supernatant as possible was sucked up and the sediment transferred into 50 mL centrifugation tubes. The container was rinsed three times with deionised water and the rinsed water transferred into the centrifugation tubes. The tubes were then centrifuged at 1450 rpm for 3 mins (**Figure 4**). The supernatant was poured off and the deposit re-suspended in 40 mL $ZnSO_4$ solution. The mixture was homogenized with a sterile spatula and centrifuged at 1450 rpm. At a specific gravity of 1.2 ($ZnSO_4$), helminth eggs float leaving other sediments at the bottom of the centrifuge tube. The $ZnSO_4$ supernatant was poured into a 2 L flask and distilled water added to the 1 L mark. This was allowed to stand overnight for the eggs to resettle. As much supernatant as possible was sucked up and the deposit re-suspended by shaking. This was then transferred into a centrifuge tube. The deposit was re-suspended in 15 mL acid/alcohol buffer solution (H_2SO_4 at 0.1 N at 35% ethanol, *i.e.*, 350 mL ethanol and 5.16 mL H_2SO_4) and about 5 mL ethyl acetate was added (**Figure 4**).

The mixture was shaken and the centrifuge tube occasionally opened to let gas out before centrifuging at 2200 rpm for 3 mins. After centrifugation, a diphasic solution (aqueous and lipophilic phase representing the acid/alcohol and ethyl acetate, respectively) was formed. With the aid of a micropipette, large volumes of the supernatant (starting from the lipophilic and then the aqueous phase) were sucked up leaving about 1 mL of the deposit. The deposit was then transferred onto a Sedgwick-Rafter

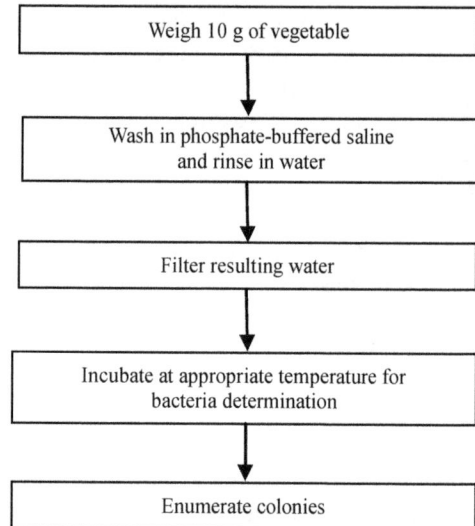

Figure 3 flowchart:
- Weigh 10 g of vegetable
- Wash in phosphate-buffered saline and rinse in water
- Filter resulting water
- Incubate at appropriate temperature for bacteria determination
- Enumerate colonies

Figure 3. Flow chart showing procedure for determination of bacteria on vegetables.

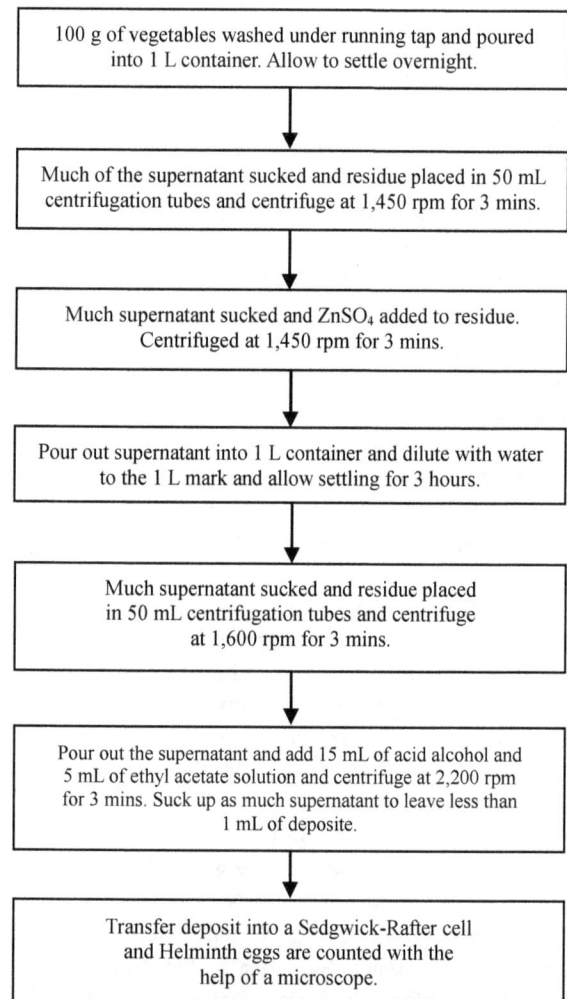

Figure 4 flowchart:
- 100 g of vegetables washed under running tap and poured into 1 L container. Allow to settle overnight.
- Much of the supernatant sucked and residue placed in 50 mL centrifugation tubes and centrifuge at 1,450 rpm for 3 mins.
- Much supernatant sucked and $ZnSO_4$ added to residue. Centrifuged at 1,450 rpm for 3 mins.
- Pour out supernatant into 1 L container and dilute with water to the 1 L mark and allow settling for 3 hours.
- Much supernatant sucked and residue placed in 50 mL centrifugation tubes and centrifuge at 1,600 rpm for 3 mins.
- Pour out the supernatant and add 15 mL of acid alcohol and 5 mL of ethyl acetate solution and centrifuge at 2,200 rpm for 3 mins. Suck up as much supernatant to leave less than 1 mL of deposite.
- Transfer deposit into a Sedgwick-Rafter cell and Helminth eggs are counted with the help of a microscope.

Figure 4. Procedure for the determination of helminth eggs on vegetables.

cell, observed under a microscope and the eggs counted (**Figure 4**). Eggs were identified by shape, size and colour. The bench aid for the diagnosis of Intestinal Parasites [18] was used for preliminary identification.

2.3. Data Analyses

Geometric means, standard deviations, minimum and maximum values were calculated using SPSS (version 16.0 for Windows). Total coliform, faecal coliform and *E. coli* loads on vegetables and wastewater were normalized by log transforming the raw data before it was analyzed. One way ANOVA was used to assess the statistical differences ($P < 0.05$) between wastewater and vegetables. Correlation analysis was done to determine the source of microbial loads.

3. Results and Discussions

The mean microbial loads on vegetables irrigated with wastewater are presented in **Table 1**. All vegetables sampled during the study period recorded high levels of total and faecal coliform bacteria. The highest level of contamination of total coliform, faecal coliform, *E. coli* and helminth eggs were recorded on lettuce. Mean levels of total coliforms (TC), faecal coliform (FC), *E. coli* and Helminthes eggs on lettuce were 4.1 ± 0.5, 3.7 ± 0.5, 3.3 ± 0.6 $\log_{10}$ CFU·g^{-1} fresh weight and two (2) helminth eggs respectively (**Table 1**).

The helminth species found in the lettuce was *Strongiloides stercoralis*. *E. coli* were detected in lettuce and amarantus, however, none was detected in cabbage during the study period. *E. coli* detected in lettuce ranged from 2.60 to 4.06 log CFU·g^{-1} fresh weight with a mean of 3.3 ± 0.6 log CFU·g^{-1} fresh weight which was higher than the International Commision on Microbiological Specifications for foods (ICMSF) recommended level of 3 $\log_{10}$ CFU·g^{-1} fresh weight. *E. coli* levels on amarantus ranged from 0.45 to 0.61 $\log_{10}$ CFU·g^{-1} fresh weight (mean 0.57 log CFU·g^{-1}). Generally all vegetables recorded faecal coliforms levels higher than ICMSFs recommended limit. Contamination of vegetables on the farm seems to emanate primarily from wastewater used in watering vegetables. According to Abdul-Ghaniyu [15] farmers in the study area usually apply water to their plots using buckets or watering cans. Correlation analysis done indicated a high positive correlation between total coliform on the vegetables and waterwater used for watering (**Table 2**). Total coliform levels on vegetables showed that they emanate from the same source.

Faecal coliform bacteria levels on vegetables showed a similar trend (**Table 3**). Faecal coliform loads exhibited a high correlation on both vegetables and wastewater. There was positive correlation between *E. coli* levels on

Table 1. Mean levels of bacteria on vegetables.

	TC	FC	E .coli	Helminth (eggs/l)
Lettuce	(3.30 - 4.73)	(2.98 - 4.37)	(2.60 - 4.06)	2
N = 12	4.1 ± 0.5	3.7 ± 0.5	3.3 ± 0.6	
Amarantus	(3.20 - 4.44)	(2.80 - 4.10)	(0.45 - 0.61)	0
N = 12	3.8 ± 0.5	3.4 ± 0.6	0.6 ± 0.07	
Cabbage	(3.00 - 4.29)	(2.50 - 3.80)	0	0
N = 12	3.4 ± 0.2	3.1 ± 0.6		

Units for total coliform (TC), faecal coliform (FC) and *E. coli* are in $\log_{10}$ CFU·g^{-1} fresh weight.

Table 2. Correlation matrix for total coliforms in vegetables and wastewater.

	Lettuce	Amarantus	Cabbage	Wastewater
Lettuce	1	0.820	0.818	0.707
Amarantus		1	0.872	0.717
Cabbage			1	0.667
Wastewater				1

Correlation significant at 0.05 level (2-tailed).

Table 3. Correlation matrix for faecal coliforms on vegetables and wastewater.

	Lettuce	Amarantus	Cabbage	Wastewater
Lettuce	1	0.806	0.588	0.573
Amarantus		1	0.571	0.703
Cabbage			1	0.455
Wastewater				1

Correlation significant at 0.05 level (2-tailed).

lettuce and amarantus ($r^2 = 0.667$), amarantus and wastewater ($r^2 = 0.660$). However, there was a weak correlation of *E. coli* levels on wastewater and amarantus ($r^2 = 0.329$), implying that the wastewater was not the only source of contamination. Other likely sources of contamination may be from "daefacation" from humans and animals on marginal lands around the farms and from the application of fresh poultry manure on farm plots [10, 19].

Bacteria counts on lettuce and amarantus were higher than that on cabbage because of the larger surface area exposed to irrigation water. This confirms work done by Amoah [10] in parts of Ghana. High levels of bacteria in vegetables from the farm gate could be a source of serious health concern to residents. Consumption of contaminated vegetables has been associated with gastrointestinal diseases like typhoid, cholera and dysentery [20-26].

The mean microbial loads in wastewater used to irrigate vegetables in the study area are presented in **Table 4**. Total coliform composition of wastewater ranged from 3.19 to 4.82 log CFU/100 ml with a mean of 4.4 log CFU/100 ml. Faecal coliform bacteria ranged from 3.36 to 4.33 log CFU/100 ml with a mean of 4.0 log CFU/100 ml. Mean FC was higher than WHO recommended limit of 3 log CFU/100 ml [27].

Equally, wastewater used for vegetable production in the study area contained *E .coli* ranging from 3.29 to 4.22 log CFU/100 ml with a mean of 3.7 log CFU/100 ml (**Table 4**). Four (4) helminthes eggs/l which were of the *Ascaris lumbricoides* were detected in wastewater during the study period. This exceeded the recommended helminth level of less than 1 helminth egg/l for unrestricted irrigation [28]. Generally coliform bacteria, *E. coli* and helminth eggs recoreded in wastewater used for watering of the vegetables were higher than that recorded on the vegetables (**Figures 5-8**). The poor quality of wastewater in the study area was due to anthropogenic activities such as the disposal of human and animal waste into drains, open defecation around farms and runoffs of fresh poultery manure used to amend soils in the vegetable farms into irrigation water. Using analysis of variance it was observed that there was no significance difference in total coliform bacteria counts in vegetables and wastewater (**Figure 5**). There were however, significant differences ($P < 0.05$) in faecal coliform counts and *E.coli* levels in vegetables and wastewater (**Figures 5** and **6**). pH and electrical conductivity of wastewater were was good for irrigation.

Helminthes eggs recorded in lettuce and wastewater should be a source of concern, since they pose significant health risk in humans. Generally results of microbial loads in wastewater and vegetables indicate that residents who patronize vegetables from these farms stand a high risk of diseases such as cholera, typhoid and intestinal worms. Data from the Regional Information Management Unit of the Ghana Health Service, Northern Region show reported cases of these diseases (**Table 5**). Though

Table 4. Physical and bacteriological analysis of wastewater used for watering vegetables.

Parameter	Min	Max	Mean	SD	[27]
Total coliform (log CFU/100 ml)	3.19	4.82	4.4	0.4	-
Faecal coliform (log CFU/100 ml)	3.36	4.33	4.0	0.4	3
E. coli (log CFU/100 ml)	3.29	4.22	3.7	0.4	
Helminthes (eggs/l)	<1	4	1	-	0
pH	7.01	7.40	7.24		
Temperature (°C)	30.9	34.1	31.8		
Conductivity (μS/cm)	240	243	242		

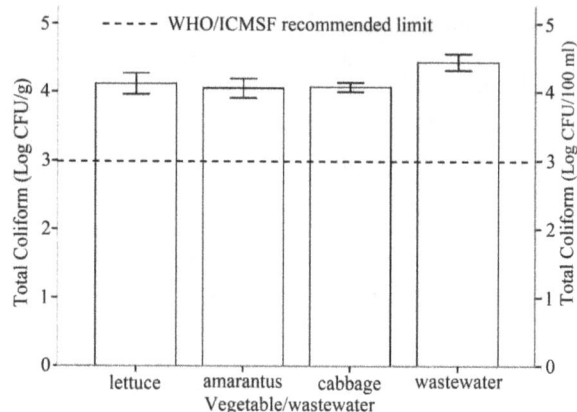

Figure 5. Mean total coliform on vegetables/wastewater.

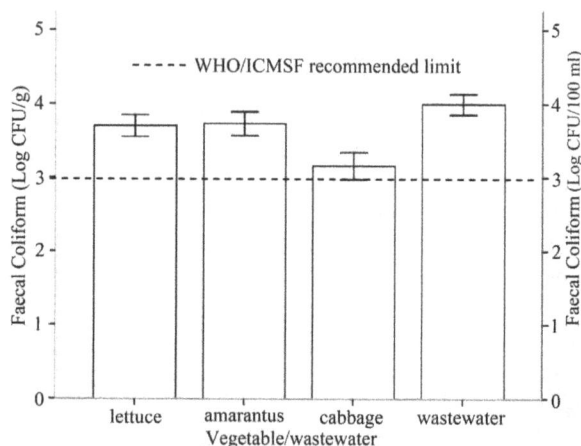

Figure 6. Mean faecal coliform on vegetable/wastewater.

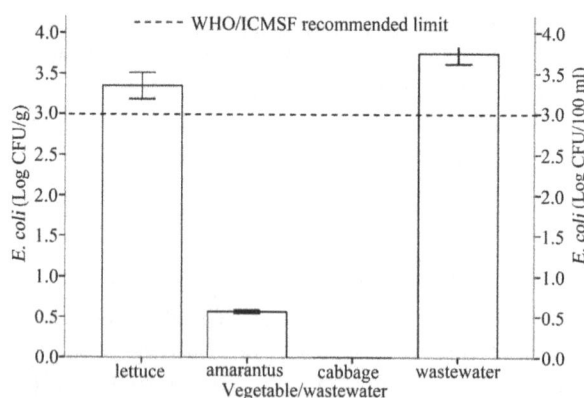

Figure 7. Mean *E. coli* counts on vegetables and wastewater.

figures may not all be as a result of the consumption of contaminated wastewater irrigated vegetables, there is still cause for concern.

4. Conclusion

The consumption of vegetables serves as a vital source of much needed minerals and vitamins for the sound deve-

Table 5. Correlation matrix for faecal coliforms on vegetables and wastewater.

Disease	Year				
	2008	2009	2010	2011	2012
Typhoid fever	673	2,740	1,799	3,023	56
Cholera	5	2	20	0	0
Diarrhoea	10,661	10,437	9,536	16,654	5,167
Intestinal worms	3,587	3,245	3,519	4,799	204

Source: Ghana Health Service, Northern Region. (Note: Data for 2012 was from Jan-Mar).

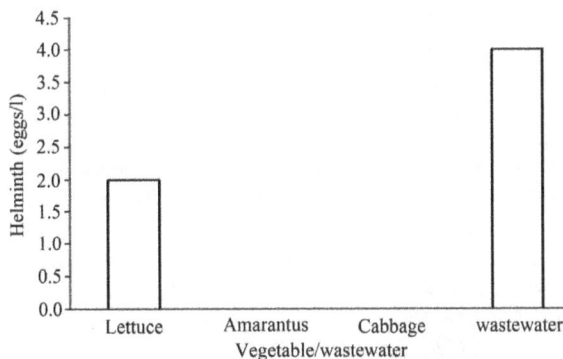

Figure 8. Helminths eggs in vegetable/wastewater.

lopment of the human body. The study shows that microbial loads on vegetables and in wastewater in this part of the Tamale Metropolis were well above ICSFM and WHOs recommended limits for vegetables and wastewater respectively. People who eat these contaminated vegetables raw, stand a high chance of contracting gastrointestinal diseases. To prevent an eminent outbreak efforts have to be made to discouraged farmers from the use of wastewater for irrigation. If they are to use continue with the usage of wastewater, then they have to be educated on good agricultural practices such as modification of the current manual irrigation system to reduce vegetables contact with the polluted irrigated water. Legislation against wastewater use for irrigation could be improved as well as education of farmers on pre-harvesting practices that will help reduce microbial loads on the vegetables. Government and non-government agencies are entreated to provide alternative water sources such as boreholes and hand dug wells on the farms for watering of the vegetables to safeguard the health of consumers.

5. Acknowledgements

The team is grateful to staff of CSIR Water Research Institute who helped in the laboratory analysis. We acknowledge the cooperation and help of vegetable farmers at the Gumbihini Waterworks area.

REFERENCES

[1] C. A. Scott, N. I. Faruqui and L. Raschid-Sally, "Wastewater Use in Irrigated Agriculture: Confronting the Livelihood and Environmental Realities," CABI, Wallingford, 2004.

[2] F. Sabiena, H. Raheela and Van der H. Wim, "Health Risks of Irrigation with Untreated Urban Wastewater in the Southern Punjab, Pakistan," Institute of Public Health (IPH) and International Water Management Institute (IWMI) Research Report No. 107, 2000.

[3] G. B. Shende and C. Chakrabarti, "Optimum Utilization of Municipal Wastewaters as a Source of Fertilizer," *Journal of Resources and Conservation*, Vol. 13, No. 2-4, 1987, pp. 281-290. doi:10.1016/0166-3097(87)90070-8

[4] Ghana Statistical Services, "2000 Population and Housing Census: Summary Report of Final Results," Ghana Statistical Services, Legon, Accra, 2002.

[5] B. Keraita, P. Drechsel, F. Huibers and L. Raschid-Sally, "Wastewater Use in Informal Irrigation in Urban and Peri-Urban Areas of Kumasi, Ghana," *Urban Agriculture Magazine*, No. 8, 2002, pp. 11-13.

[6] P. Amoah, P. Drechsel and R. C. Abaidoo, "Irrigated Urban Vegetable Production in Ghana: Sources of Pathogen Contamination and Health Risk Reduction," *Irrigation Drainage*, Vol. 54, No. S1, 2005, pp. 49-61. doi:10.1002/ird.185

[7] R. C. Abaidoo, B. Keraita, P. Amoah, P. Drechsel, J. Bakang, G. Kranjac-Berisavljevic, F. Konradsen, W. Agyekum and A. Klutse, "Safeguarding Public Health Concerns, Livelihoods and Productivity in Wastewater Irrigated Urban and Peri-Urban Vegetable Farming," CPWF PN 38 Project Report, Kumasi, 2009.

[8] M. A. S. McMaho and I. G. Wilson, "The Occurrence of Enteric Pathogens and Aeromonas Species in Organic Vegetables," *International Journal of Food Microbiology*, Vol. 70, No. 1-2, 2001, pp. 155-162. doi:10.1016/S0168-1605(01)00535-9

[9] L. R. Beuchat, "Food Safety Issues: Surface Decontamination of Fruits and Vegetables Eaten Raw: A Review," Food Safety Unit, WHO, Geneva, 1999.

[10] P. Amoah, P. Drechsel, R. C. Abaidoo and M. Henseler, "Irrigated Urban Vegetable Production in Ghana: Microbiological Contamination in Farms and Markets and Associated Consumer Risk Groups," *Journal for Water and Health*, Vol. 5, No. 3, 2007, pp. 455-466. doi:10.2166/wh.2007.041

[11] D. D. Mara, P. A. Sleigh, U. J. Blumenthaland and R. M. Carr, "Health Risks in Wastewater Irrigation: Comparing Estimates from Quantitative Microbial Risk Analyses and Epidemiological Studies," *Journal of Water and Health*, Vol. 5, No. 1, 2007, pp. 39-50. doi:10.2166/wh.2006.055

[12] J. H. J. Ensink, "Water Quality and the Risk of Hookworm Infection in Pakistani and Indian Sewage Farmers," Ph.D. Thesis, University of London, London, 2006.

[13] J. A. Frost, M. B. McEvoy, C. A. Bently, Y. Andersson and B. Rowe, "An Outbreak of *Shigella sonnei* Infection Associated with Consumption of Iceberg," *Emerging In-*

fectious Disease, Vol. 1, No. 1, 1995, pp. 26-28. doi:10.3201/eid0101.950105

[14] K. Takeuchi, A. N. Hassan and J. F. Frank, "Penetration of *Escherichia coli* O157:H7 into Lettuce as Influence by Modified Atmosphere and Temperature," *Journal of Food Protection*, Vol. 64, No. 11, 2001, pp. 1820-1823.

[15] S. Abdul-Ghaniyu, G. Kranjac-Berisavljevic, I. B. Yakubu and B. Keraita, "Sources and Quality of Water Used in Urban Vegetable Production in Tamale Municipality, Ghana," AU Magazine no.8-Wastewater Reuse in Urban Agriculture, 2005.

[16] APHA, AWWA, WEF, "Standard Methods for the Examination of Water and Wastewater," 20th Edition, American Water Works Association, Denver, 1998.

[17] J. Schwartzbrod, "Methods of Analysis of Helminth Eggs and Cysts in Wastewater, Sludge, Soil and Crops," University Henry Poincare, Nancy, 1998.

[18] World Health Organization, "Guidelines for Drinking Water Quality," First Addendum to 3rd Edition, Recommendations, Vol. 1, 1993, p. 444.

[19] N. O. B. Ackerson and E. Awuah, "Microbial Risk Assessment of Urban Agricultural Farming: A Case Study on Kwame Nkrumah University of Science and Technology Campus, Kumasi, Ghana," *International Journal of Science and Technology*, Vol. 1, No. 3, 2012, pp. 118-125.

[20] M. A. Halablab, I. H. Sheet and H. M. Holail, "Microbiological Quality of Raw Vegetables Grown in Bekaa Valley, Lebanon," *American Journal of Food Technology*, Vol. 6, No. 2, 2011, pp. 129-139. doi:10.3923/ajft.2011.129.139

[21] L. R. Beuchat, "Ecological Factors Influencing Survival and Growth of Human Pathogens on Raw Fruits and Vegetables," *Microbial Infections*, Vol. 4, No. 4, 2002, pp. 413-423. doi:10.1016/S1286-4579(02)01555-1

[22] J. Spencer, H. R. Smith and H. Chart, "Characterization of Enteroaggregative *Escherichia coli* Isolated from Outbreaks of Diarrhoeal Disease in England," *Epidemiology and Infection*, Vol. 123, No. 3, 1999, pp. 413-421. doi:10.1017/S0950268899002976

[23] S. M. Faruque, M. J. Albert and J. J. Mekalanos, "Epidemiology, Genetics and Ecology of Toxigenic Vibrio Cholerae," *Microbiology and Molecular Biology Reviews*, Vol. 62, No. 4, 1998, pp. 1301-1314.

[24] P. S. Mead and P. M. Griffin, "*Escherichia coli* O157:H7," *Lancet*, Vol. 352, No. 9135, 1998, pp. 1207-1212. doi:10.1016/S0140-6736(98)01267-7

[25] H. Michino, K. Araki, S. Minami, *et al.*, "Massive outbreak of *Escherichia coli* O157:H7 Infection in School Children, Sakai City, Japan, Associated with Consumption of White Radish Sprouts" *American Journal Epidemiology*, Vol. 150, No. 8, 1999, pp. 787-796. doi:10.1093/oxfordjournals.aje.a010082

[26] H. I. Shuval, "Health Guidelines and Standards for Wastewater Reuse in Agriculture: Historical Perspectives," *Water Science and Technology*, Vol. 23, No. 10-12, 1991, pp. 2037-2080.

[27] World Health Organization, "Health Guidelines for the Use of Wastewater in Agriculture and Aquaculture," Technical Report Series No. 778. WHO Scientific Group, Geneva, 1989.

[28] World Health Organisation, "Guidelines for the Use of Wastewater Excreta and Grey Water: Wastewater Use in Agriculture," Vol. 1, 2, 4, WHO, Geneva, 2006.

19

Effect of Treated Waste Water Irrigation on Plant Growth and Soil Properties in Gaza Strip, Palestine

Yasser El-Nahhal[1,2]**, Khalil Tubail**[1]**, Mohamad Safi**[1]**, Jamal Safi**[1,3*]

[1]Environmental Protection and Research Institute, Gaza, Palestine; [2]The Islamic University of Gaza, Gaza, Palestine; [3]Al-Azhar University of Gaza, Gaza, Palestine.
Email: *eprigaza@palnet.com

ABSTRACT

This study investigated the effect of treated wastewater (TWW) irrigation on growth of Chinese cabbage and corn and on soil properties in Gaza Strip, Palestine. Chinese cabbage and corn were planted in winter and summer seasons respectively in a sandy soil. The experimental design was a randomized complete block using 2 treatments with 4 replicates. Soil samples were collected from 0.0 - 120 cm depths from all plots and analyzed for pH, electric conductivity (EC) and nutrient contents. The plants were irrigated with either TWW or fresh water (FW) fortified with NPK, while control used drip irrigation system. The biomass (total fresh weight of the plants) was used as an indicator of the plant yields. Concentration of heavy metals on plant leaves was determined by Inductive Coupled Plasma Analyzer (ICP) and was taken as an indicator of plant quality. Biomass of Chinese cabbage and corn grown in plots irrigated with TWW was higher than those grown in plots irrigated with FW. These results indicate the ability of TWW supplying the necessary nutrients for plant growth. Heavy metal content in plant leaves in all treatments (TWW and FW) was nearly similar and below EPA standards, indicating high quality of plants. Soil analysis showed great changes in soil properties due to irrigation with TWW. The interesting outcome of this study is that TWW is an effective source for plant nutrients. It is encouraging to reuse TWW in agricultural system after full treatment.

Keywords: TWW; FW; Corn; Chinese Cabbage; Heavy Metals

1. Introduction

The quantity of treated wastewater (TWW) in Gaza is about 111,900 m^3/day [1]. This quantity is expected to increase during the coming years due to population growth. Guidelines for safe and effective reuse of TWW for agricultural purposes are not yet approved by the Palestinian Authority, or any Research Institute. Regardless of this fact, this high quantity of TWW should be reused in the agricultural or industrial sectors to solve the problem of water quality in Gaza Strip Palestine. Although TWW in Gaza did not meet the international standards, several trials have been made by international institutions in Gaza for the reuse of TWW in agriculture. For instance, effluent from the existing Gaza waste water treatment plants (WWTP) is currently being used by farmers through pilot projects funded by the Spanish and French governments. In the Spanish project the trial includes irrigation of citrus and olive trees in Gaza area

(Stawi Farm in Al Zeitoon area, 100 dunums), and in the French project the trial includes irrigation of forage crops in North area (beit Lahia, 40 dunums). Obviously, these trials are not based on scientific research. They used the TWW without looking into the quality of the products or the effect of TWW in soil properties.

Nevertheless, elsewhere several investigations have been made to evaluate the quality of TWW for possible reuse in agricultural sectors. For instance, Evett et al. [2] evaluated the feasibility of using TWW irrigation strategies based on 1) water use of different tree species, 2) weather conditions in different climate zones of Egypt, 3) soil types and available irrigation systems, and 4) the requirement to avoid deep percolation losses that could lead to groundwater contamination. They concluded that drip irrigation systems are preferred to achieve several small irrigations per day in order to avoid deep percolation losses. Mendoza-Espinosa et al. [3] evaluated the effect of treated wastewater on the growth of cabernet sauvignon and merlot grapes from the Guadalupe Valley,

---

*Corresponding author.

Mexico. They reported that the number of leaves per shoot and the overall biomass increased in plants irrigated with wastewater and grape production per plant was 20% higher and the concentration of carbohydrates, organic acids and pH were similar in grapes from vines irrigated with wastewater to those irrigated with groundwater. Oron *et al.* [4] evaluated the influence of TWW on sustainable agricultural production and safe groundwater recharge using filed experiment. They reported that ultrafiltration stage is efficient in the removal of the pathogens and suspended organic matter while the successive Reverse Osmosis (RO) stage provides safe removal of the dissolved solids (salinity). Best agricultural yields were obtained when applying effluent had minimal content of dissolved solids (after the RO stage) as compared with secondary effluent without any further treatment and extended storage. Mosse *et al.* [5] investigated the effect of application of winery wastewaters to physicochemical properties of soils. They concluded that long-term application of winery wastewaters had significant impacts on soil respiration, nitrogen cycling and microbial community structure, but the treated wastewater application showed no significant differences in wetting alone. Singh *et al.* [6] investigated the effect of land application of sewage sludge on the physicochemical properties of soils. They concluded that amending soil with sewage sludge modified the physicochemical properties of soils, and might contaminate ground water, stock ponds, or produce food chain contamination from eating food grown in sludge-treated land. Castro *et al.* [7] studied the effects of wastewater irrigation on turfgrass growth, and reported that plants irrigated with treated wastewater had the highest sodium content. Pritchard *et al.* [8] investigated the risks of the environment and food crops that may come from land application of sewage sludge in Australia. They reported that the attention was given to researches related to plant nutrient uptake, particularly nitrogen and phosphorus (including reduced phosphorus uptake in alum sludge-amended soil); the risk of heavy metal uptake by plants, specifically cadmium, copper and zinc; the risk of pathogen contamination in soil and grain products; change of soil pH. Belyaeva *et al.* [9] investigated the effects of adding biosolids to a green waste feedstock (100% green waste, 25% v/v biosolids or 50% biosolids) on the properties of composted products. They found that addition of biosolids to the feedstock increased total N, EC, extractable NH(4), NO(3) and P but lowered pH, macroporosity, water holding capacity, microbial biomass C and basal respiration in composts. This paper investigates the effect of TWW irrigation on Chinese cabbage and corn growth and quality and on soil properties in Gaza Soil.

2. Materials and Methods

2.1. Environmental Background of the Study Area

Gaza Strip is a semi-arid region of roughly 365 km^2 which lies on the Mediterranean Sea. On this narrow strip, almost 1.625 million of the Palestinian people live and work [10]. The ground water is used for irrigation, industrial and domestic purposes. A "Catastrophic" water shortage, water pollution with high salinity and micropollutants, lack of sewage and solid waste treatment, maritime pollution, overcrowding, poverty and uncontrolled use of pesticides are the most pressing environmental problems in the Gaza Strip. Internationally suspended, banned and canceled pesticides which considered mutagenic and carcinogenic are still used in the agricultural environment. The wastewater sector in the Gaza Strip is characterized by poor sanitation, insufficient treatment and unsafe disposal. Currently, there are four wastewater treatment plants in operation in the Gaza Strip namely: Beit Lahia, Gaza, Khan Younis and Rafah Wastewater Treatment Plants (WWTP's) receiving about 24 Million m^3 of raw sewage per year.

2.2. Experimental Design

The experimental site was selected as sandy soil in the north zone of the Gaza Strip. Two treatments were selected: one irrigated with TWW and the other with fresh water (FW). Each treatment contains 4 replicates (plots). Plot dimensions are 5 × 4 m. Each plot is divided into 5 rows for planting (12 plants in each row). The experimental design was established as a randomized complete block design. Chinese cabbage seedlings were planted for first season on 28/11/2010 and harvested on 9/3/2011 and for the 2nd season on 11/11/2011 and harvested on 21/2/2012. Corn seeds were sown for the 1st season on 21/3/2011 and harvested on 25/6/2011 and for 2nd season 21/3/2012 and harvested on 25/6/2012.

2.3. Soil Analysis

Soil samples from depths of 0.0 - 30, 30 - 60, 60 - 90 and 90 - 120 cm depth were collected from eight soil profiles dug in all plots. The soil samples were air dried, sieved through 2 mm mesh and kept in plastic bags in the laboratory for pH, EC and nutrient contents using the standard methods.

2.4. Irrigation Water and Analysis

Treated wastewater (TWW) from Beit Lahia wastewater treatment plant and fresh water (FW) from local well were used for irrigation. The irrigated plants are Chinese cabbage as winter crop and corn (Zea maiz, Variety

Merit) as summer crop was grown on a sandy soil. Irrigation was managed by a drip irrigation system with discharge of 4 L/plant/h according to the standard water requirements [11].

Samples of TWW and FW were analyzed for physico-chemical properties following the procedure described in the standard method [12].

2.5. Harvesting, Plant Sampling and Preparation

Plants were harvested after four months of planting day or by the end of each season. The biomass of Chinese cabbage and corn were collected and weighed and used as growth indicator.

Chinese cabbage and corn leaves were sampled randomly from several plants of each plot on the harvesting day at the end of winter season, whereas, corn leaves were sampled at the end of summer season. Leave samples were collected from several plants of each plot. The samples were washed with tap water to remove atmospheric dust sand and then washed with distilled water. Leave samples were then oven dried at 65°C for 48 hours, ground and kept in well sealed plastic bags and stored at room temperature for elemental analysis.

2.6. Elemental Analysis

About 0.5 g of oven dried leave sample as mentioned above was digested with concentrated nitric acid in glass tube at 80°C for 48 - 72 h then heated up to 120°C for 4 - 8 h to have clear solution as previously described [13]. Samples then were cooled and diluted with distilled water up to 25 ml, filtered using small glass or plastic funnels pre-washed with sulfuric acid. Elements concentration in the filtrate was determined using ICP. Two determinations were conducted per replicate

2.7. Data Analysis

The data were statistically analyzed using mean and standard deviations. Analysis of variance between treatments was conducted using T-Test. P-values associated with T-test were taken as an indicator of significant differences among the treatments. P-values are presented below Tables. P-values less than 0.05 are considered significant.

3. Results and Discussion

3.1. Soil Analysis

Soil components and soil texture of the field plots are shown in **Table 1**. It can be seen the clay fraction of soil ranged between 2% - 5% in different depths of the soil profiles. Accordingly soil can be classified as sandy soil.

Furthermore, the sand fraction of soils in all depths ranged between 91% - 94% indicating sand texture of soil.

Biological, physical and chemical properties of irrigation water are shown in **Tables 2**.

BOD, COD, TSS, and EC stand for: biological oxygen demand, chemical oxygen demand, total suspended solids and electric conductivity, respectively.

It can be seen that BOD, COD and TSS are nil in FW whereas high values are observed in TWW. In addition, nitrate level in FW is higher than in TWW. The explanation of these results is that FW is nitrogen phosphorus

Table 1. Soil fractions and texture.

Depth (cm)	Sand %	Silt %	Clay %	Soil Texture
0 - 30	94	4	2	Sandy
30 - 60	92	5	3	Sandy
60 - 90	91	4	5	Sandy
90 - 120	92	4	4	Sandy

Table 2. biological and chemical properties of irrigation water, 2011.

Properties	FW	TWW
BOD (mg/L)	-	95.8
COD (mg/L)	-	242.3
TSS (mg/L)	-	108.7
pH	8.22	8.41
EC (dsim/m)	2.39	2.2
$N-NO_3$ (mg/L)	38.9	1.6
$N-NH_4$ (mg/L)	-	51.6
K mg/l	3.8	21.7
Na (mg/L)	115	159
Ca mg/L	215	112
Mg	36	41
SAR	1.9	3.3
S mg/L	34	20
P mg/L	0.07	4.89
B (ppm)	0.07	0.17
Cl (ppm)	505	351
Cr (ppm)	0.005	0.001
Cu (ppm)	0.002	0.003
Fe (ppm)	0.002	0.009
Mn (ppm)	0.001	0.002
Ni (ppm)	0.016	0.018

potassium (NPK) fortified whereas in TWW the nitrate level is being reduced to ammonium hydroxide due to an aerobic condition. Accordingly low level of nitrate is available in TWW. Sodium and Potassium are several times higher in TWW than in FW. Calcium concentration is higher in FW than in TWW whereas Magnesium has opposite direction. Sulfur concentration is higher in FW than TWW due to possible transformation of sulfate to hydrogen sulfide in TWW due to an aerobic conditions. Phosphorus and Barium are higher in TWW than in FW. Chloride is higher in FW than in TWW due to chlorination process in drinking water. Heavy metal contents ranged from 0.001 to 0.018 ppm in FW and TWW indicating low contents. Comparison with EPA standards shows that the properties of the used water are within the range. Accordingly, the current water situation may be used for agricultural irrigation. The following: Ag, As, Bi, Cd, Co, Hg, Mo, Pb, Se and Sn were not detected in the used water.

Table 3 shows pH and EC values for the soil profiles. It is obvious that pH values ranged between 7.69 ± 0.21 to 8.05 ± 0.15 in FW and from 7.75 ± 0.1 to 8.1 ± 0.01 in TWW plots.

It is obvious that soil is more acidic at the top layers 0.0 - 30 cm depth and less acidic at deeper depths in FW and TWW plots. EC values are high in the top soil layers (**Table 3**) and several times lower in deeper depths (90 - 120 cm) in both FW and TWW plots. This may be due to accumulation of less soluble salts in the tope soil layer and possible formation of organic acids due to biodegradation of organic compounds in soils. These results are in accord with Belyaeva et al. [9] who found lower pH in the top soil due to addition of biosolids.

Statistical analysis for comparison between pH, and EC values in soil profiles of 2011 does not show significant differences in pH and EC values in TWW and FW. P-values are 0.18 and 0.43 respectively.

Our results agree with Castro et al. [7] who investigated the effects of wastewater irrigation on soil properties and turfgrass growth and concluded that there were no negative effects with respect to changes in soil pH but a significant increase in electrical conductivity and sodium content was observed in wastewater-irrigated soil.

Table 3. pH, and EC (dS/m) of soil profile, 2011.

Depth (cm)	FW		TWW	
	pH	EC (dS/m)	pH	EC (dS/m)
0 - 30	7.69 ± 0.21	0.57 ± 0.25	7.73 ± 0.1	0.52 ± 0.21
30 - 60	7.88 ± 0.21	0.27 ± 0.06	8.08 ± 0.14	0.27 ± 0.07
60 - 90	7.89 ± 0.12	$0.18 (0.02)$	8.05 ± 0.16	0.15 ± 0.02
90 - 120	8.05 ± 0.15	0.15 ± 0.01	8.10 ± 0.14	0.14 ± 0.02

Nitrate concentrations (**Table 4**) decreased from the top 0 - 30 to deeper depths in both treatments (TWW and FW). Nitrate concentration of the soil 0 - 30 of FW-profile is lower than TWW-profile as well as higher than other depths.

Chloride concentrations of soil profile are higher in the FW-samples than in the TWW-treated samples. Statistical analysis for comparison between N-NO3, and Cl⁻ values in soil profiles of 2011 does not show significant differences in N-NO3, and Cl⁻ values in TWW and FW. P-values are 0.2 and 0.34 respectively. Our results agree with Boruah and Hazarika [14] who concluded that available N, K, S and exchangeable and water soluble Na, K, Ca, Mg were highest in effluent irrigated soil.

Regardless to the highest values of organic matter that found at depth 30 - 60 cm (**Table 5**), the mount of organic carbon decreased from the top to deeper depths in both treatments (FW and TWW). Statistical analysis shows no differences between treatments, P-value equals to 0.45.

Our results agree with Adrover et al. [15] who investigated the chemical properties and biological activity in soils of Mallorca following twenty years of treated wastewater irrigation and did not observe negative effects on cation exchange capacity, pH, calcium carbonate equivalent, and soil organic matter.

Macronutrients, micronutrients and heavy metal concentrations in soil profiles are presented in **Tables 6-8**.

It can be seen that the concentrations of P (**Table 6**) are ranged between 0.19 - 0.58 ppm in all depths of FW and TWW treatments. This low value is due to low solubility of P in soil solution due to high soil pH (**Table 3**). These results agree with Metson et al. [16] who found

Table 4. N-NO3 and Cl⁻ concentrations (mmol) in soil water extract (1:1).

Depth (cm)	FW		TWW	
	N-NO3	Cl⁻	N-NO3	Cl⁻
0 - 30	0.76 ± 0.22	3.0 ± 1.78	1.01 ± 0.43	2.8 ± 1.35
30 - 60	0.46 ± 0.26	1.09 ± 0.56	0.64 ± 0.38	1.1 ± 0.44
60 - 90	0.41 ± 0.05	0.62 ± 0.26	0.49 ± 0.15	0.34 ± 0.21
90 - 120	0.32 ± 0.02	$0.43' \pm 0.35$	0.41 ± 0.23	0.39 ± 0.11

Table 5. Total organic carbon (ppm) of soil profile, 2011.

Depth (cm)	FW	TWW
0 - 30	26.2 ± 2.75	27.7 ± 10.9
30 - 60	32.1 ± 23.7	27.5 ± 13.6
60 - 90	15.6 ± 3.48	14.76 ± 4.33
90 - 120	12.0 ± 1.23	12.61 ± 2.95

Table 6. Macronutrients concentration (ppm) in soil profile, 2011.

Element	FW				TWW			
	0 - 30	30 - 60	60 - 90	90 - 120	0 - 30	30 - 60	60 - 90	90 - 120
P	0.22 ± 0.1	0.24 ± 0.1	0.43 ± 0.3	0.19 ± 0.06	0.17 ± 0.06	0.21 ± 0.07	0.58 ± 0.32	0.57 ± 0.46
K	6.5 ± 2.2	1.70 ± 0.6	1.80 ± 0.4	2.50 ± 1.2	5.9 ± 3.30	2.2 ± 0.70	2.2 ± 0.40	2.5 ± 1.0
Ca	60.7 ± 4.13	29.2 ± 7.0	19.5 ± 2.20	18.5 ± 2.70	48.5 ± 15.6	32.7 ± 4.10	20.1 ± 6.6	25.7 ± 11.0
Mg	9.25 ± 4.13	3.95 ± 0.88	2.58 ± 0.19	2.18 ± 0.25	3.24 ± 0.33	3.55 ± 1.66	2.81 ± 0.50	2.71 ± 0.63
S	13.4 ± 6.90	5.60 ± 1.10	4.10 ± 1.10	3.30 ± 0.60	2.19 ± 0.27	2.57 ± 2.82	1.15 ± 0.42	1.0 ± 0.57
Na	41.1 ± 20.8	20.8 ± 3.80	13.2 ± 2.30	13.5 ± 4.70	2.53 ± 1.07	1.38 ± 0.64	1.07 ± 0.30	1.1 ± 0.21

Table 7. Micronutrients concentration (ppm) in soil profile, 2011.

Element	FW				TWW			
	0 - 30	30 - 60	60 - 90	90 - 120	0 - 30	30 - 60	60 - 90	90 - 120
Fe	0.104 ± 0.11	0.320 ± 0.32	0.622 ± 0.33	0.833 ± 0.37	0.05 ± 0.02	0.35 ± 0.28	1.44 ± 0.58	1.69 ± 1.19
Zn	0.031 ± 0.01	0.037 ± 0.02	0.034 ± 0.02	0.022 ± 0.01	0.023 ± 0.01	0.026 ± 0.01	0.032 ± 0.02	0.028 ± 0.01
Mn	0.020 ± 0.02	0.07 ± 0.09	0.03 ± 0.02	0.05 ± 0.02	0.01 ± 0,01	0.04 ± 0.02	0.14 ± 0.03	0.10 ± 0.08
Cu	0.011 ± 0.00	0.01 ± 0.00	0.01 ± 0.00	0.01 ± 0.00	0.011 ± 0.00	0.01 ± 0.00	0.01 ± 0.00	0.01 ± 0.00
B	0.04 ± 0.01	0.04 ± 0.02	0.03 ± 0.00	0.03 ± 0.01	0.05 ± 0.01	0.041 ± 0.02	0.03 ± 0.01	0.03 ± 0.01

Table 8. Heavy metals concentration (ppm) in soil profile, 2011.

Element	FW				TWW			
	0 - 30	30 - 60	60 - 90	90 - 120	0 - 30	30 - 60	60 - 90	90 - 120
Co	0.002 ± 0.00	0.000 ± 0.00	0.000 ± 0.00	0.000 ± 0.00	0.002 ± 0.00	0.001 ± 0.00	0.001 ± 0.00	0.001 ± 0.00
Cr	0.001 ± 0.00	0.002 ± 0.00	0.003 ± 0.00	0.004 ± 0.00	0.002 ± 0.00	0.005 ± 0.00	0.004 ± 0.00	0.007 ± 0.00
Ni	0.013 ± 0.00	0.011 ± 0.00	0.013 ± 0.00	0.012 ± 0.00	0.011 ± 0.00	0.012 ± 0.00	0.017 ± 0.00	0.017 ± 0.00
Pb	0.109 ± 0.11	0.333 ± 0.33	0.648 ± 0.34	0.879 ± 0.37	0.047 ± 0.02	0.36 ± 0.28	1.54 ± 0.62	1.77 ± 1.26
Al	0.123 ± 0.13	0.398 ± 0.39	0.823 ± 0.43	1.161 ± 0.47	0.053 ± 0.03	0.47 ± 0.39	2.120 ± 0.80	2.378 ± 1.54

similar results of P in an urban ecosystem. Concentrations of K, Ca, Mg, S and Na are higher in the top soil layers than deeper depths in both treatments. Concentration of micronutrients in soil profile are shown in **Table 7**. It can be seen that except Fe, concentrations of Zn, Mn, Cu, and B are below 0.15 ppm indicating poor nutrient conditions. Furthermore, it can be seen that concentration of Fe is increased from top soil layer to deeper depths, indicating leaching of iron in Gaza Soils. However, the poor concentration of micronutrients in soil is in agreement with the general concept of sandy soils.

Concentrations of heavy metals are shown in **Table 8**. It is obvious that concentrations of Co, Cr, and Ni are below 0.02 ppm in both treatment (FW and TWW). These values indicate low contents of heavy metals in soil.

Similar results were recently observed [17], who made a geochemical survey in Italy and revealed the presence of huge volumes of composite wastes which accumulated up to a thickness of 25.6 m.

Furthermore, levels of Pb and Al are increasing gradually as increasing soil depth in both treatment (FW and TWW) indicating leaching of these metals in Gaza soils.

An interesting conclusion of these results is that Fe, Pb and Al pose threat to groundwater in Gaza.

3.2. Effect of TWW on Biomass

The total biomass of Chinese cabbage and corn are presented in **Tables 9**. Generally, it is obvious that there is an increase in both plant growth from year 2011 to year 2012. Furthermore, the fresh weight of Chinese cabbage in plots irrigated with TWW is higher than those irrigated with FW in year 2011 and 2012.

Statistical analysis showed a significant difference between the average bio-mass of the two treatments (TWW and FW). P-value equals to 0.015. Moreover, similar trend is observed for corn plants (**Table 9**) indicating high yields.

Statistical analysis showed significant differences between FW and TWW treatments. This suggests that TWW can supply enough nutrients the same as the NPK fortified FW treatment equivalent to the nutrient contents of treated wastewater. This suggestion is supported by the data in **Table 2** (water analysis). In addition, our results agree with resent reports [7,18-20] who analyzed the long term effects of two gradients: spatial (relative distance from the water channel) and land use intensity (cropping frequency) and addition of organic amendment on soil properties and model crop (barley) response. They demonstrated the clear and consistent patterns in soil properties and plant response along the gradients and points out the probable long-term environmental trends in a "would be" scenario for agricultural use of similar polluted soils.

Comparison between the biomass 2011 and 2012 shows a great increase in the biomass in year 2012. The explanation of these results is that application of TWW may enrich the soil with necessary nutrients that enabled plant growth. Beside the fact that TWW contains some bacteria as shown from the high BOD value (**Table 2**) that participate in the degradation or organic matter that maintain soil fertility. This explanation is supported by Mousavi *et al.* [21] who showed that irrigation with TWW had a significant positive impact on all characters of quality of maize.

3.3. Determination of Micronutrients and Heavy Metals in Plant Leaves

Levels of micronutrients and heavy metals in Chinese cabbage leaves are shown in **Tables 10** and **11** respectively. The levels of micronutrients in Chinese cabbage leaves in 2010-2011 (**Table 10**) indicate that Fe levels are high in Chinese cabbage leaves in both treatments. Its concentration did not exceed 192.35 ± 62.81 in both FW

Table 9. Average weight of Chinese cabbage and corn (Kg/plot).

Treatment	Chinese cabbage		Corn	
	2011	2012	2011	2012
FW	33.9 ± 2.8	42 ± 9.1	32.9 ± 3.93	55.17 ± 12.9
TWW	38.6 ± 3.1	47 ± 10.4	46.73 ± 6.6	52.93 ± 10.73

Fresh Weight $\pm$ SD, P value between 2011 treatments = 0.015 for Chinese cabbage; P value between 2011 treatments = 0.04.

Table 10. Micronutrients level (mg/kg) in Chinese cabbage leaves 2010-2011.

Element	FW		TWW	
	2010	2011	2010	2011
Fe	164.3 ± 33.9	192.35 ± 62.81	166.2 ± 25.78	148.27 ± 22.89
Cu	3.54 ± 0.49	3.79 ± 0.86	3.28 ± 0.63	4.38 ± 0.62
Zn	30.7 ± 5.3	38.16 ± 0.81	27.99 ± 2.65	38.33 ± 7.15
Mn	32.53 ± 6.44	42.37 ± 8.17	36.57 ± 8.02	35.15 ± 4.17
B	nd	26.5 ± 5.4	nd	30.61 ± 7.05

Table 11. Heavy metals level in Chinese cabbage leaves (Mean ± SD).

Element	FW		TWW	
	2010	2011	2010	2011
Cr	0.673 ± 0.12	0.55 ± 0.25	0.58 ± 0.1	0.54 ± 0.16
Ni	1.42 ± 0.49	0.77 ± 0.24	1.06 ± 0.34	0.81 ± 0.21
Sn	0.58 ± 0.33	4.34 ± 1.47	0.82 ± 0.33	6.87 ± 3.95
Cd	0.08 ± 0.01	0.12 ± 0.08	0.07 ± 0.01	0.06 ± 0.00
Co	nd	0.18 ± 0.10	nd	0.14 ± 0.02

and TWW treatments.

Concentrations of Cu did not exceed 4.38 ± 0.62 mg/kg in both treatments (FW and TWW) during the 2 growing seasons. Concentrations of Zn reached 38.33 ± 7.15 mg/kg indicating elevated levels. Concentrations of Mn are nearly similar in both TWW and FW but in year 2011 the levels reached to 42.37 ± 8.17 mg/kg indicating high concentrations. Concentration levels of B are high in year 2011 and not detected in year 2010 in both treatments. Concentrations of heavy metals in Chinese cabbage leaves are shown in **Table 11**. It can be seen that concentrations of Cr, Ni, Sn and Cd did not exceed 2 mg/kg in year 2010 whereas only Sn exceeded 2 mg/kg in year 2011 and reached 4.34 and 6.87 mg/kg in FW and TWW respectively. These elevated levels indicating high contamination levels. These results agree with Ferrara *et al.* [17] who revealed that levels of As, Cd, Cr, Cu, Hg, Pb, Sn, Tl and Zn exceeding the intervention legal limits when irrigated with TWW.

Levels of micronutrients and heavy metals in corn leaves are shown in **Tables 12** and **13** respectively.

Concentrations of some micronutrients in corn leaves are shown in **Table 12**.

It can be seen that concentrations of Fe ranged between 79.4 ± 11.19 to 128.41 ± 16.00 mg/kg indicating wide variations. Concentrations of Cu ranged between 6.18 ± 0.83 to 7.63 ± 1.22 mg/kg indication similarity in both treatments in the 2 growing season.

Table 12. Micronutrients level (mg/kg) in corn leaves 2010-2011 (Mean ± SD).

Element	FW		TWW	
	2010	2011	2010	2011
Fe	104.6 ± 30.86	128.41± 16.00	79.4 ± 11.19	124.8 ± 23.4
Cu	7.63 ± 1.22	6.18 ± 0.83	6.23 ± 1.34	6.21 ± 1.43
Zn	83.48 ± 32.05	63.09 ± 12.24	57.04 ± 28.31	50.8 ± 19.9
Mn	44.03 ± 7.34	48.24 ± 6.42	45.53 ± 9.38	55.6 ± 9.8
B	nd	12.75 ± 2.74	nd	16.3 ± 6.5

Statistical analysis did not detect any significant difference at $\alpha = 0.05$.

Table 13. Heavy metals level (mg/kg) in corn leaves 2010-2011 (average ± SD).

Element	FW		TWW	
	2010	2011	2010	2011
Cr	2.07 ± 0.63	1.97 ± 0.32	2.07 ± 0.41	1.14 ± 0.31
Ni	1.16 ± 0.27	1.07 ± 0.14	1.77 ± 0.68	0.91 ± 0.33
Sn	53.91 ± 3.82	26.48 ± 14.24	39.55 ± 12.25	18.6 ± 6.1

Statistical analysis detect significant difference at $\alpha = 0.05$ only for Cr at year 2011, P-value is Cr = 0.01.

Concentrations of Zn ranged between 50.8 ± 19.9 and 83.48 ± 32.05 mg/kg indicating wide variations among the treatments.

Concentrations of Mn ranged between 44.03 ± 7.34 and 55.6 ± 9.8 mg/kg. This range is not as wide as in Zn indicating similarity among the treatments. Concentrations of B are detected only in Year 2011. Statistical analysis did not detect any significant difference at $\alpha = 0.05$ level.

Concentrations of heavy metal in corn leaves are presented in **Table 13**. It can be seen that concentration of Cr ranged between 1.14 ± 0.31 and 2.07 ± 0.63 mg/kg in both treatment indicating low concentrations and variations. Concentrations of Ni ranged between 0.91 ± 0.33 and 1.77 ± 0.68 mg/kg in both treatments (FW and TWW). Concentration of Sn ranged between 18.6 ± 6.1 and 53.91 ± 3.82 mg/kg indicating wide variations and high concentrations.

Concentrations of Hg, Pb, As, and Se are under ICP detection limit. These high levels of heavy metals may be attributed to the irrigation with TWW. Our suggestion agrees with Pritchard *et al.* [8] who concluded that attention must be given to heavy metal uptake by plant due to irrigation with TWW.

4. Conclusions

The rational of this study emerges from the fact that the country suffers from arid and semi arid conditions. Accordingly, the use of treated waste water is an option to save water recourses for domestic uses. Our results demonstrated that FW and TWW have physicochemical properties that allow for a safe use. Irrigation with TWW demonstrates the effectiveness to increase the biomass of Chinese cabbage and corn. Analytical results of soil profile indicate leaching of Fe, Al, and Pb from the top soil and accumulation in deeper depths. This situation may pose health risk to groundwater.

Micronutrient and heavy metal contents in the plant leaves are not extremely high and can be within the range of local standards.

Although it is still too early to recommend the use of TWW as an alternative option for irrigation, the presented results are promising and encouraging. Further research work is needed before recommending TWW as an alternative source of fresh water irrigation for vegetables. The future research may include the impact of long term application of TWW on human health and environment in terms of heavy metals and pathogens.

5. Acknowledgements

This research was funded by DFG Grant no. GZ:MA 1830/9-2.

Special thanks to Prof Dr Bernd Marschner, Bochum University, Germny, for his suggestions during the research activity.

REFERENCES

[1] Palestinian Central Bureau of Statistics, "Palestinian National Authority, Yearly Report," 2012.

[2] R. S. Evett Jr., S. R. Zalesny, F. N. Kandil, A. J. Stanturf and C. Soriano, "Opportunities for Woody Crop Production Using Treated Wastewater in Egypt. II. Irrigation Strategies," *International Journal of Phytoremediation*, Vol. 13, No. 11, 2011, pp. 122-139.

[3] G. L. Mendoza-Espinosa, A. Cabello-Pasini, V. Macias-Carranza, W. Daessle-Heuser, V. M. Orozco-Borbón and L. A. Quintanilla-Montoya, "The Effect of Reclaimed Wastewater on the Quality and Growth of Grapevines," *Water Science and Technology*, Vol. 57, No. 9, 2008, pp. 1445-1450. doi:10.2166/wst.2008.242

[4] G. Oron, L. Gillerman, A. Bick, Y. Manor, N. Buriakovsky and J. Hagin, "Membrane Technology for Sustainable Treated Wastewater Reuse: Agricultural, Environmental and Hydrological Considerations," *Water Science and Technology*, Vol. 57, No. 9, 2008, pp. 1383-1388. doi:10.2166/wst.2008.243

[5] P. K. Mosse, F. A. Patti, J. R. Smernik, W. E. Christen and R. T. Cavagnaro, "Physicochemical and Microbiological Effects of Long- and Short-Term Winery Wastewater Application to Soils," *Journal of Hazardous Material*, Vol. 201, No. 202, 2012, pp. 219-228.

doi:10.1016/j.jhazmat.2011.11.071

[6]   P. R. Singh, P. Singh, H. M. Ibrahim and R. Hashim,
      "Land Application of Sewage Sludge: Physicochemical
      and Microbial Response," *Review of Environmental Con-
      tamination and Toxicology*, Vol. 214, 2011, pp. 41-61.
      doi:10.1007/978-1-4614-0668-6_3

[7]   E. Castro, P. M. Mañas and J. De Las Heras, "Effects of
      Wastewater Irrigation on Soil Properties and Turfgrass
      Growth," *Water Science and Technology*, Vol. 63, No. 8,
      2011, pp. 1678-1688. doi:10.2166/wst.2011.335

[8]   L. D. Pritchard, N. Penney, J. M. McLaughlin, H. Rigby
      and K. Schwarz, "Land Application of Sewage Sludge
      (Biosolids) in Australia: Risks to the Environment and
      Food Crops," *Water Science and Technology*, Vol. 62, No.
      1, 2010, pp. 48-57. doi:10.2166/wst.2010.274

[9]   N. O. Belyaeva, J. R. Haynes and C. E. Sturm, "Chemical,
      Physical and Microbial Properties and Microbial Diver-
      sity in Manufactured Soils Produced from Co-Compost-
      ing Green Waste and Biosolids," *Waste Management*,
      2012, in Press. doi:10.1016/j.wasman.2012.05.034

[10] Palestinian Central Bureau of Statistics, "Palestinian Na-
      tional Authority, Yearly Report," 2007.

[11] FAW, CROP WAT, VERSION 8, Food Agriculture Or-
      ganization, Rome, 2007.

[12] A. D. Eaton, L. S. Clesceri and A. E. Greenberg, "Stan-
      dard Methods for Wastewater Analysis," American Pub-
      lic Health Association, Washington DC, 20005.

[13] B. J. Jones and V. W. Case, "Sampling, Handling and
      Analyzing Plant Tissue Samples," In: R. L. Westerman,
      Ed., *Soil Testing and Plant Analysis*, 3rd Edition, Soil
      Science Society of America, Madison, 1990, pp. 389-427.

[14] D. Boruah and S. Hazarika, "Normal Water Irrigation as
      an Alternative to Effluent Irrigation in Improving Rice
      Grain Yield and Properties of a Paper Mill Effluent Af-
      fected Soil," *Journal of Environmental Science and En-

gineering*, Vol. 52, No. 3, 2010, pp. 221-228.

[15] M. Adrover, E. Farrús, G. Moyà and J. Vadell, "Chemical
      Properties and Biological Activity in Soils of Mallorca
      Following Twenty Years of Treated Wastewater Irriga-
      tion," *Journal of Environmental Management*, Vol. 95,
      2012, pp. 188-192. doi:10.1016/j.jenvman.2010.08.017

[16] S. G. Metson, L. R. Hale, M. D. Iwaniec, M. E. Cook, R.
      J. Corman, S. C. Galletti and L. D. Childers, "Phosphorus
      in Phoenix: A Budget and Spatial Representation of Phos-
      phorus in an Urban Ecosystem," *Ecological Applications*,
      Vol. 22, No. 2, 2012, pp. 705-721. doi:10.1890/11-0865.1

[17] L. Ferrara, M. Iannace, M. A. Patelli and M. Arienzo,
      "Geochemical Survey of an Illegal Waste Disposal Site
      under a Waste Emergency Scenario (Northwest Naples,
      Italy)," *Environmental Monitoring and Assessment*, 2012,
      in Press.

[18] A. S. Hashemi, "The Investigation of Irrigation with
      Wastewater on Trees (*Populus deltoids* L.)," *Toxicology
      and Industrial Health*, 2012, in Press.
      doi:10.1177/0748233712442738

[19] N. Nikolic and M. Nikolic, "Gradient Analysis Reveals a
      Copper Paradox on Floodplain Soils under Long-Term
      Pollution by Mining Waste," *Science of the Total En-
      vironment*, Vol. 15, No. 425, 2012, pp. 146-154.
      doi:10.1016/j.scitotenv.2012.02.076

[20] E. B. Jones, J. R. Haynes and R. I. Phillips, "Addition of
      an Organic Amendment and/or Residue Mud to Bauxite
      Residue Sand in Order to Improve Its Properties as a
      Growth Medium," *Journal of Environmental Manage-
      ment*, Vol. 95, No. 1, 2012, pp. 29-38.

[21] R. S. Mousavi, M. Galavi and H. Eskandari, "Effects of
      Treated Municipal Wastewater on Fluctuation Trend of
      Leaf Area Index and Quality of Maize (*Zea mays*)," *Wa-
      ter Science and Technology*, Vol. 67, No. 4, 2013, pp.
      797-802. doi:10.2166/wst.2012.624

20

Hydrogeochemical Characteristics and Assessment of Water Resources in Beni Suef Governorate, Egypt

Ahmed Melegy[1], Ahmed El-Kammar[2], Mohamed Mokhtar Yehia[3], Ghadir Miro[2]
[1]Department of Geological Sciences, National Research Centre, Dokki, Egypt
[2]Department of Geology, Faculty of Science, Cairo University, Giza, Egypt
[3]Central Laboratory for Environmental Quality Monitoring, National Water
Research Center, Kanater El-Khairia, Egypt.
Email: amelegy@yahoo.com, amkammar@hotmail.com, saliha.achi@yahoo.fr, ghadeer.miro@gmail.com

ABSTRACT

The main target of this research is to investigate the hydrogeochemical characteristics of surface and groundwater resources in Beni Suef Governorate, Egypt. A combination of major and heavy metals has been used to characterize surface and groundwater in Beni Suef Governorate, Egypt. Twenty water samples were collected from the water resources: River Nile, El-Ibrahimia canal, Bahr Youssef, irrigation and drainage channels and Quaternary aquifer. The collected water samples were analyzed for major and heavy metals. Surface and groundwater in the study area are considered potable for drink and irrigation based on the TDS and major ions concentrations. Regarding the MPL quoted by the WHO [1] and FAO [2], water resources in the study area are polluted by Al, Cd, Fe, Mn, Co and Cu. The highest concentrations of Al, Cd, Fe, Mn, Co and Cu recorded in the study area were 2545, 400, 1415, 2158, 239 and 1080 µg/l, respectively. Both surface and groundwater in Beni Suef suffer from pollution due to the impact of unsupervised anthropogenic activities including: fertilizers and pesticide, waste disposal and industrial waste, seepage from septic tanks, construction of water pipes, wastewater from El-Moheet drainage and evaporation processes during flood irrigation.

KEYWORDS

Beni Suef; Surface Water; Groundwater; Hydrogeochemical Characteristics; Heavy Metals

1. Introduction

Pollution of the environment is one of the major concerns throughout the universe, which could originate by several ways such as continuous discharging of the large variety of toxic inorganic and organic chemicals into the environment. This causes severe water, air and soil pollutions [3]. Contamination of water supplies by heavy metals has steadily been increased over the last decades, as a result of over population and expansion of industrial activities [4,5]. Sources of heavy metals in water comprise natural sources including eroded minerals within sediments, leaching of ore deposits and volcanic materials and anthropogenic sources such as solid waste disposals, industrial or municipal effluents and wharf channel dredging. Heavy metals in nature are not usually hazardous to the environment and human health as the amounts of them are not significant; furthermore some heavy metals are required at low concentrations as catalysts for enzyme activities in human body (Co, Cu, Fe, Mn and others). However, if the levels of these metals are elevated to higher than the normal ranges, they can cause malfunction and toxicity to human body [6].

Beni Suef Governorate within the Nile Valley occupies an area of approximately 10950 km[2] bounded by the Eastern and Western Deserts. About 85% of the total populated area is agriculture land. Surface water, which is mostly represented by River Nile, El-Ibrahimia Canal and Bahr Youssef, is considered the major source in securing water requirements (drinking, irrigation and industry). 66% of drinking water is provided from surface water and 34% from groundwater.

Providing suitable water to Beni Suef Governorate is considered as the basic environmental issue that worries the population. That is associated with the different types of pollutants that affect the surface and ground water quality. Water resources (mainly surface water) at Beni Suef are currently threatened by contamination from municipal, industrial and agricultural pesticides. Most of the sewage water is discharged in irrigation canals and drains. The domestic sewage water in governorate is estimated about 163,000 m^3 daily. Agricultural pesticides and wastewater enforce the major impact on degradation of groundwater quality in the study area [7].

The present work aims to assess the quality of water resources in Beni Suef in terms of major ions and heavy metals chemical analysis data with respect to human uses and irrigation.

2. Study Area

Beni Suef Governorate is located between latitudes 28°45' and 29°25'N and longitudes 30°45' and 31°15'E, occupying a part of the lower Nile Valley. It includes seven towns namely; Beni Suef (the capital), El-Wasta, Naser, Beba, Ihnasia, Somosta and El-Fashn (**Figure 1**). The surface on Beni Suef Governorate is built of sedimentary rocks and sediments; the exposed sequence is formed of Upper and Middle Eocene, Pliocene, Pleisto-

cene and Holocene rock units. The Eocene rocks are mainly composed of limestone rocks which are karstified in some places. The exposed Pliocene formations are made up of alternating conglomerate consisting of limestone, chert and quartz sand grains bounded together by a clay matrix and sandstone beds. The Pleistocene and Holocene deposits in the Nile Valley composed of gravel, sand, pebbly sandstone, sandy gravel beds and loose to fairly well indurated deposits of sand, silt and mud [8,9].

Beni Suef Governorate occurs within the arid to semiarid desert belt of Egypt. Its climate is characterized by a hot dry summer and mild very short winter with scarce rainfall. Dry winds with dust, traditionally known as Khamassen wind storms, are blown on Egypt from southwest.

As well known, the *River Nile* is the main perennial stream, which flows all over the year from south to north. It has the widest course in Beni Suef Governorate (100 to 500 m). Beni Suef Governorate uses 2 billion m^3 of the River Nile water, 92% for irrigation, 3% for industry and 5% for drinking. As well as, two irrigation canals, namely; El-Ibrahimia canal and Bahr Youssef, run parallel to River Nile, provide the water for irrigation. The two canals were constructed along the western flood plain of the Nile Valley [7,10].

The yield of groundwater from the Eocene waterbearing rocks along the western slopes of Beni Suef area is very limited and the obtained water is saline with TDS ranging from 4500 to 6600 ppm [10].

Many wells were drilled and penetrated a considerable thickness of the Quaternary section. They provide the main supply of the groundwater for different uses in the area. The flow regime and the fluctuation regime indicate that the main sources of recharge is the seepage from the Nile water in the irrigation canals and drainage water in the drains and the direct infiltration of return flow after irrigation. While the River Nile acts as an effluent stream especially after the construction of the High Dam in 1966. The uncontrolled pumping of groundwater for drinking and irrigation purposes acts as main factor for decline and discharge beside the subsurface seepage into the drains and the River Nile.

3. Methodology

To get an idea on the water quality of Beni Suef and its suitability for drinking and irrigation, twenty samples were collected from the study area during one period on March 2013 (**Figure 1**). Eighteen samples were collected from different sources of surface water represented by; River Nile, El-Ibrahimia canal, Bahr Youssef as well as the irrigation and drainage channels (**Table 1**). That considered the main source in obtaining water supplies in the study area. Two groundwater samples were collected from wells on depths among 20 - 30 m. The boreholes

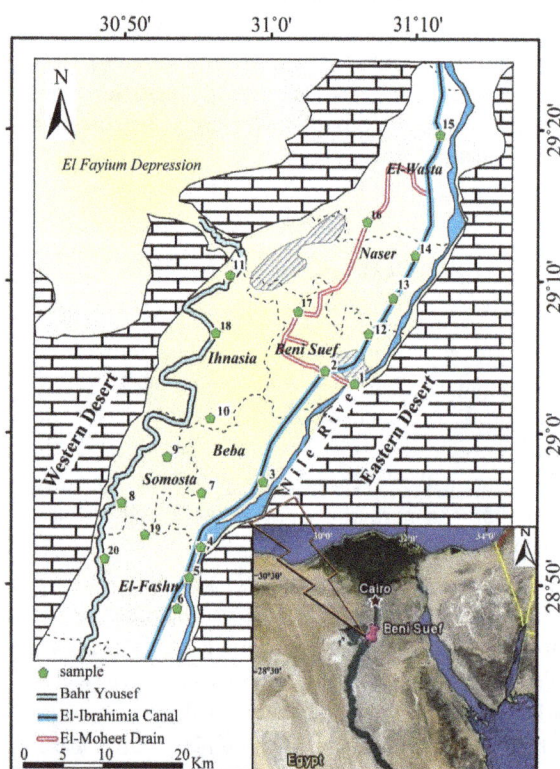

Figure 1. Location map of Beni Suef Governorate.

Table 1. Location of surface and ground water samples.

Sample No.	Location	Latitude (N)	Longitude (E)	Elevation, m (a.s.l.)
1	River Nile (Beni Suef)	29 03 32 97	31 05 48 00	46
2	El-Ibrahimia canal (Beni Suef)	29 04 1515	31 03 59 16	43
3	El-Ibrahimia canal (Fabriqet Beba)	28 56 35 92	30 59 23 14	47
*G 4	Harab Shant (Beba)	28 52 35 27	30 55 14 55	52
*G 5	El- Shrahna (El-Fashn)	28 51 04 06	30 54 50 51	57
6	El-Ibrahimia canal (El-Fashn)	28 48 42 62	30 53 43 15	48
7	Irrigation drainage (Between Helyah and Badhal)	28 55 52 39	30 55 15 80	50
8	Bahr Youssef (Somosta)	28 55 02 74	30 49 36 86	48
9	Irrigation channel (Bani Hallah)	28 57 55 10	30 53 08 89	56
10	Drainage channel (Ihnasia)	29 00 55 65	30 56 01 54	49
11	Bahr Youssef (Ihnasia)	29 10 25 37	30 57 11 69	47
12	El-Ibrahimia canal (El-Jnabia)	29 06 26 56	31 06 58 92	47
13	El-Ibrahimia canal (Naser)	29 08 34 96	31 08 11 77	41
14	El-Ibrahimia canal (Bani Adi)	29 11 41 47	31 09 59 45	36
15	El-Ibrahimia canal (El-Wasta)	29 19 46 44	31 54 15 00	40
16	El-Moheet drainage (El-Wasta)	29 12 41 50	31 05 47 10	48
17	El-Moheet drainage (Beni Suef)	29 08 30 96	31 03 59 10	48
18	Bahr Youssef (Ihnasia)	29 05 1510	30 55 15 70	47
19	Irrigation drainage (El-Fashn)	28 53 52 39	30 53 07 89	50
20	Bahr Youssef (El-Fashn)	28 51 52 39	30 49 35 86	48

*G: groundwater sample.

were pumped for about 5 minutes to purge the aquifer of stagnant water to acquire fresh aquifer samples for analysis. In addition, two soil samples were combined from positions 5 (El-Fashn) and 8 (Somosta) (**Figure 1**). A well-constrained Global Positioning System (GPS) was used for navigation to record the sampling sites accurately.

Water samples were taken in duplicate for major ions and heavy metals analyses. They were collected in 1L plastic bottles rinsed three times and filled to the brim, before sealing tightly to include as little air as possible in the top of the bottle. Few drops of nitric acid were added only to the samples analyzed for heavy metals. Physical examination of the water samples including temperature, pH, and electrical conductivity was done in the field. The samples were preserved in a refrigerator until they were transported to the laboratory for chemical analysis. Chemical analysis was carried out in the Central Laboratory for Environmental Quality Monitoring, Kanater El-Khairia, Egypt. The Inductively Coupled Plasma-Mass Spectrometry (ICP-MS) using to analyzed heavy metals (Al, As, Ba, Co, Cd, Cr, Cu, F, Fe, Mn, Ni, Pb, Sb, Se, Sn, V and Zn) and major cations (Ca^{+2}, Mg^{+2}, Na^+ and K^+). Major anions concentrations (HCO_3^- SO_4^{2-}, Cl^- and NO_3^-) were determined using Ion Chromatography (IC). Gravimeteric method is being used to determine

TDS. A well-mixed sample is filtered through a standard glass fiber filter. The filtrate is evaporated and dried to constant weight at 103°C - 1050°C. The GIS mapping technique was adopted to highlight the spatial distribution pattern of major ions and heavy metals concentrations in the surface water by using ArcGIS (version 9.2). Descriptive statistics, spearman correlation and factor analysis among all the parameters were calculated using software SPSS 16.

4. Results and Discussion

4.1. Surface Water

The surface water in the study area is generally alkaline with pH ranging between 7 and 7.9 (**Table 2**). The optimum range of pH, as proposed by WHO [1], is 6.5 - 8.5. The lowest value of pH (7) is recorded at River Nile mainly deduced as a result of organic matter oxidation; whereas pH values of 7.7, or higher, dominate in the other locations.

The EC values range from 319 to 1473 μS/cm (**Table 2**) and it is mutually distributed with TDS where the highest values dominate in the north and the intermediate values prevail in the west. The palatability of water with a TDS level of less than about 600 mg/l is generally considered to be good; drinking-water becomes significantly

Table 2. Descriptive statistics of the obtained data for surface water (n = 18). Note that major elements are given in mg/l while heavy metals in µg/l.

Element	Min	Max	Mean	Standard Deviation
pH	7	7.93	7.79	0.2
EC	319	1473	570	356
TDS	204	943	365	228
K	2.5	16	7.3	4.4
Na	18	204	47	58
Ca	34	79	48	16
Mg	9	24	14	4
Cl	21	274	61	78
SO$_4$	20	118	44	31
NO$_3$	0.3	16	7	5
HCO$_3$	128	297	174	50
Al	1608	2545	2116	325
Cd	<1	400	98	121
Co	<5	239	55	59
Cu	<1	1080	404	368
Fe	13	1415	624	332
Mn	37	713	137	154
Ba	20	225	134	67.6
V	<5	187	57	63.4
Zn	<1	1700	331	426.5

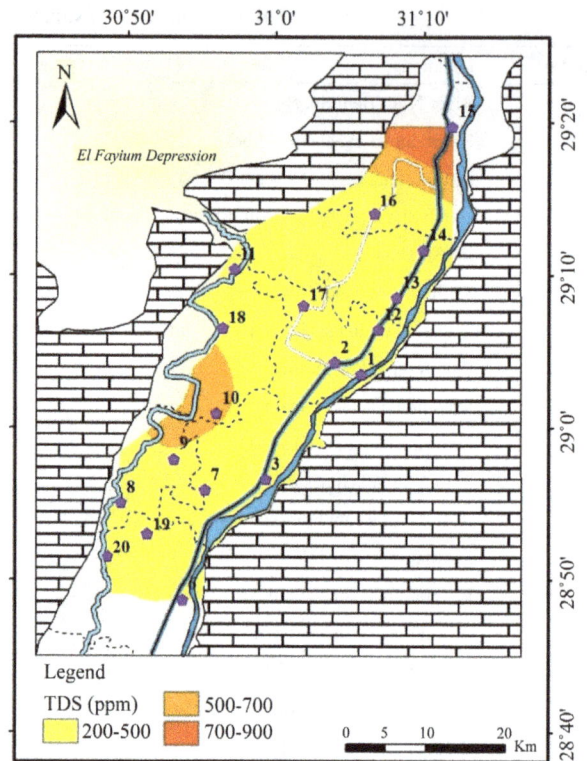

Figure 2. Geochemical map of TDS in the surface water in the study area.

and increasingly unpalatable at TDS levels greater than about 1000 mg/l. The concentration of TDS ranges between 204 and 943 mg/l, averaging 365 mg/l (**Table 2**). High contents exceeding 600 mg/l are only recorded in the northern part of the study area in El-Ibrahimia canal at El-Wasta center (**Figure 2**). This concentration agrees with the highest concentrations of major ions (**Figures 3-10**). The TDS values are strongly correlated with the major ions; Ca^{2+}, K$^+$, Mg^{2+}, Na$^+$, Cl$^-$, SO$_4^{2-}$ and HCO$_3^-$, where r ranges from 0.76 with Mg^{2+} to 0.98 with SO$_4^{2-}$ (**Table 3**) due to the role of lithologic impact and evaporation.

The main source of such exceeding's can be interpreted to the wastewater coming from El-Moheet drainage. It is the main drainage in the governorate and associated with high number of sub-drainages. It receives high amount of untreated sewage water from treatment plant in Beni Suef city, as well as the wastewater from agricultural and industrial activities. In spite of this relative increase in the TDS content, the surface water in study area is generally considered to be good for drink (<1000 mg/l) and irrigation (<2000 mg/l).

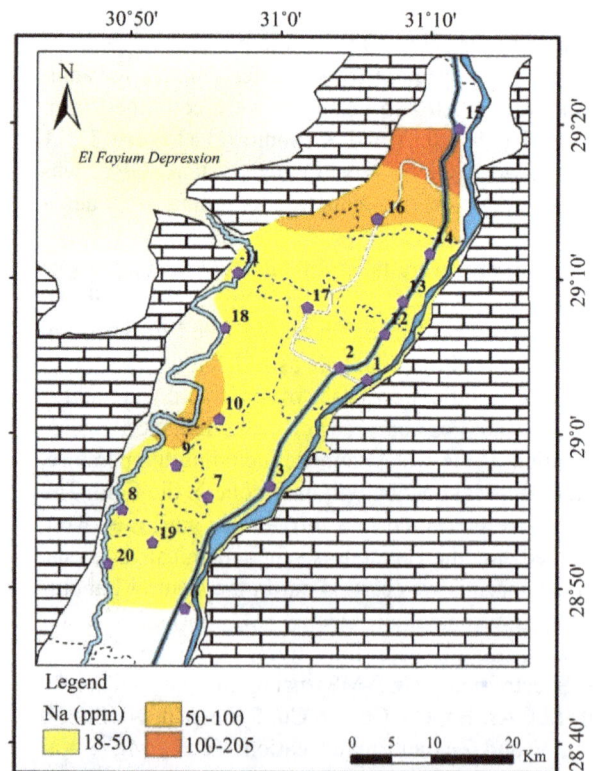

Figure 3. Geochemical map of Na in the surface water in the study area.

Table 3. Correlation coefficients among major parameters of surface water.

	pH	EC	TDS	Ca	K	Mg	Na	Cl	NO$_3$	SO$_4$	HCO$_3$
pH	1										
EC	0.10	1									
TDS	0.20	1	1								
Ca	0.27	0.87	0.88	1							
K	0.22	0.90	0.90	0.84	1						
Mg	0.21	0.76	0.76	0.96	0.80	1					
Na	0.14	0.97	0.97	0.73	0.82	0.58	1				
Cl	0.14	0.96	0.97	0.73	0.82	0.57	1.00	1			
NO$_3$	0.24	0.76	0.76	0.64	0.64	0.59	0.75	0.75	1		
SO$_4$	0.19	0.98	0.98	0.90	0.93	0.82	0.93	0.92	0.77	1	
HCO$_3$	0.24	0.80	0.80	0.98	0.80	0.97	0.62	0.62	0.52	0.83	1

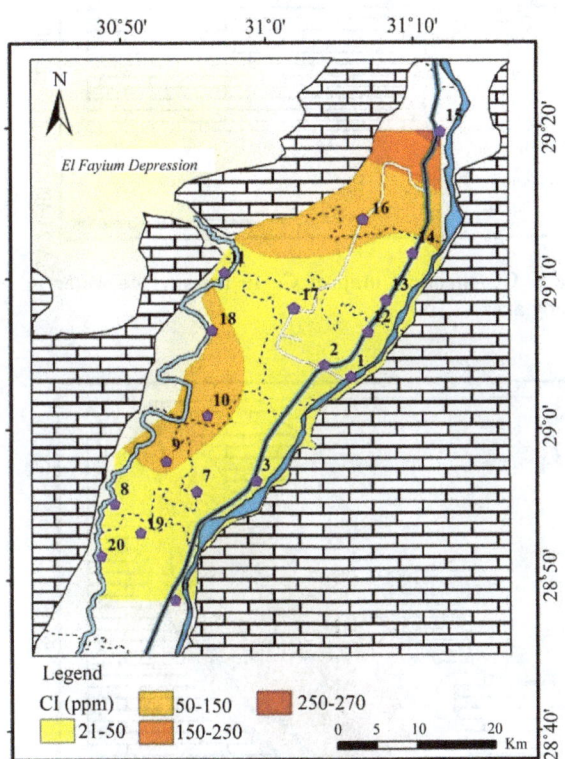

Figure 4. Geochemical map of Cl in the surface water in the study area.

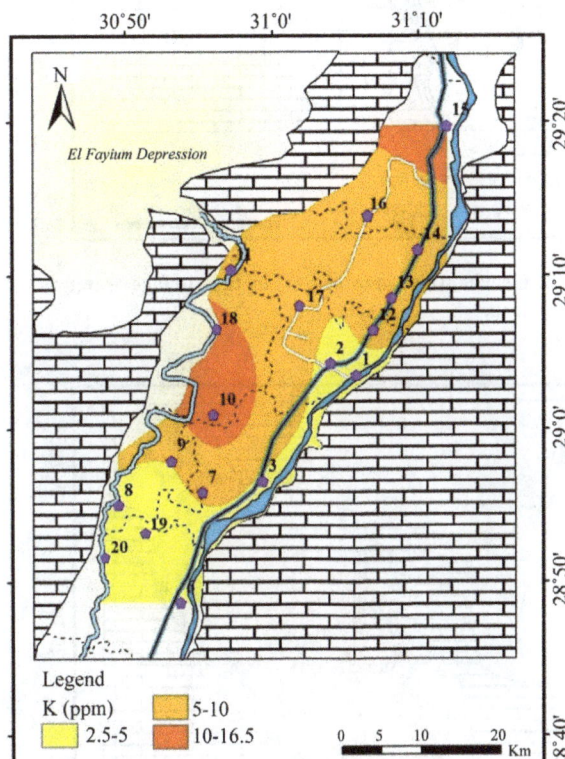

Figure 5. Geochemical map of K in the surface water in the study area.

4.2. Major Ions

The quantity of major ions (cations; Ca^{2+}, Mg^{2+}, Na^+ and K^+, and the anions; Cl^-, SO_4^{2-}, HCO_3^- and NO_3^-) in water depends primarily on the type of rocks or soil with which the water has been in contact and the length of time of contact. Industrial effluents, irrigation drainage, fertilizers, pesticide, septic tank, and other sources that result from the anthropogenic activities are considered the additional sources of elements [11-13]. Relative to the distribution of the highest concentration of major ions

in the study area, they could be classification into groups, namely:

1) The first group includes the ions; Na^+, K^+, Cl^-, SO_4^{2-} and NO_3^-, where the highest concentrations are recorded in the north at El-Wasta center (El-Ibrahimia canal). Na^+ concentration ranges from 18 to 204 mg/l averaging 47 mg/l (**Table 2** and **Figure 3**). The concentration of Cl^- ranges from 21 to 274 mg/l with an average of 61 mg/l (**Table 2**, **Figure 4**). The perfect correlation between Na^+ and Cl^- (r = 1) as well as between TDS and

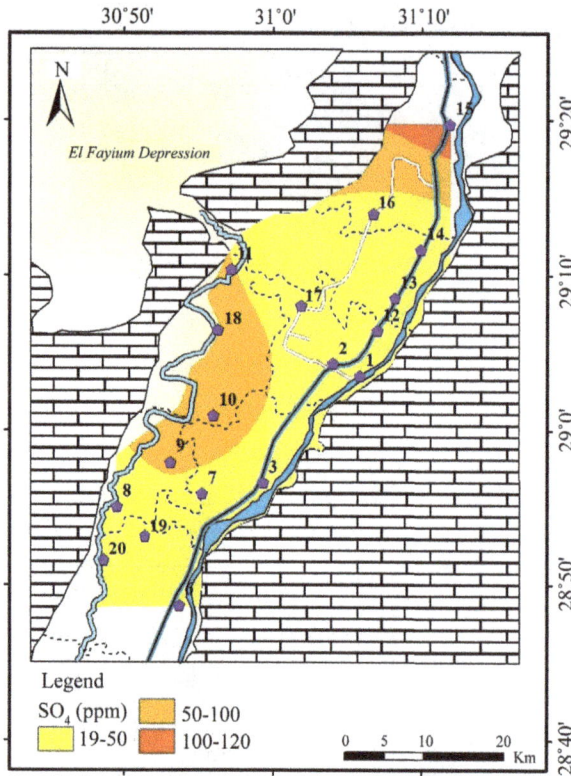

Figure 6. Geochemical map of SO₄ in the surface water in the study area.

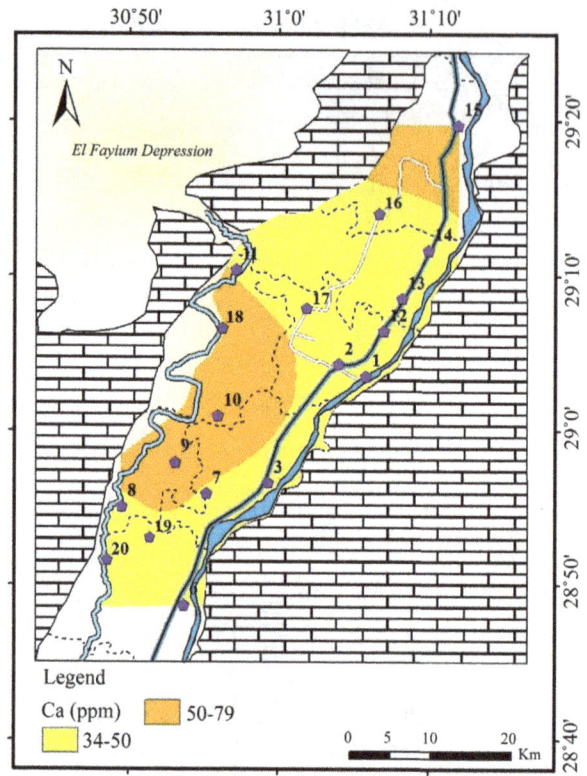

Figure 8. Geochemical map of Ca in the surface water in the study area.

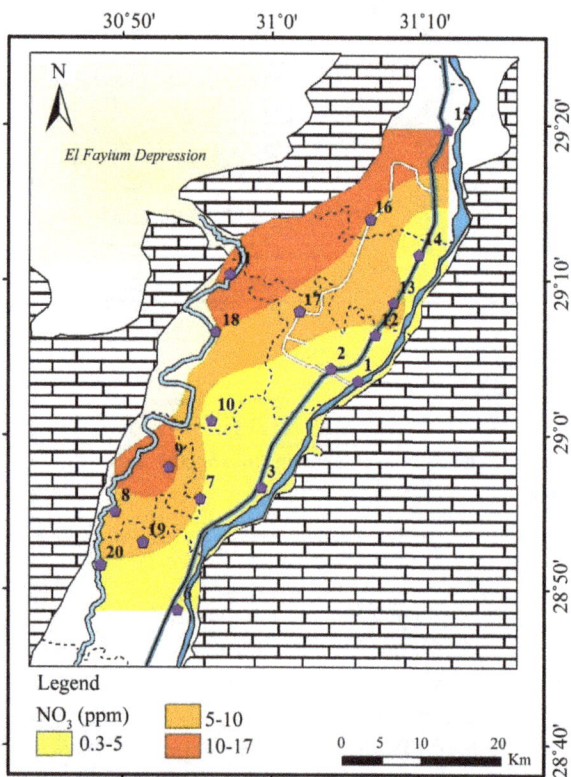

Figure 7. Geochemical map of NO₃ in the surface water in the study area.

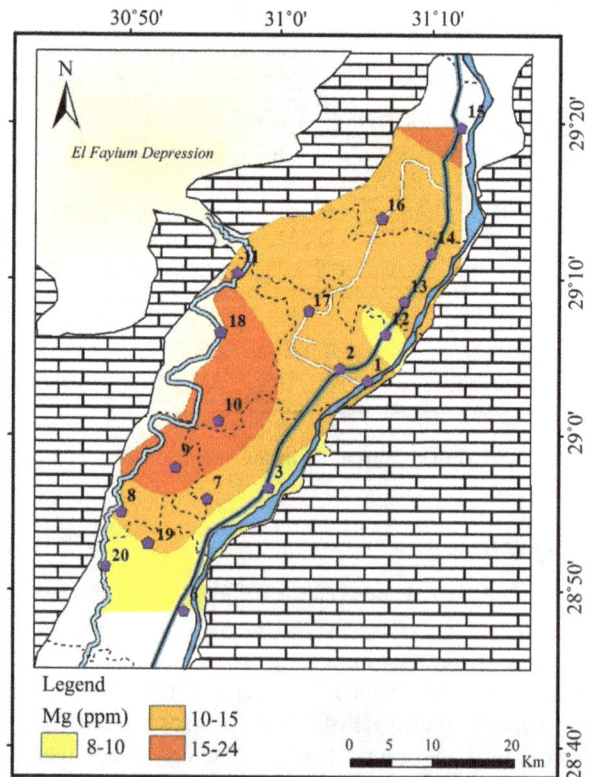

Figure 9. Geochemical map of Mg in the surface water in the study area.

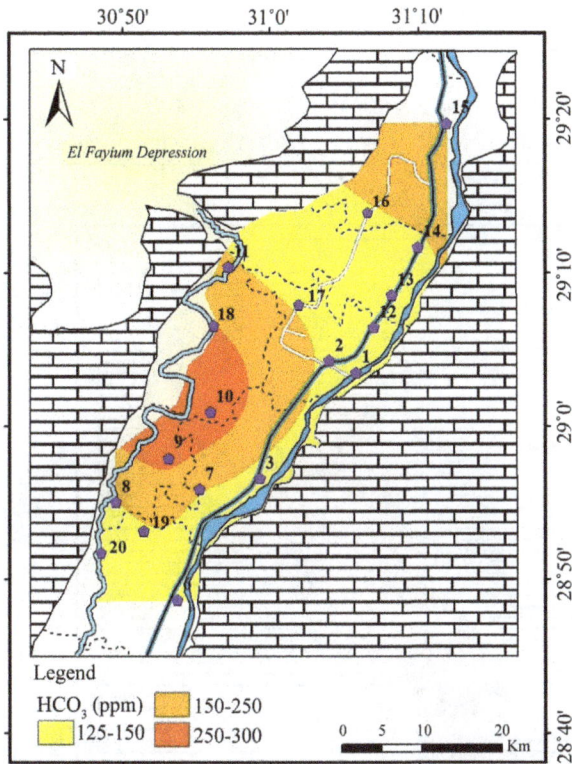

Figure 10. Geochemical map of HCO_3^- in the surface water in the study area.

both Na^+ and Cl^- (r = 0.97) point to the role of NaCl as main contributor to salinity of surface water (**Table 3**).

Leaching from clay minerals and sediments is one of the sources of Na^+ and Cl^-. **Table 4** shows the concentration of major ions in soil samples that collected from positions 5 and 8 in study area (**Figure 1**). Agricultural output (irrigation drains, fertilizer and pesticide) and sewage effluent are other possible sources.

The concentration of K^+ ranges between 2.5 and 16 mg/l with a mean value of 7.3 mg/l. SO_4^{2-} varies from 20 to 118 mg/l with an average of 44 mg/l (**Table 2** and **Figures 5** and **6**). The concentration of NO_3^- ranges from 0.3 to 16 mg/l with an average of 7 mg/l (**Table 2**, **Figure 7**).

Potassium is primarily from leaching of silicate minerals and in small amounts from evaporation minerals, fertilizers and rain water. Sulfate minerals, fertilizers, insecticide, irrigation water and sewage water are probable additional sources of sulfate in water. Sources of excess nitrate in water include fertilizers, septic systems, wastewater treatment ponds, animal wastes, industrial wastes and food processing wastes.

Strong correlations are perceived between SO_4^{2-} and K^+, Na^+ (r = 0.93), Cl^- (r = 0.92) and NO_3^- (r = 0.68), (**Table 3**). The highest concentrations of these ions in El-Ibrahimia canals in the north can mainly be attributed

to discharging of untreated sewage water from El-Moheet drainage and sewage water from treatment plant in Beni Suef city. In addition to the impact of fertilizers, pesticide and insecticide used in agriculture.

2) The second group involves Ca^{2+}, Mg^{2+} and HCO_3^-. The highest concentrations of ions are noticed in the irrigation drainages in the west.

The concentration of Ca^{2+} ranges from 34 to 79 mg/l with an average of 48 mg/l while Mg^{2+} from 9 to 24 mg/l averaging 14 mg/l (**Table 2** and **Figures 8** and **9**). The concentration of HCO_3^- ranges between 128 and 297 mg/l with a mean value of 174 mg/l (**Table 2** and **Figure 10**). The strong correlations between each of Ca^{2+} and Mg^{2+} with HCO_3^- (r = 0.98 and r = 0.97, respectively) indicate possible derivation from dissolution of carbonate minerals (**Table 3**). The significant positive correlations between Ca^{2+} and Mg^{2+} (r = 0.97) and among Na^+, K^+, Cl^- and SO_4^{2-} (r ranges from 0.57 to 0.90) propose leaching of Ca^{2+} and Mg^{2+} from clays of the Pleistocene sediments is essential source of these ions (**Table 3**).

The highest concentration of Ca^{2+}, Mg^{2+} and HCO_3^- in irrigation canals in the west is attributed to the effect of irrigation effluent, fertilizers and pesticide and sewage water that used in irrigation. In addition, the cement industry in Beba, can be considered as source of Ca^{2+} and HCO_3^-. No health-based guideline values (GV) are proposed for ions in drinking and irrigation water. However, guideline values have been established for some ions that may cause taste or odor in drinking water such as; Na^+, Cl^-, SO_4^{2-} and NO_3^- [1]. In general, the concentrations of ions in the study area remain below the GV, except Na^+ and Cl^- ions which exceed the limit (200 and 250 mg/l, respectively) only in El-Wasta due north of the study area. The increase in sodium and chloride is accompanied with mutual increase in the TDS content that is generally backed to the effect of wastewater coming from El-Moheet drainage. High concentrations of Na^+ and Cl^- give a salty taste to water and beverages.

4.3. Heavy Metals

In the surface water samples, the concentrations of many heavy metals; e.g., Cd, Fe, Mn, Co and Cu, exceed the maximum permissible limits for drinking and irrigation water (**Table 2**).

Cadmium (Cd): It may enter aquatic systems through weathering and erosion of soils and bedrock, atmospheric deposition direct discharge from industrial operations, leakage from landfills and contaminated sites, and the dispersive use of sludge and fertilizers in agriculture [14]. The concentration of Cd ranges from below detection limit (<1) to 400 μg/l with standard deviation (σ) equals 121 (**Table 2**). High concentrations of Cd have been recorded in most of the study area, whereas the low concentration

Table 4. Chemical analysis of soil samples (Major ions and heavy metals mg/kg).

Sample position	Na	K	Ca	Mg	Cl	HCO₃	Al	Cd	Cu	Fe	Pb	Mn
5	40	32	489	96	4033	9302	31650	174	2902	110200	619	3543
8	49	88	1051	274	5041	7442	19470	166	609	56420	1510	2047

of Cd occupies rather limited extent (**Figure 11**). The MPL recommended by WHO [1] for Cd for potable water is 3 µg/l, while FAO [2] quoted 10 µg/l for irrigation water.

Except that positions where the value of Cd below detection limit (<1 µg/l), the concentration of Cd in the study area exceeds the MPL quoted by the WHO and FAO in all samples and up to 135 fold. The kidney is the main target organ for cadmium toxicity. It accumulates primarily in the kidneys and has a long biological half-life in humans of 10 - 35 years [1]. The strong correlations between Cd and both Na^+ and Cl^- (r = 0.91) and its moderate correlations with TDS and SO_4^{2-} (r = 0.84 and 0.81, respectively, **Table 5**) confirms that leaching from soil and anthropogenic activities are the main sources of Cd in the study area. High concentration of Cd has also been recorded in the analyzed soil samples (174 and 166 mg/l, **Table 4**). The industrial wastes and sewage water, which are discharged in the surface water without being treated, are possible candidates for Cd pollution in the study area. There are limited locations in the study area that do not have sewage system. They discharge their wastes through septic tank. Moreover, the use of pesticides and phosphate fertilizer in agriculture can be considered as a potential source of Cd pollution [6,15-17].

Iron (Fe) and Manganese (Mn): Both Fe and Mn are commonly found in water and are essential elements required in small amounts by all living organisms. The presence of iron or manganese in drinking water can affect the taste, smell, or color of the water [18]. In the study area, the concentration of Fe varies between 13 and 1415 µg/l with σ equals 332 (**Table 2**). The concentration of Fe in the southwest increases toward the west in the irrigation canals, whereas in the northeast it increases toward the east at El-Ibrahimia canal (**Figure 12**). The WHO and FAO have estimated the MPL for Fe to be 300 µg/l for drinking purposes and 5000 µg/l for irrigation uses. The concentration of Fe surpasses the MPL for drink water in the most samples and still within the limit for irrigation water.

The concentration of Mn ranges from 37 to 713 µg/l with σ equals 154 (**Table 2**). **Figure 13** shows the distribution of Mn in Beni Suef region. It is similar to the distribution of Fe, where it increases to the east in the northeast and toward west in the southwest. The highest value of Mn in the study area is recorded in the River Nile that exceeds the MPL for drinking water as given by

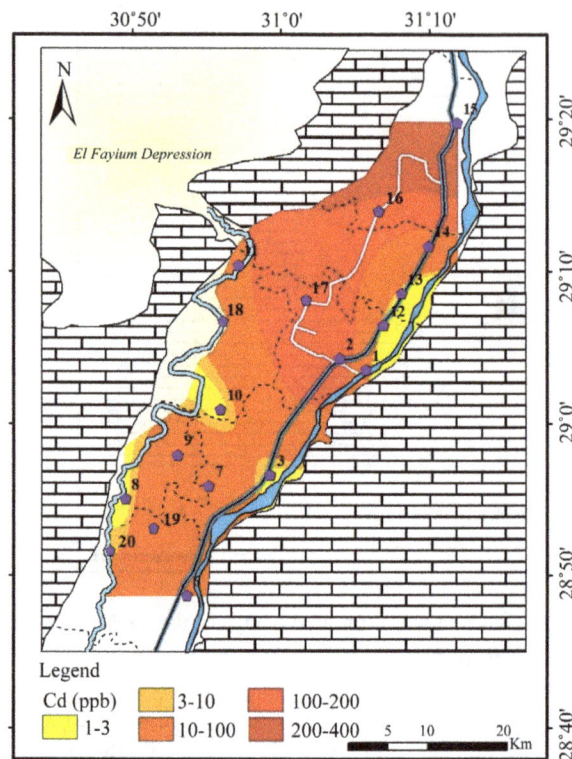

Figure 11. Geochemical map of Cd in the surface water in the study area.

WHO (400 µg/l), whereas the other values remain below the limit. The highest concentration of Mn in the River Nile can be attributed to low value of pH, where metals tend to be more soluble. This is supported by the strong negative correlation between Mn and pH (r = −0.86, **Table 5**). The value of Mn exceeds the MPL given for irrigation water by FAO [2] (200 µg/l) in irrigation canals.

The high concentrations of Fe and Mn are possibly backed to many sources including; interaction with sediments and soil, using of fertilizers, irrigation by untreated water, industrial and domestic waste and the metallic pipes that are used to transport water [1,17,19].

Copper (Cu): Copper can be released into the environment by both natural sources and human activities where it is widely used in numerous industrial and agricultural applications (mining, metal production, wood production and phosphate fertilizer production). It is often found near mines, industrial settings, landfills and waste disposals. Copper works its way into the water by dissolving from copper pipes in the household plumbing. Usually water soluble copper compounds occur in the

Table 5. Correlation matrix of major parameters and heavy metals of surface water.

	Parameters										
	pH	TDS	Ca	K	Mg	Na	Cl	NO_3	SO_4	HCO_3	Cd
Al	0.33	0.14	0.29	0.23	0.34	0.02	0.03	0.31	0.15	0.32	−0.13
Cd	0.15	0.84	0.55	0.69	0.41	0.91	0.91	0.67	0.81	0.43	1
Co	0.27	0.23	0.57	0.46	0.66	−0.03	−0.04	−0.01	0.34	0.65	−0.26
Cu	−0.02	0.43	0.30	0.45	0.29	0.46	0.46	0.28	0.44	0.26	0.622
Fe	0.27	0.02	0.33	−0.05	0.34	−0.13	−0.13	−0.05	−0.01	0.39	−0.30
Mn	−0.86	0.06	0.16	−0.01	0.21	0.02	0.02	−0.08	0.07	0.21	−0.08

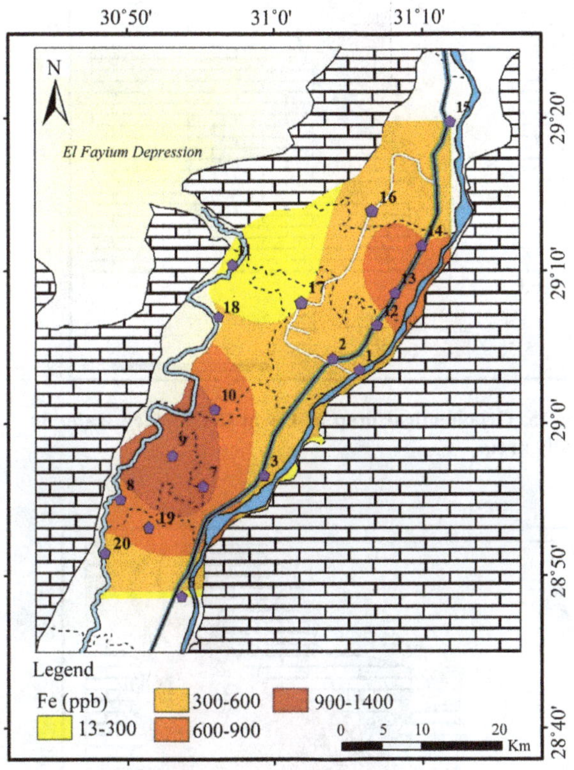

Figure 12. Geochemical map of Fe in the surface water in the study area.

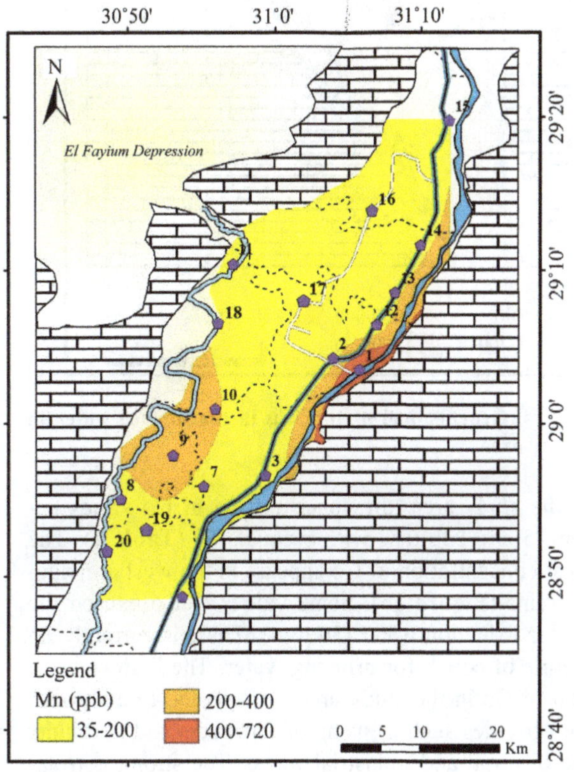

Figure 13. Geochemical map of Mn in the surface water in the study area.

environment after release through fertilizer application in agriculture [20].

It is an essential element for living organisms, including humans, and in small amounts necessary in our diet to ensure good health. However, too much copper can cause adverse health effects, including vomiting, diarrhea, stomach cramps, nausea, liver damage and kidney disease [21].

In the study area, the concentration of Cu in surface water ranges from <1 to 1080 μg/l with σ equals 368 (**Table 2**). High values of Cu are recorded in El-Ibrahimia canal at two positions; El-Wasta and Beni Suef (**Figure 14**). Cu was found to be less than 2000 μg/l (MPL in drinking water WHO [1] and higher than 200 μg/l (MPL in irrigation water FAO [2]. Moderate correlation between Cu and Cd (r = 0.62, **Table 5**). High concentrations of Cu in surface water are attributed to several factors; agricultural and industrial activities, waste disposal and wastewater from El-Moheet drainage.

***Cobalt* (*Co*):** Cobalt is a relatively rare element of the earth's crust. It is an essential micronutrient required for the formation of vitamin B_{12} and for its function in enzymatic processes. Small amounts of cobalt are present naturally in rock, soil, water, plants, animals and air. Natural sources of cobalt to the environment include volcanic eruptions, seawater spray and forest fires. Anthropogenic sources to the environment contain burning of fossil fuels, sewage sludge, phosphate fertilizers, mining and smelting of cobalt-containing ores and industrial processes that use cobalt compounds [22,23].

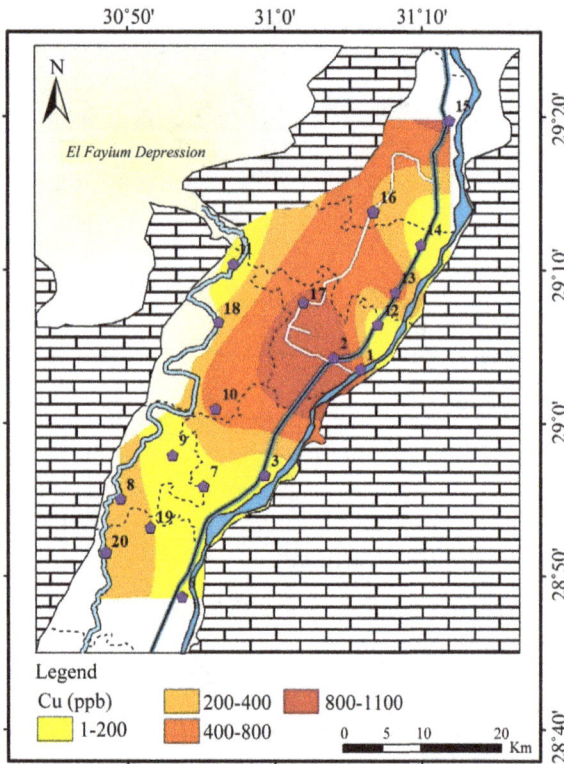

Figure 14. Geochemical map of Cu in the surface water in the study area.

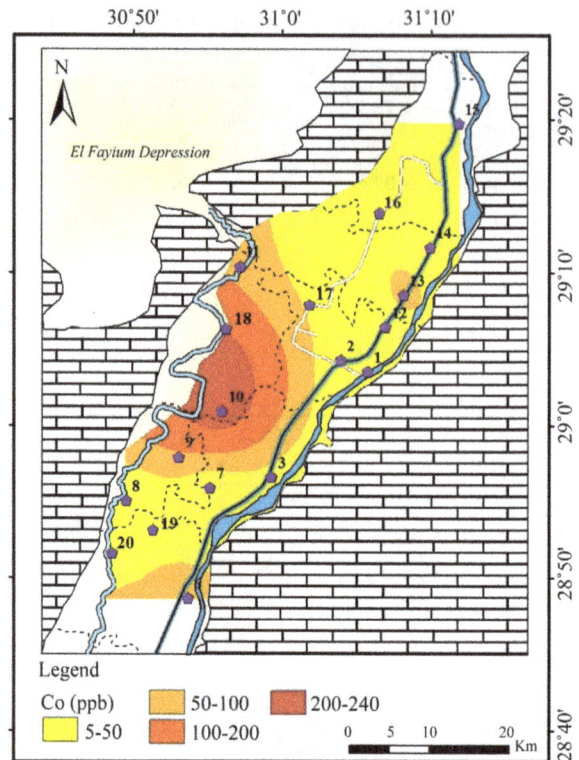

Figure 15. Geochemical map of Co in the surface water in the study area.

In the study area, the concentration of Co varies between <5 and 239 µg/l with σ equals 59 (**Table 2**). The highest concentration of Co appears in the west at irrigation drains (**Figure 15**). Some values of Co exceed the MPL for irrigation water (50 µg/l). There is currently no guideline of cobalt for drinking water. The high concentration of Co in the study area is attributed to anthropogenic activities such as using of fertilizers and pesticide, waste disposal and industrial waste. The strong correlation between Co and each of Mg^{2+}, HCO_3^- and Ca^{2+} (**Table 5**) confirms the anthropogenic contribution.

Aluminum (Al): The concentration of Al varies from 1608 to 2545 µg/l with σ equals 325 (**Table 2**). **Figure 16** shows the distribution of Al in Beni Suef governorate where the highest concentrations are recorded in the southwest in irrigation canals and Bahr Youssef.

FAO organization [2] has recommended a MPL of Al to be 5000 µg/l for irrigation water. However, for drinking water a MPL of 200 µg/l was established for Al by WHO [1], based on aesthetic considerations. Al is not considered to be a heavy metal like lead, but it can be toxic in excessive amounts and even in small amounts if it is deposited in the brain. Many of the symptoms of Al toxicity mimic those of Alzheimer's disease and osteoporosis [24].

In the study area, the concentration of Al is less than the MPL for irrigation. However, it exceeds the MPL for

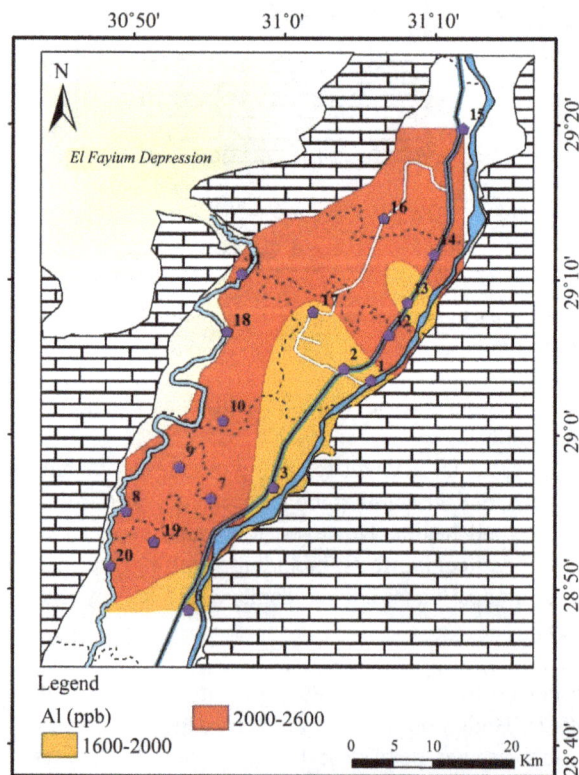

Figure 16. Geochemical map of Al in the surface water in the study area.

drinking water in all surface water samples (ten-fold the GV limit). Aluminum is the most abundant metallic element and constitutes about 8% of Earth's crust. It occurs naturally in the environment as silicates, oxides, and hydroxides, combined with other elements, such as sodium and fluoride, and as complexes with organic matter. In addition to the natural sources of Al from soil (**Table 4**), this serious water pollution by Al (up to 2500 µg/l) is most probably provoked by anthropogenic activity, especially because there is no constrains on using Al for manufacturing household utensils in Egypt as well as most other Arab States. It is used in the transportation machinery, construction, a wide range of household items; from cooking utensils to watches; electrical conductors, explosives, pigments, paints, and in the coagulation process on water treatment to reduce organic matter, color, turbidity and microorganism levels. It is also used in food processing (baking powders, bleached flour, grated cheese, and table salt) and in aluminum cans [1,19,25].

Barium (Ba) and Zinc (Zn): the concentration of Ba and Zn range from 20 to 225 µg/l and <1 and 1700 µg/l, respectively (**Table 2**). The highest concentrations of Ba are noticed in the northeast and southwest (**Figure 17**). Whereas, the highest concentration of Zn is recorded in the southwest (**Figure 18**). Ba and Zn values were below the MPL for drinking and irrigation water [1,2].

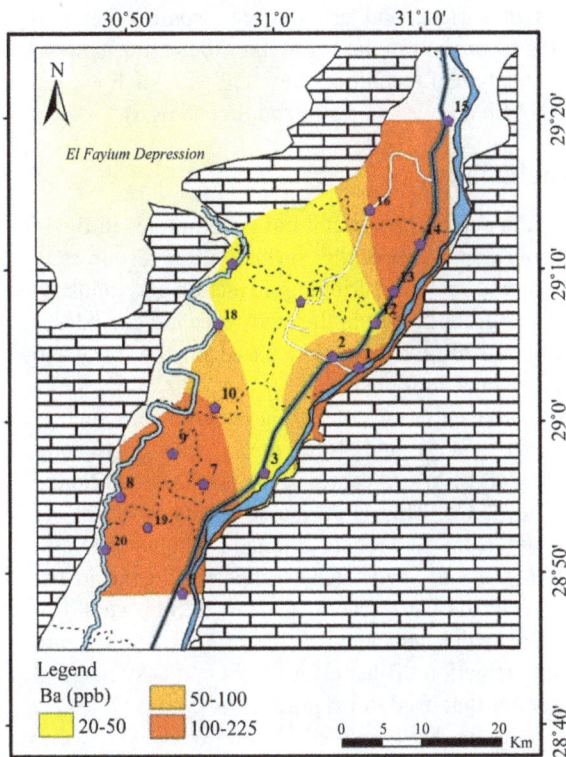

Figure 18. Geochemical map of Zn in the surface water in the study area.

Vanadium (V): The concentration of V varies between <5 to 187 µg/l, (**Table 2**) and highest value appears in the northeast (**Figure 19**). The concentration of vanadium in drinking-water depends significantly on geographical location.

To summarize the information in a smaller set of factors or components for prediction purposes, factor analysis (a multivariate statistical technique) has been utilized to examine the underlying patterns or relationships for the major ions and heavy metals in surface water. Five factors were extracted (**Figure 20**).

Factor 1 is highly positive loading with EC, TDS, Cd, Cu, V and major ions (**Figure 20**). It is attributed to influence of lithogenic and anthropogenic activities. It can be named lithogenic and anthropogenic factor. Factor 2 is highly positive loading with Co, Fe, Mg, Ca and HCO_3 (**Figure 20**) being of common source (agriculture activities). It can be named agriculture factor. Factor 3 is highly positive loading with Mn and negatively with pH (**Figure 20**). It is caused by the impact of pH. Factor 4 is highly positive loading with Al and Fe and negatively with Cu (**Figure 20**). It may be attributed to the tools and material which made by aluminum that human used especially. Factor 5 is highly positive loading with Ba and Zn (**Figure 20**) which occur in water from industrial wastes.

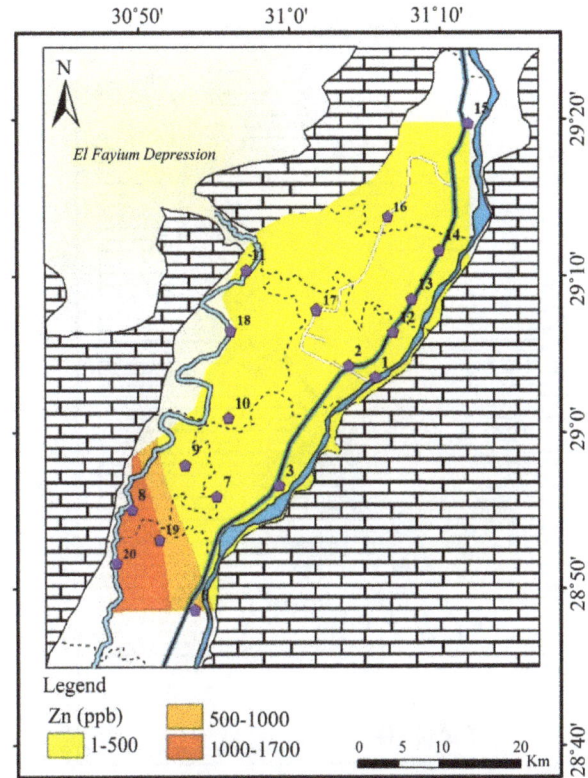

Figure 17. Geochemical map of Ba in the surface water in the study area.

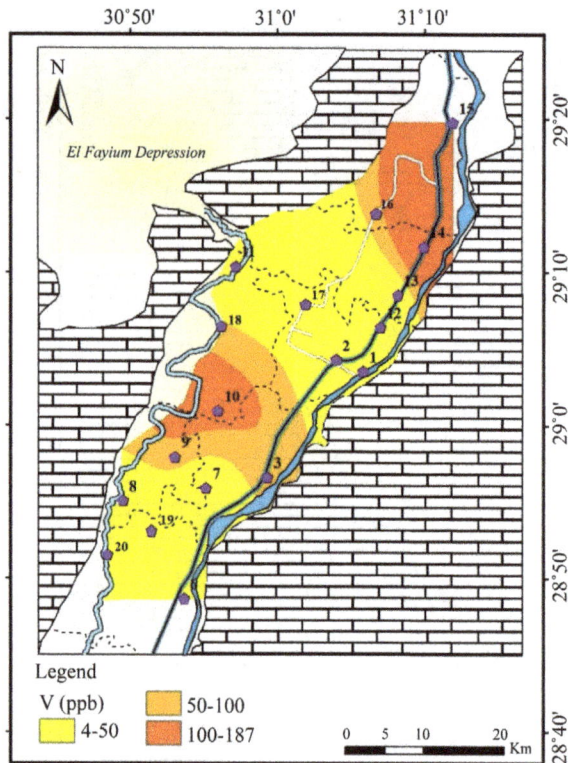

Figure 19. Geochemical map of V in the surface water in the study area.

Table 6. Chemical analysis of groundwater samples. Note that major elements are given in mg/l while heavy metals in µg/l.

Elements	G4	G5
pH	7.96	7.72
EC	1417	493
TDS	907	315
Ca	99	52
K	29	6
Mg	32	16
Na	136	18
Cl	124	23
SO_4	94	23
HCO_3	507	222
NO_3	19	0.2
Al	1526	1397
Mn	1188	2158
Cd	<1	28
Fe	48	599
Pb	<1	159
Ba	341	203
V	191	95
Zn	47	<1

4.4. Groundwater

The chemical analyses of two groundwater samples collected from Beni Suef Governorate (G4 and G5) at Beba and El-Fashn centers (**Figure 1** and **Table 6**) suggest:

High values of pH, Ec, TDS and major ions (Ca^{2+}, Mg^{2+}, Na^+, K^+, Cl^-, SO_4^{2-}, HCO_3^- and NO_3^-) are noticed in G4 sample at Beba center (**Table 6**). In spite of this increase, the values are still below the MPL given by WHO [1]. As for the heavy metals, high concentrations of Al and Mn are noticed in both samples (G4 and G5), higher than MPL given by WHO [1] (200 and 400 µg/l, respectively). Besides, Cd, Fe and Pb concentrations exceed the MPL for drinking water (3, 300, 10 µg/l, respectively) in G5 sample at El-Fashn center (**Table 6**).

The TDS and major ions in groundwater can mainly be attributed to leaching processes through the facies of the Quaternary water bearing formations (Holocene and Pleistocene deposits). In addition to the interaction with sediments and soil, the direct infiltration of surface water (irrigation water and wastewater from drains), seepage from septic tanks, construction of water pipes, use of fertilizers and pesticides and evaporation processes during flood irrigation contribute to the heavy metals content in groundwater especially in G5, being an agriculture area [11,13,17,19].

In addition to the previously mentioned hazardous elements in surface and groundwater, some other elements have concentrations below their detection limit of the used analytical technique, *i.e.*, ICP-MS, such as; As, Cr, Pb, Ni, Sb and Se (<1 µg/l), and Sn (<5 µg/l).

5. Conclusions

Beni Suef is one of the agricultural governorates in Egypt. It is mainly dependent on the surface water resources to supply its requirements (drinking, irrigation and industry). Surface and groundwater in the study area are considered potable for drink and irrigation based on the TDS and major ions concentrations.

Highest concentrations of TDS and major ions such as Na^+, K^+, Cl^-, SO_4^{2-} and NO_3^- are recorded in El-Ibrahimia canal at El-Wasta center in the north. In addition to high value of Cd. This is attributed to the effect of sewage water from El-Moheet drainage. High concentrations of Mg^{2+}, Ca^{2+} and HCO_3^- appeared in the irrigation drains in the west accompanied with high values of Fe and Co, which attributed to the effect of agriculture activities (irrigation effluent, fertilizers and pesticide and sewage water that used in irrigation).

Regarding the MPL quoted by the WHO [1], water resources in the study area are polluted by Al, Cd and Fe and by Co and Cu. High concentration (>MPL) of Mn was recorded in River Nile and groundwater samples.

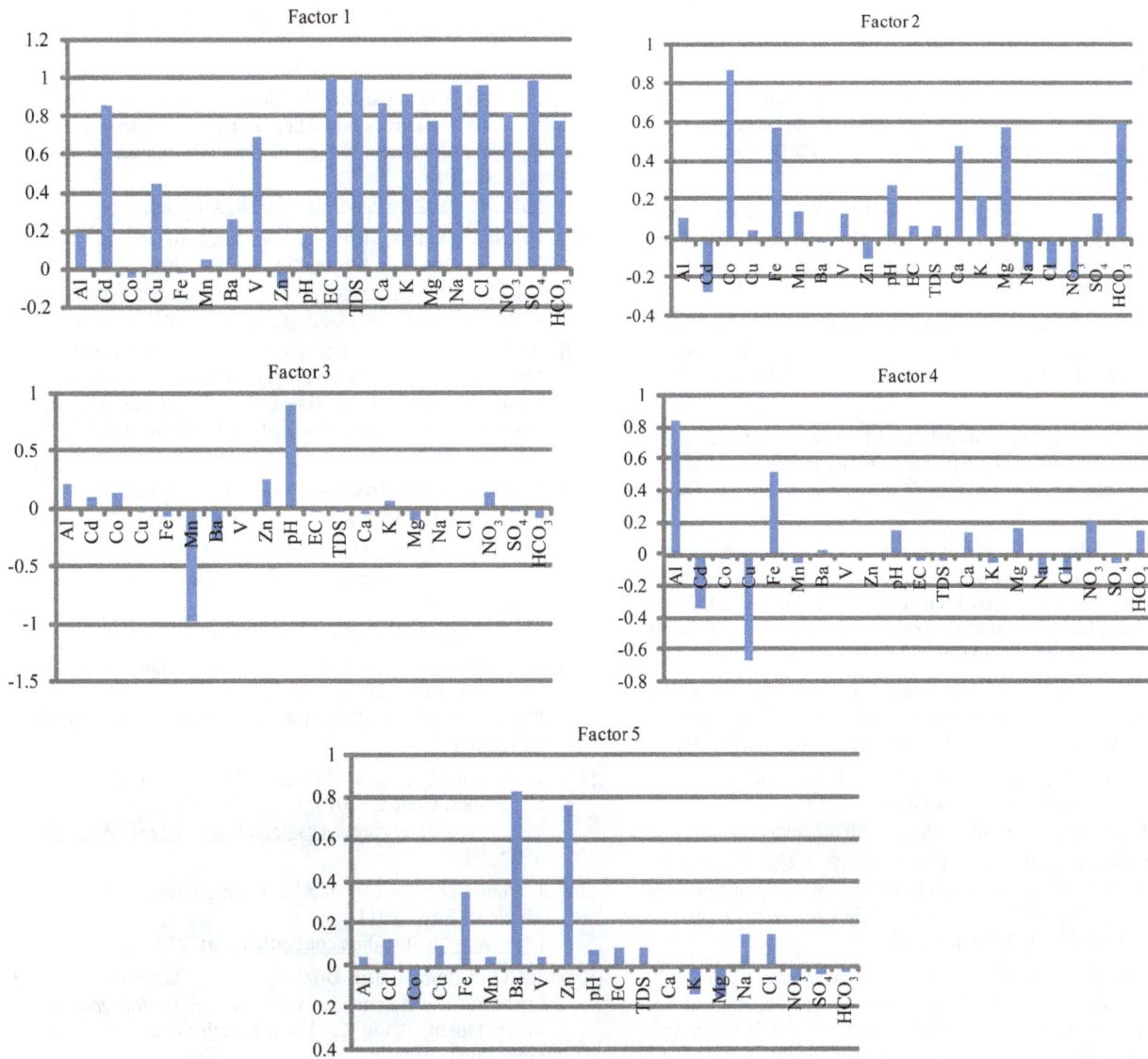

Figure 20. Factor analyses of surface water in study area.

The surface and groundwater in Beni Suef suffer from pollution due to the impact of unsupervised anthropogenic activities including excessive usage of fertilizers and pesticide, waste disposal and industrial waste, seepage from septic tanks, construction of water pipes, wastewater from El-Moheet drainage and evaporation processes during flood irrigation.

REFERENCES

[1] WHO, "Guidelines for Drinking Water Quality," 4th Edition, World Health Organization, Geneva, 2011.

[2] FAO, "Water Quality for Agriculture," Food and Agriculture Organization, Rome, 1994. http://www.fao.org/docrep/003/t0234e/t0234e00.HTM

[3] S. K. Das, M. K. Bhunia and A. Bhaumik, "Solvothermal Synthesis of Mesoporous Aluminophosphate for Polluted Water Remediation," *Microporous and Mesoporous Materials*, Vol. 155, 2012, pp. 258-264. http://dx.doi.org/10.1016/j.micromeso.2012.01.034

[4] A. Melegy, "Adsorption of Lead (II) and Zinc (II) from Aqueous Solution by Bituminous Coal," *Geotechnical and Geological Engineering*, Vol. 28, No. 5, 2010, pp. 549-558. http://dx.doi.org/10.1007/s10706-010-9309-5

[5] S. Wang and Y. Peng, "Natural Zeolites as Effective Adsorbents in Water and Wastewater Treatment," *Chemical Engineering Journal*, Vol. 156, No. 1, 2010, pp. 11-24. http://dx.doi.org/10.1016/j.cej.2009.10.029

[6] M. Pirsaheb, T. Khosravi, K. Sharafi, L. Babajani and M. Rezaei, "Measurement of Heavy Metals Concentration in Drinking Water from Source to Consumption Site in Kermanshah—Iran," *World Applied Sciences Journal*, Vol. 21, No. 3, 2013, pp. 416-423.

[7] Egyptian Environmental Affairs Agency, "Environmental characterization of Beni Suef Governorate," 2003. (In

Arabic)
http://www.eeaa.gov.eg/arabic/info/report_gov_profiles.a
sp

[8]   M. Abdel-Rahman, "Pliocene and Quaternary Deposits of
       Beni Suef—East Fayum and Their Relation to the Geo-
       logical Evaluation of the River Nile," Ph.D. Thesis, Cairo
       University, Giza, 1992.

[9]   H. S. Hassan, "Geological Studies of the Pliocene-Qua-
       ternary Sediments in the Saff-Sukkara Area and Their
       Uses as Building Materials," Ph.D. Thesis, Cairo Univer-
       sity, Giza, 1994.

[10] G. Abdel-Mageed, "Hydrogeological and Hydrogeoche-
       mical Studies of the Nile Valley Area, Beni Suef Gover-
       norate, Egypt," M.Sc. Thesis, Cairo University, Giza,
       1997.

[11] M. El Kashouty, "Modeling of Limestone Aquifer in the
       Western Part of the River Nile between Beni Suef and El
       Minia," *Arabian Journal of Geosciences*, Vol. 6, No. 1,
       2013, pp. 55-76.

[12] N. S. Magesh, S. Krishnakumar, N. Chandrasekar and J.
       P. Soundranayagam, "Groundwater Quality Assessment
       Using WQI and GIS Techniques, Dindigul District, Tamil
       Nadu, India," *Arabian Journal of Geosciences*, Vol. 6, No.
       11, 2013, pp. 4179-4189.

[13] S. Salman, "Geochemical and Environmental Studies on
       the Territories West River Nile, Sohag Governorate—
       Egypt," Ph.D. Thesis, Al-Azhar University, Cairo, 2013

[14] ICdA (The International Cadmium Association), "Level
       of Cadmium in the Environment," 2013.
       http://www.cadmium.org/pg_n.php?id_menu=6

[15] J. Pan, J. A. Plant, N. Voulvoulis, C. J. Oates and C. Ih-
       lenfeld, "Cadmium Levels in Europe: Implications for
       Human Health," *Environmental Geochemistry and Health*,
       Vol. 32, No. 1, 2009, pp. 1-12.

[16] J. M. R. S. Bandara, H. V. P. Wijewardena, Y. M. A. Y.
       Bandara, R. G. P. T. Jayasooriya and H. Rajapaksha,
       "Pollution of River Mahaweli and Farmlands under Irri-
       gation by Cadmium from Agricultural Inputs Leading to a
       Chronic Renal Failure Epidemic among Farmers in NCP,

Sri Lanka," *Environmental Geochemistry and Health*,
Vol. 33, No. 5, 2010, pp. 439-453.

[17] A. Melegy, A. M. Shaban, M. M. Hassaan and S. A. Sal-
       man, "Geochemical Mobilization of Some Heavy Metals
       in Water Resources and Their Impact on Human Health
       in Sohag Governorate, Egypt," *Arabian Journal of Geos-
       ciences*, 2013.
       http://dx.doi.org/10.1007/s12517-013-1095-y

[18] British Columbia, "Iron and Manganese in Groundwater,"
       Water Stewardship Information Series, 2007.
       http://www.env.gov.bc.ca/wsd/plan_protect_sustain/groundwa
       ter/library/ground_fact_sheets/pdfs/fe_mg(020715)_fin2.pdf

[19] A. El-Kammar, M. El-Kashouty, M. M. Yehia and Gh.
       Miro, "Environmental Hydrogeochemistry of Groundwa-
       ter in Central Damascus Basin, Syria," *Journal of the Se-
       dimentological Society of Egypt*, Vol. 20, No. 1110-2527,
       2012, pp. 73-81.

[20] Lenntech (Water Treatment Solution), "Copper," 2013.
       http://www.lenntech.com/periodic/elements/cu.htm

[21] MDH Minnesota Department of Health, "Copper in
       Drinking Water, Health Effects and How to Reduce Ex-
       posure," 2005.
       http://www.health.state.mn.us/divs/eh/water

[22] N. K. Nagpal, "Technical Report—Water Quality Guide-
       lines for Cobalt," 2004.
       http://www.env.gov.bc.ca/wat/wq/BCguidelines/cobalt/co
       balt_tech.pdf

[23] Environment Canada, "Federal Environmental Quality
       Guidelines, Cobalt," 2013.
       http://www.ec.gc.ca/ese-ees/default.asp?lang=En&n=92F
       47C5D-1

[24] L. Paul, "The Dangers of Aluminum Toxicity. Alternative
       Medicine Site," 2011.
       http://www.bellaonline.com/articles/art7739.asp

[25] WHO, "Aluminum in Drinking Water: Background Doc-
       ument for Development of WHO Guidelines for Drinking-
       Water Quality," Vol. 2, World Health Organization, Ge-
       neva, 2003.

21

Development of El-Salam Canal Automation System

Noha Samir Donia
Environmental Studies and Researches Institute, Ain Shams University, Cairo, Egypt
Email: ndonia@gmail.com

ABSTRACT

In Egypt irrigation water is becoming more scarcer with the continuously increasing demand for agriculture, domestic and industrial purposes. To face this increasing irrigation demand, the available water supply in Egypt is supplemented by the reuse of agricultural drainage water as in El-Salam Canal that do not satisfy water quality standards defined for the canal. This paper introduces an automation system for El-Salam Canal to control the flow of the fresh water and drainage water supplied to the canal. This automatic control system (ACS) is able to process data of various flows and water quality data along the canal. This control system is represented by a canal computer model. This system computes the required control actions at the Damietta branch and the feeding drains. It is also able to generate optimum solutions for the canal to satisfy the pre-defined canal conditions and standards.

Keywords: Water Quality; Automatic Control; Modeling

1. Introduction

As water is becoming more and more a scarce resource all over the world, proper management of the available water is essential. For an optimal use of the available water resources, water management strategies have to be developed. A water management strategy is based on a water control system. The two main factors that determine the designated water use are the water quality and water quantity of a water system. Controlling the quality and quantity of a water system is done using monitoring devices, water gates, pump stations, power stations and other operational devices. There are different types of controlling a water system. However, the use of automatic control has lately proven to have more advantages over other types. Automatic control provides accuracy, reliability, time-saving and man-power saving. It also enhances flexibility and saves water and improves production.

Many researches have been conducted for implementtation of automatic control water systems. [1] studied the real-time control of combined surface water quantity and quality for polder flushing. [2] studied the Elements of a decision support system for real-time management of dissolved oxygen in the San Joaquin River Deep Water Ship Channel. In Thailand, on the Kamphaengsaen Irrigation Canal, the canal's automation system has been developed and tested during October 2006 to July 2008. The canal automation system consists of the master station and six remote terminal units (RTU) which communicate by VHF radio. The six RTUs installed in the canal

irrigation system are for monitoring and controlling of water levels and discharges in the canal system, monitoring rainfall, air temperature and relative humidity. The system has provided flexible, accurate and reliable control of irrigation water supply [3]. In Arizona USA, on the Salt River Project Canal system an automatic control system was proposed. This system automates and enhances functions already performed by operators. Some of these functions are control of water levels and flow control at check structures. The proposed system consists of three separate controllers with a configuration that makes control actions computed independently of gate hydraulics. The controllers are centrally operated, that is monitoring and determining control actions is done from a remote site. The control system has proven to be a stable and robust system [4]. In Australia, on the Coleambally Canal Network, an automation system has been introduced, with the objective of reducing the operating cost of the canal system, reducing conveyance losses and improving the ability of the supply system to respond to irrigation demands. There is an ability to remotely monitor and regulate the main canal which results in a much improved standard of service to the secondary canal off-takes. Gates are being automated and a software system controls the opening and closing of the gates automatically. The control system assists irrigators to improve the efficiency of water use [5].

This study focuses on introducing El-Salam Canal control system that consists mainly of an automatic monitoring system and an automatic control system which is represented by a computer control model based on a data driven model.

2. Study Area Description

El-Salam Canal is located in the North East of Egypt where it supplies water for the reclamation of new lands in that part of the country. These areas are originally parts of the sedimentary formation of the ancient Nile branches in that area. The canal intake is on the right bank of Damietta Branch at Km 219, 3.0 Km upstream the Faraskur Dam. The canal passes through five governorates: Damietta, Dakahliya, Sharkiya, PortSaid and North Sinai [6], the total length of the canal is about 277 Km and is divided into two main parts. The first part is West of Suez Canal, it is about 86 Km long and the second part lies east of Suez Canal and is about 191 Km long. The western part of the canal is known as El-Salam Canal. It starts from the intake at Damietta Branch (Nile River) runs in a south-eastern direction and crosses the Suez Canal through a siphon, it continues after the siphon and the eastern part of the canal is known as El-Sheikh Gaber Canal. A layout of El-Salam Canal is shown in **Figure 1**. El-Salam Canal was designed to supply the irrigation water to a total area of 620,000 feddans consisting of 220 thousand feddans on the western side of the Suez Canal and 400 thousand feddans east of the Suez Canal in Sinai. The canal was planned to convey a discharge of 4.45 billion m^3/year. About 2.2 billion m^3/year would be fresh water supplied from the Nile and

transferred through the canal at its intake. And about 2.25 billion m^3/year is to be supplied from two drains called Bahr Hadous and Lower Serw drains. The water quality represented by salinity was also a concern when designing the canal.

Salinity should not exceed 1250 ppm generally in the canal. Many structures are constructed along El-Salam Canal. The first group of these structures is for water regulation purposes, consisting of pump stations and regulators. The second group of structures is crossing structures such as siphons and bridges.

Some of the objectives and benefits that are gained from implementing El-Salam Canal are: redistributing population in Egypt, protecting the eastern borders of the country, strengthening the Egyptian agricultural policy through increasing the cultivated areas and agricultural yield, increasing agricultural and national production and thus increasing exporting vegetables and fruits while decreasing food import, benefiting and making good use of agricultural drainage water as an important water resource, creating work opportunities for the youth and establishing tourism, industrial and mining projects.

Therefore, careful investigation and prediction of the quality of water throughout the canal is crucial. Many studies have been carried for assessment of the water quality of Bahr Hadous and El-Serw drains, [7-11], also many studies have been conducted about the agriculture development of El-Salam water [12-16], and few studies were conducted to study the water quality along El-Salam

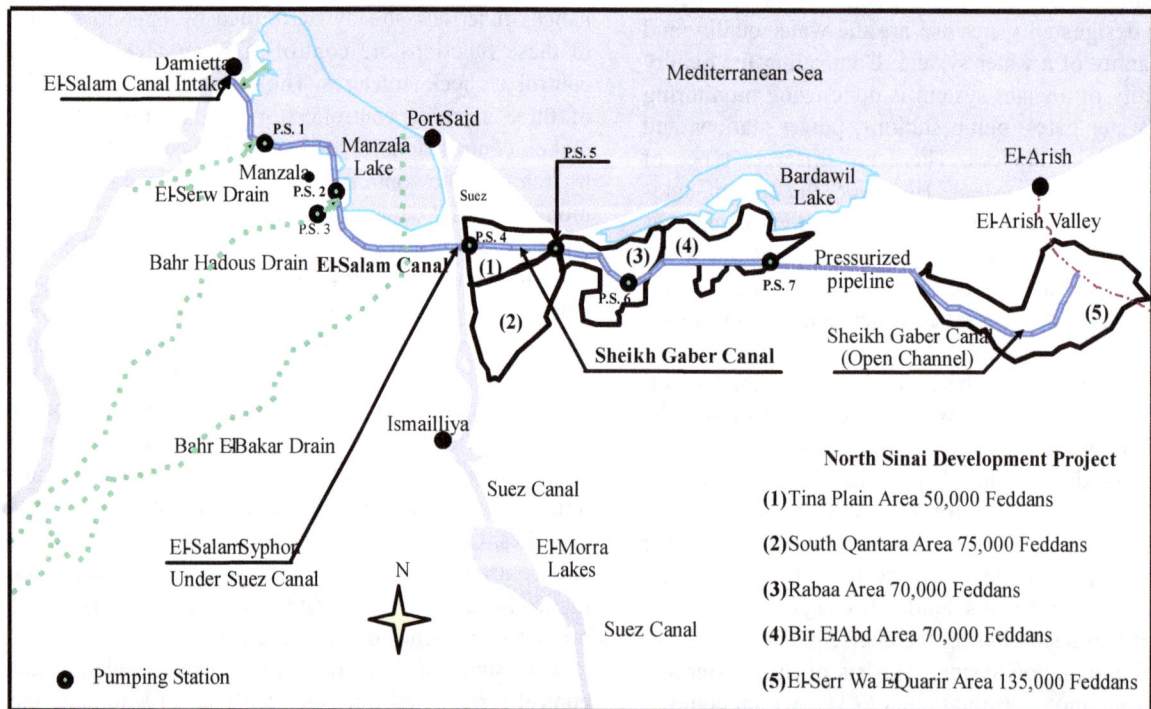

Figure 1. Layout of El-Salam Canal project.

Canal, [17-22] developed a decision support system (DSS) to choose the required treatment option of discharging drains in order to satisfy with these guidelines but little attention has been for real time operational water quality management of the canal [23,24].

3. Computer-Aided Control System for El-Salam Canal

The Control System on El-Salam Canal integrates the water quality monitoring and the water quality control policy using:
- An automatic monitoring system (AMS), which is capable of collecting data of different flows and water quality along the canal.
- An automatic control system (ACS), which is able to process data of various flows and water quality data along the canal. This control system is represented by a computer model designed for the canal.

This computer model is able to generate optimum solutions for the canal to satisfy the pre-defined canal conditions and standards. The model can also compute the required control actions at the Damietta branch and the feeding drains which supply the canal with its water. It calculates the gate opening required for each mixing drain.

3.1. The Automatic Monitoring System (AMS)

The type of automatic monitoring system used consists of a Data Acquisition System (DAS) which runs a data software collection platform (DCP). This DAS includes at each local station:
- a) A Data Collection Unit (DCU)
- b) A Data Terminal Unit (DTU)
- c) Computer Control Model

The DCU collects data from sensors and is triggered by the DTU, whereas the DTU is the part that triggers the DCU and sends data to the computer control model at the main station [7,8]. The communication equipment is installed at each DTU and at the main station. The communication system also supports voice communication between any two stations. The facilities of the voice communication system include telephone, earpiece and mouthpiece. To fulfill web communication, a web-enabled software is introduced to the control system at the main station to support remote monitoring and viewing of databases for station details, historical and actual data through the internet. In case of failure of the automatic system that sends the control actions from the main station to all the DTUs of all stations, the data communication system delivers the control actions to the concerned stations in the form of messages. These messages are displayed on the DTU for the managing of the station manager and the operators. Upon the reception of a mes-

sage, alerting devices like a horn and a flashing light are automatically activated through digital signals delivered to the DTU. All electrical devices are connected with cables to deliver power and to transport signals and data. Cable guidance tubes, ducts and similar connections are used to give the cables proper protection.

3.2. Description of the Automatic Real-Time Control System (ARTCS)

- The supply, transport and distribution of the irrigation water are managed through real-time control of the structures on El-Salam Canal. The structures which we consider in this study are:
- The head regulator at Damietta Branch admitting fresh water from the Nile.
- The regulators at the Lower Serw drain admitting drainage water from the agricultural drain.
- Pump station No. 3 lifting water from Bahr Hadous drain to El-Salam Canal.

3.2.1. Automatic Real-Time Control System Features
The ARTCS system is based on:
- Full utilization of the available fresh Nile water with a water quantity control at the rest of the intakes to El-Salam Canal.
- Presence of instantaneous information available on the actual flow of the drains and of Damietta Branch feeding El-Salam Canal.
- Presence of instantaneous information available on the salinity of the drains and of Damietta Branch feeding El-Salam Canal.
- The difference between the actual value (measured) and the setpoint (desired output response) is checked every suggested period (e.g. 30 minutes) and control actions are calculated by the controller. Those actions are automatically communicated and act on the actuators that execute the control actions physically causing the operation of the gates and pump stations as desired.

Thus the automatic real-time control system fulfills the following functions:
- Receiving the measured data once every 30 minutes.
- Processing data and comparing it with setpoint values
- Computing required actions by pump stations and gates.
- Communicating these actions to the needed gates and pump stations and operating them as desired.

3.2.2. Control Method Description
The computer model is installed at the main station. It includes the software that receives the monitored data from the DTU and makes all the necessary computations (processing of data). It then gives an output of control

actions that are sent back to the DTUs of all stations .In the computer control model, the control method that is applied is called the Master-Slave controller. The Master controller determines the flows that need to be applied at the control structures (Damietta Branch, El-Serw drain and Bahr Hadous drain), while the Slave controller of each structure converts the flow to a local setting of the structure. As the Slave controller receives information from the Master controller about the flow change that the concerned structure has to implement, it converts this flow change to a change in the opening height of the gates or in a change of the pump flow by the following relationship (Equation (1)):

$$U = f(Q) \tag{1}$$

where:

U: structure setting (gate opening or pump flow)

Q: flow through the structure

Slave controllers use upstream and downstream water levels (h) around the structure in this formula. A detailed explanation of this formula is given earlier in chapter three.

3.3. The Automatic Control System (ACS)

The type of control system used is the "multivariable closed-loop water management control system with disturbance and feed forward monitoring". This control system is a combination of feedback control and feed forward control methods. Parts of the automatic control system are shown in **Figure 2** [8].

The computer control model represents the automatic control system used. This computer model is based on a data driven model. The data measured along El-Salam Canal over the years 2006 to 2008 are being used in this model.

3.3.1. Mathematical Background of the Computer Control Model

The basic equations governing El-Salam Canal are :

- **Mass Balance Equations: Equations (2) and (3)**

$$Q_t = Q_{dam} + Q_{serw} + Q_{hadous} \tag{2}$$

$$Q_t * TDS_t = Q_{dam} * TDS_{dam} + Q_{serw} * TDS_{serw} + Q_{hadous} * TDS_{hadous} \tag{3}$$

- **Data Driven Equations: Equations (4) and (5)**

$$Q_{serw}/Q_{hadous} = R \tag{4}$$

$$OMR = (Q_{serw} + Q_{hadous})/Q_{dam} \tag{5}$$

where:

Q_t = output discharge of El-Salam Canal (million m³/day)

TDS_t = salinity at the output discharge of El-Salam Canal (ppm)

Q_{dam} = flow of Nile water at Damietta Intake (million m³/day)

TDS_{dam} = salinity of Nile water at Damietta Intake (ppm)

Q_{serw} = discharge of El-Serw drain (million m³/day)

TDS_{serw} = salinity of El-Serw drain (ppm)

Q_{hadous} = discharge of Bahr Hadous drain (million m³/day)

TDS_{hadous} = salinity of Bahr Hadous drain (ppm)

R = measured ratio between discharge of El-Serw drain and discharge of Bahr Hadous drain

OMR = optimum mixing ratio of fresh water and drainage water

- **Flow-Gate Equation: Equation (6)**

From Bernoulli equation the following flow-gate Equation (6) is derived:

$$Q = c_d \cdot A\sqrt{2 \cdot g \cdot (h_1 - h_2)} \tag{6}$$

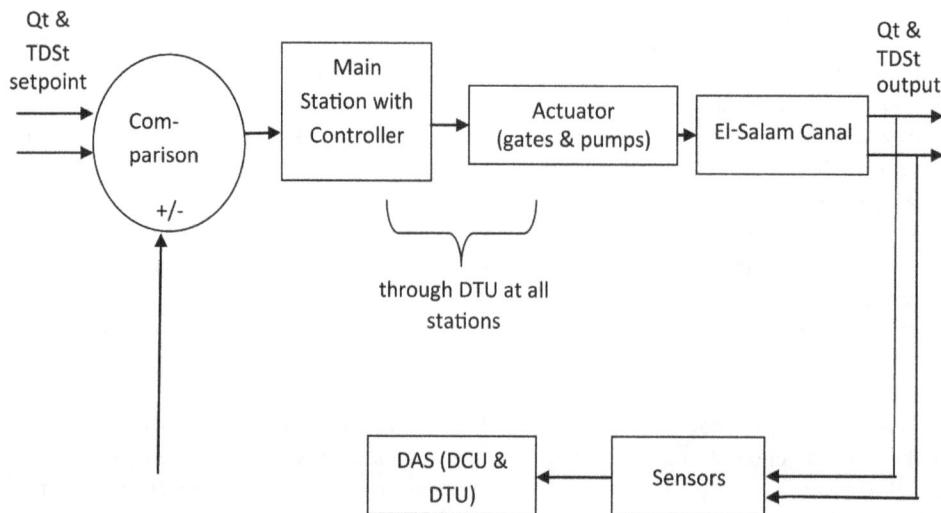

Figure 2. Design of the suggested automatic monitoring and control system for El-Salam Canal.

with

$$A = W_g \cdot Go$$

where:

Q = Discharge through the gated structure (m³/s)

c_d = Overall discharge coefficient

A = Wetted area (m²)

Wg = Gate width (m)

Go = Gate Opening height (m)

g = Gravity acceleration (m/s²)

h_1 = Upstream water level (m)

h_2 = Downstream water level (m)

Constants: c_d = 0.6 – 0.65, W_g = 25 m for Damietta intake & 12 m for El-Serw drain, g = 9.81.

- **Flow-Pump Equation: Equation (7)**

$$NOP = Q/COP \qquad (7)$$

where:

NOP = No. of Pumps

Q = Discharge needed to be pumped (m³/s)

COP = Capacity of Pump (m³/s)

Constant: COP = 16.5

It has been found from the data measured over the years 2006 to 2008, that the best scenario to be used to satisfy the specified conditions for El-Salam Canal is fully utilizing the available fresh Nile water (Damietta Branch) together with the optimum discharge of the available drains feeding El-Salam Canal (El-Serw drain and Bahr Hadous drain). Both fresh and drainage waters are mixed with an optimum mixing ratio. It has also been concluded that if the available fresh water (Damietta Branch) is greater or equal to half the required discharge of El-Salam Canal, then both fresh and drainage waters are mixed with mixing ratio 1:1 as designed and in that case this would be the optimum mixing ratio.

To satisfy the quantity and quality standards defined for El-Salam Canal, we have to calculate an optimum value of the drains discharges and an optimum mixing ratio between fresh water and drainage water. To do so, Equations (1)-(4) are solved in a numerical method. After the optimum values are calculated, control actions are computed using Equations (5) and (6).

4. Automatic Control System Implementation

In order to represent the optimum values of the feeding drains discharge, the optimum mixing ratios and the suitable control actions which satisfy the standards defined for El-Salam Canal, the model is run under different input discharges and different values of input water quality parameter (TDS) from Damietta Branch, El-Serw drain and Bahr Hadous drain. Data obtained through the years 2006 to 2008 represent the different scenarios that are chosen by the model.

4.1. Scenarios Analysis

Input values of discharge and TDS at the Damietta intake, El-Serw drain and Bahr Hadous drain are shown in **Figure 3**, and input values of upstream and downstream water levels at Damietta intake and El-Serw drain are shown in **Figure 4**. Iinput values of constants are shown in **Figure 5**. The input values are used by the model to define the control actions of water levels of the drains

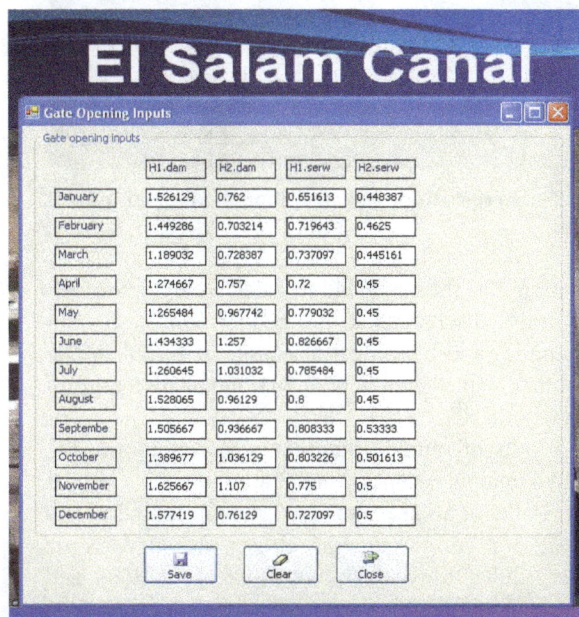

Figure 3. Screen displaying the input discharge and TDS at feeding points along El-Salam Canal year 2007.

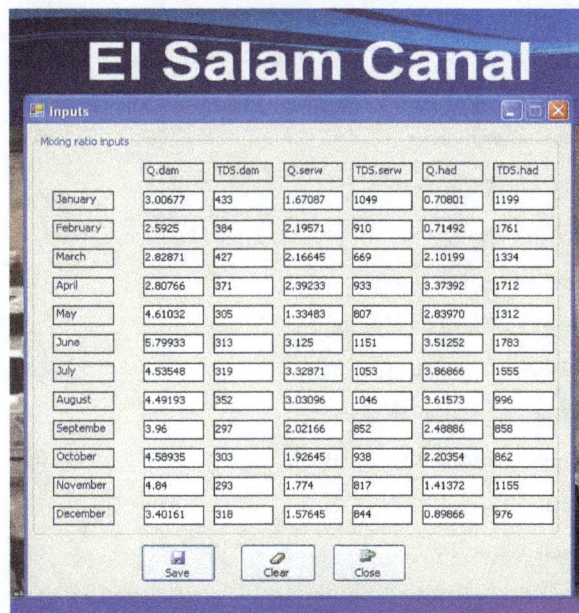

Figure 4. Screen displaying the input values of levels upstream and downstream water along El-Salam Canal year 2007.

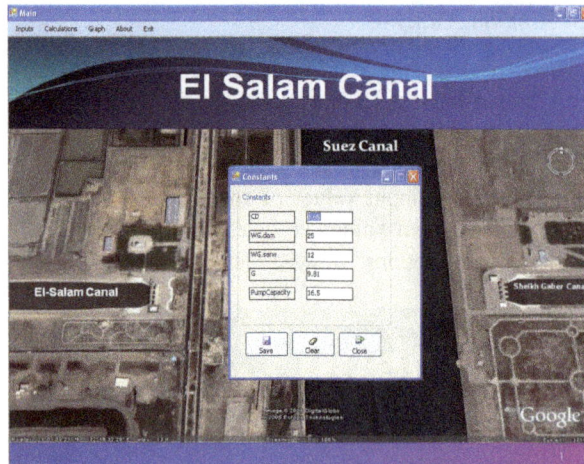

Figure 5. Screen displaying the input values of pumps constants.

discharging into the canal and to calculate optimum values of drains discharges at the feeding points, an optimum mixing ratio the output discharge of El-Salam Canal together with the salinity at the output discharge of the canal.

The results of running different scenarios by the implemented computer control model are shown in **Table 1**. Output results of all scenarios presented in this study are displayed for certain months chosen as an example (February 2006, May 2006, July 2006, June 2008 and one assumed month). The table shows the control actions taken at Damietta Branch, El-Serw Drain and Bahr

Hadous Drain concerning the gate opening and number of pumps are under different scenarios.

In **Figure 6**, the results of a run of the model for the selected month January 2007 chosen as an example are displayed. Values entered shown in **Figure 6** are used to compute the control actions that are required at the Damietta intake and the feeding drains. In **Figure 7**, the calculated control actions for the selected month January 2007 chosen as an example are displayed.

4.2. Analysis of Scenarios Outputs

In scenario 1 (June 2008), it is concluded that when the available fresh water (Damietta Branch) is greater or equal to half the required discharge of El-Salam Canal and the salinity at the output discharge of El-Salam Canal is within canal's standards, then the optimum mixing ratio between fresh and drainage waters will be 1:1 as designed. This will increase the discharge of El-Salam Canal to the required discharge (improve) and will maintain the salinity within the canal's standards 1).

In scenario 2 (January 2007), salinity at the output discharge of El-Salam Canal and the required discharge of the canal are within the canal's standards, thus the optimum mixing ratio between fresh and drainage waters will continue to be as measured.

In scenario 3 (July 2006), it is concluded that salinity at the output discharge of El-Salam Canal is within canal's standards and the output discharge of El-Salam Canal is increased to the required discharge (improve).

Table 1. Measured and calculated Data (GO & No. of Pumps) under different scenarios.

Chosen months	June 2008	Jan. 2007	July 2006	Feb. 2006	March 2007	Nov. 2007
Scenario name	Scen. 1	Scen. 2	Scen. 3	Scen. 4	Scen. 5	Scen. 6
GOdam[a] (original) meter	2.75	1.73	1.09	0.54	2.34	0.27
GOdam (calculated) meter	2.55	1.73	1.09	0.54	2.34	1.39
GOserw[b] (original) meter	0.34	2.4	1.08	0.61	1.34	1
GOserw[b] (calculated) meter	0.41	2.4	1.62	1.6	1.04	1.39
FChadous[c] (original) No. of pumps	3	2	2	1	2	1
PChadous[c] (original) No. of pumps	0.17	0.95	0.43	0.3	0.46	0.91
FChadous[d] (calculated) No. of pumps	3	2	3	3	1	2
PChadous[d] (calculated) No. of pumps	0.79	0.95	0.64	0.4	0.91	0.68
TDS total (measured)	841	994	1016	975	1622	1473
TDS total (calculated)	871	994	1142	975	1250	1099
Mixing ratio (measured)	0.77	1.54	1.2	1.14	3.73	3.73
Mixing ratio (calculated)	1	1.54	1.8	2.99	3.12	1

Where: [a]GOdam = gate opening at Damietta Branch; [b]GOserw = gate opening at El-Serw Drain; [c]FChadous = full capacity of pumps at Bahr Hadous Drain; [d]PChadous = partial capacity of pumps at Bahr Hadous Drain.

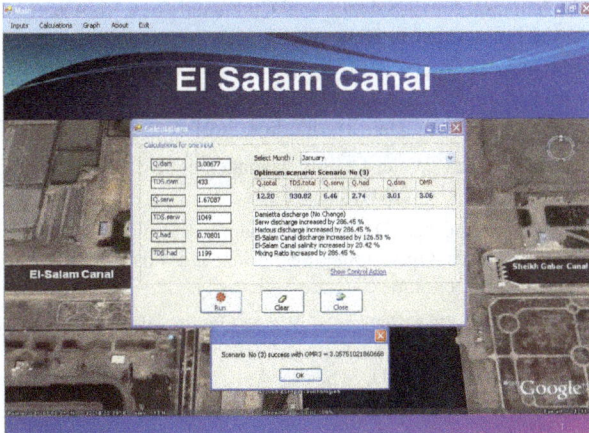

Figure 6. Screen displaying the original and calculated control actions at the feeding points along El-Salam Canal for January 2007.

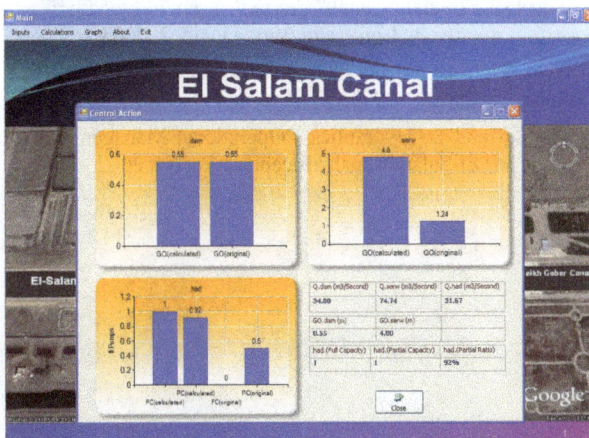

Figure 7. Screen displaying the original and calculated control actions at the feeding points along El-Salam Canal for January 2007.

In scenario 4 (February 2006), it is concluded that salinity at the output discharge of El-Salam Canal is decreased (improve) and the output discharge of El-Salam Canal is increased although not reaching the required discharge (improve).

In scenario 5 (March 2007), it is concluded that salinity at the output discharge of El-Salam Canal is decreased to the standard value (improve) and the output discharge of El-Salam Canal does not increase but may decrease, thus sacrifice with the discharge for the sake of the improved salinity of the canal.

In scenario 6 (November 2007), it is concluded that salinity at the output discharge of El-Salam Canal is decreased to the standard value (improve) and the output discharge of El-Salam Canal does not increase but may decrease, thus sacrifice with the discharge for the sake of the improved salinity of the canal.

In all cases, control actions are taken at the Damietta Branch, El-Serw drain and Bahr Hadous drain to fulfill all scenarios. On El-Salam Canal the gated intakes are at Damietta Branch and at El-Serw drain. The pumped intake is at Bahr Hadous drain. Thus the gated intakes use Equation (5) to calculate the control action needed (gate opening height) and the pumped intake uses Equation (6) to calculate the control action needed (no. of pump units required to operate).

5. Conclusion

Based on the results of this work, the following may be concluded that the computer-aided control system proposed in this paper could successfully monitor and control the flow of the fresh and drainage waters supplied to El-Salam Canal allowing variable mixing ratios. Also, mixing the fresh and drainage waters at the designed ratio 1:1 does not improve the value of the total output discharge except when using fresh water as half the required discharge of El-Salam Canal. Finally, fully utilizing the available fresh water together with optimum discharge of drainage water has improved the total output discharge of El-Salam Canal and the salinity at the output discharge of the canal.

REFERENCES

[1] M. Xu, P. J. van Overloop, N. C. van de Giesen and G. S. Stelling, "Real-Time Control of Combined Surface Water Quantity and Quality: Polder Flushing," *Water Science & Technology*, Vol. 61, No. 4, 2010, pp. 869-878. doi:10.2166/wst.2010.847

[2] N. W. T. Quinn, K. Jacobs, C. W. Chen and W. T. Stringfellow, "Elements of a Decision Support System for Real-Time Management of Dissolved Oxygen in the San Joaquin River Deep Water Ship Channel," *Environmental Modelling & Software*, Vol. 20, No. 12, 2005, pp. 1495-1504. doi:10.1016/j.envsoft.2004.08.014

[3] V. Vudhivanich and V. Sriwongsa, "Development of Kamphaengsaen Canal Automation System," *The 6th Regional Symposium on Infrastructure Development*, Bangkok, 8-10 December 2008, p. 89.

[4] E. Bautista, A. J. Clemmens and R. J. Strand, "River Project Canal Automation Pilot Project: Simulation Tests," *Journal of Irrigation and Drainage Engineering*, Vol. 132, No. 2, 2006, pp. 143-152. doi:10.1061/(ASCE)0733-9437(2006)132:2(143)

[5] M. Nayar and S. Murray, "Improving Water Use Efficiency," The Coleambally Irrigation Area Modernization Project, 2007.

[6] T. Emam and D. Hydraulics, "Operational Management System for El-Salam Canal," Inception Report, 2000.

[7] A. M. Mostafa, S. T. Gawad and S. M. Gawad, "Development of Water Quality Indicators for Egyptian Drains," *18th International Congress on Irrigation and Drainage*, Montereal, 2012, pp. 1-20.

[8] A. M. Mostafa, "Development of Water Quality Indicators and Atlas of Drainage Water Quality Using GIS

tools," Technical Report, NAWQAM Project, 2002.

[9]   A. M. Mostafa, A. Abdelsatar, S. T. Gawad and S. M. Gawad, "A New Technique for the Estimation of BOD/DO from Unmonitored/Non-Point Sources of Pollution," *Journal of Engineering and Applied Science*, Vol. 51, No. 3, 2003, p. 483.

[10] A. M. El-Degwi, A. Abdelsatar, S. T. Gawad and S. M. Gawad, "Variation of BOD Pollution Rate within Hadous Drain Catchments of Egypt," *2nd ICID Asian Conference on Irrigation and Drainage*, Moama, 14-17 March 2004, pp. 14-26.

[11] R. M. S. El Kholy and M. I. Kandil, "Trend Analysis for Irrigation Water Quality in Egypt," *Emirates Journal for Engineering Research*, Vol. 9, No. 1, 2004, pp. 35-49.

[12] Etkens, "Feasibility Studies for North Sinai Project," Ministry of Water Resources and Irrigation, 1989.

[13] JICA, "The Integrated Development of North Sinai Governorate," Ministry of Water Resources and Irrigation, 1989.

[14] FAO, "The Development Agricultural Projects of North Sinai," Ministry of Water Resources and Irrigation, 1989.

[15] Shata, "Structural Development of Sinai Peninsuls, Egypt," Bulletin MWRI-HXS, 1995.

[16] G. G. Refae, A. M. El Jawary and S. Yehia, "Saline Soil Reclamation for El-Salam Canal Command Area," NAWQAM Project, Cairo, 2006.

[17] T. Emam, "Operational Management System for El-Salam Canal, Egypt," Hydraulics Research Institute, Drainage Research Institute, Delft Hydraulics, 2001.

[18] M. G. Ahmed, "Water Quality Management for El-Salam canal," Cairo University, Giza, 2003.

[19] F. M. Eweida, "BOD Variations along El-Salam Canal under Various Operational Scenarios and Possible Enhancement Techniques," Cairo University, Giza, 2006.

[20] M. I. Kandil, "Evaluation of Water Quality of El-Salam Canal and Prediction of Its Effect on Soil and Plant Characteristics," Ph.D. Thesis, Ain Shams University, Cairo, 2006.

[21] R. M. S. El Khouly, "Drainage Water Reuse for Land Reclamation: Risks and Opportunities (Case Study El-Salam Canal—Egypt), NAWQAM Project, NWRC, 2004.

[22] N. S. Donia, "Decision Support System for Water Quality Control of El-Salam Canal," *Journal of Faculty of Engineering, Ain Shams University*, Vol. 43, No. 2, 2008.

[23] DHON (Delft Hydraulics of Netherlands), "El-Salam OMS Tender Document," 2002.

[24] T. Emam and D. Hydraulics, "Operational Management System for El-Salam Canal," Inception Report, 2002.

22

Design and Evaluation of Dadu Canal Lining for Sustainable Water Saving

Ashfaque A. Memon[1], Khalifa Q. Leghari[1], Agha F. H. Pathan[1], Kanya L. Khatri[2], Sadiq A. Shah[2], Kanwal K. Pinjani[3], Rabia Soomro[2], Kameran Ansari[1]

[1]Department of Civil Engineering, Mehran UET, Jamshoro, Sindh, Pakistan
[2]Department of Civil Engineering, Mehran UET Khairpur Campus, Sindh, Pakistan
[3]Water Resources Division, National Engineering Services, Lahore, Pakistan
Email: rajaln@yahoo.com, ashfaquememon144@gmail.com

ABSTRACT

Pakistan livelihood depends on agriculture and so for this on irrigation system. The irrigation system in Sindh province depends on three barrages. The canals off taking from these three barrages irrigate 5.5 million hectares of agriculture land. Sukkur Barrage, which is the oldest one, irrigates more than 2.0 million hectares of land. The Dadu Canal off taking from Sukkur barrage is an earthen canal. A huge amount of irrigation water is lost from the canal in the form of seepage from banks and bed. It is estimated that 40 to 50 per cent of water is lost between the canal head works to the farm-gate. The seepage from the canal creates twin problems of salinity and water logging consequently a large agriculture land has gone out of use, and this process is continued particularly in Sindh. Lining of Canals is considered an effective solution to this problem. But lining of canals in Sindh is a great issue as canals will need to be closed long enough to deprive the farmers at least one crop season and the farmers are unable to pay this price for canal. Therefore, in this study, the Dadu Canal is proposed to be redesigned as an adjacent lined canal which involves design of cross section for various lining options at locations where changes in the hydraulic conditions occur at cross regulators and fall structures. The proposed lining is preferred to be plain cement concrete lining which is selected after investigating local conditions. Quantity and cost estimation at selected RDs (Reduced Distance) proved feasible and significant in long term functioning of Dadu Canal.

Keywords: Dadu Canal; Concrete Lining; Canal Design; Lining Evaluation; Sustainable Water Saving

1. Introduction

Indus basin irrigation system (IBIS), being one of the world's largest irrigation systems, encompasses 12,676 km, 33,884 km and 122,268 km length of canals at Primary, Secondary and Tertiary levels, respectively [1]. Along with a number of operational problems in IBIS, high rate of seepage occurs while conveyance of approximately 0.123 Million Hectare Meter (MHM) annually [2-6]. Due to excess seepage losses through canals along a lot of agricultural land water table has turned into within one meter depth. Such a serious scenario has resulted into considerable decline in yields of all crops except that of rice [7]. Furthermore, joint adverse effect of water logging and consequent salinity is more substantial than that of water logging only [7]. Hence, it is stated that in Pakistan there is large and cost-effective capacity to enhance water deliveries to agricultural farms by reducing conveyance losses while flow through canal system [4,8].

As a usual trend along river banks, irrigated agriculture indicated the era of development of human civilization in Sindh also. Early agriculture involving mainly food production changed slowly to modern agriculture through a continuous evolution of agriculture technologies. The present irrigation system in Sindh depends on three barrages namely Sukkur, Kotri and Guddu constructed in 1932, 1955 and 1962 respectively. Except Akram wah (canal) off taking from Kotri Barrage, all other canals are earthen canals. Large amount of irrigation water is lost from these canals in the form of seepage from banks and bed. It is estimated that 40 to 50 per cent of water is lost between the canal head works to the croplands [9].

Prabhata et al. [10] recommend that canals of new irrigation project in arid or semi-arid regions must be lined to avoid chances of abundant seepage in case of dry soil and deep groundwater table. Lining of canals is shown a good solution to this problem worldwide. Lining of canals in Sindh is a great matter as canals will need to be closed long enough to and farmers may lose enough crop production which is imperative for their livelihood and nobody is ready to give them subsidy of this loss. The irrigation application rates within the farms are also high because of reliance on the conventional flood irrigation. With the passage of time, water as a commodity is becoming more and more precious. Above all it is a finite source. This high percentage of wastage, therefore, cannot be afforded for long time. Wastage of water through poor infrastructure or poor water management constitutes a major issue related to the water resources of Pakistan.

Another aspect of this issue is the productivity of the farms against per cusec of irrigation water. Pakistan has a much lower rate of production. The irrigation efficiency, therefore, needs to be enhanced. In view of the huge (around 45 to 50 per cent) losses of irrigation water between canal-heads and the farm gate, water conservation should be accorded a high priority. Lining of Canals and water courses should be taken in hand more vigorously.

2. Dadu Canal

2.1. Water Availability

Dadu canal has been designed for a normal discharge of 89.2 cumec at the head. Main crops are cotton, sugarcane, wheat, rice, fruits and vegetables. Water availability of the Dadu Canal is highly variable due to stochastic nature of flow. Average annual availability on the basis of 5-year canal diversion at the canal head and at farm gate is 0.174 MHM and 0.092 MHM respectively, which shows a significant water loss of 0.082 MHM [11].

2.2. Crop Water Requirements

The net crop water requirements of 0.4, 0.85, 1.8, 2.0 and 0.5 m were used for wheat, cotton, rice, sugarcane and fodder crops respectively [9]. These five major crops cover more than 80% of the total cropped area of the canal and consume 81% of total water needed for total crop consumptive requirement of the cropped area. This includes: wheat (12.8% of total), cotton (15.9% of total), rice (28.3% of total), sugarcane (15% of the total) and fodder (6.8% of the total). Total water requirements for Kharif crops and Rabi crops are 176 HM (Hectare·Meter) [9].

2.3. Alignment

Total length of Dadu Canal is 212 km. The alignment of Dadu Canal has a regime section. It originates from the Sukkur barrage, along right bank of river Indus, passes through the districts of Sukkur, Larkana and Dadu running parallel to the Indus highway N55. Main cities/

towns near the canal are Bagarji, Madeji, Naudero, Larkana, Dokri, Badah village, Mehar, Radhan village, Sita road village and Dadu. The canal is accessible through the district road bridges (DRBs) and village road bridges (VRBs) from most of the above mentioned towns. Moreover, there are many hydraulic structures and road bridges on canal which provide interconnectivity to both the embankments of the canal. However at some places there can be constraints for passage of larger vehicles. Topography of the canal sites is characterized by flat lands, waterlogged in most of the reaches. Basic parameters are given in **Table 1**.

3. Lining and Remodeling

According to Streeter [12] the prime parameters of hydraulic efficiency, practicability, and economy should be considered while using uniform flow formula to design a lined canal. The lining of an existing canal invariably involves remodeling of the section. The reasons are to maintain flow levels and command due to difference in Manning's coefficient between unlined and lined material to improve deteriorated canal sections before proceeding with lining and to provide a firm sub-base for application of lining.

Different lining materials are listed by Sharma and Chawla [13]. Amongst various conventional types of lining Cement concrete lining are considered as most suitable because of its long life, durability, structural stability, resistance to erosion, high permissible velocity, etc. Besides the high construction cost of Cement concrete lining, there occurs some seepage due to expansion/contraction of joints. Cement concrete lining, the most commonly used type of lining, consists a layer of cement concrete placed on a well prepared and compacted sub-base. The recommended thickness of lining as per IS: 3873-19192 is given in **Table 2**.

4. Design Methodology

The design methodology involves design of cross section

Table 1. Basic parameters of Dadu canal.

Sr.No.	Parameters	Description
1	Type	Perennial
2	Length (km)	212
3	Gross Command Area (hectares)	222.967
4	CCA (Hectares)	201.810
5	Design Discharge (cumec)	89.2
6	Actual Discharge (cumec)	124
8	No. of Channels	112
9	No. of Outlets	2145

Source: operation and maintenance manual, ipd, government of Sindh.

Table 2. Thickness of cement concrete lining.

Canal discharge (cumec)		Water depth (m)		Lining thickness (cm)	
Up to 5.0		Up to 1.0		5	
5.0	50.0	1.0	2.5	5	7.6
50.0	199.8	2.5	4.5	7.6	10.2
199.8	299.6	4.5	6.6	10.2	12
299.6	699.4	6.6	9.1	12	15.2

at the locations where changes in the hydraulic conditions occur, e.g. at cross regulators/fall structures.

4.1. Design Discharge

Dadu Canal was originally designed in 1932 for a discharge of 89.2 cumec (**Table 3**). Over the years, the flows have been increased to 124 cumec and the estimated data show that the design discharges were proposed up to 152 cumec (**Table 4**).

To fix appropriate value of discharge in Dadu canal, data were collected from following two sources:

1) Sindh water Resources Development and Management Investment Program-Program Report, "Irrigation".

2) Prepared Design Statement of Dadu Main Canal (**Table 5**).

To avoid discrepancies in these data cumulative dis-

Table 3. Design statement of selected reaches of Dadu canal (Design Data, 1932).

S. No	Name of Regulator	Name of Channel	MRD	Discharge (cumec)	Bed Width B (m)	Depth D (m)	Mean Velocity V (m/sec)
1	212th km X.R	d/s	1002.4		5.49	1.07	0.57
		u/s			8.84	1.52	
	(Tail Regulator)	Bhambha Distry	1002.4	4.93			
		Daim Branch	1002.4	11.04			
		Bilawal pur Distry	1002.4	1.95			
		Dubi Minor	952.5	0.47			
		Lower Form Minor	937.3	0.80			
		Dircet Outlets (D.Os)		0.82			
				20.10			
2	198th km X.R	d/s	932.4		9.30	1.52	0.57
		u/s			11.28	1.83	0.65
		Phaka Distry	932.4	5.65			
		Upper Form Minor	932.4	1.35			
		Jahir Minor	932.4	2.62			
		Lower Noor Wah Minor	883.9	1.41			
		Direct Outlets (D.Os)		1.14			
				12.17			
3	175th km X.R	d/s	829.9		12.19	1.83	0.66
		u/s			13.87	1.98	0.70
		Upper Noor Wah Minor	829.9	1.14			
		Pir Gunio Disty	829.9	7.00			
		Pat Minor	795.4	0.69			
		Pukan Minor	795.4	0.43			
		Werang Channel	784.9	0.36			
		Direct Outlets (D.Os)		3.35			
				12.97			

Table 4. Proposed design statement of dadu main canal.

S. No.	Name of Regulator	Name of Channel	MRD	Q (cumec)	H.G. 1 in	Q_{XR} (cumec)	Qp (cumec)	B (m)	FSD D (m)	A (m²)	V_m (m/sec)	Qc (cumec)
1	212th km X.R	d/s	1002.42			10.45	20.02	13.72	1.98	31.22	0.65	20.27
		u/s										
	(Tail Regulator)	Bhambha Distry	1002.42	4.93								
		Daim Branch	1002.42	11.04								
		Bilawal pur Distry	1002.42	1.95								
		Dubi Minor	952.51	0.47								
		Lower Form Minor	937.27	0.80								
		Direct Outlets (D.Os)		0.82								
				20.10								
2	198th km X.R	d/s	932.44		9448	10.53	20.02	13.72	1.98	31.22	0.65	20.27
		u/s				20.56	32.93	17.07	2.44	47.28	0.70	33.27
		Phaka Distry	932.44	5.65								
		Upper Form Minor	932.44	1.35								
		Jahir Minor	932.44	2.62								
		Lower Noor Wah Minor	883.93	1.41								
		Direct Outlets (D.Os)		1.14								
				12.17								
3	175th km X.R	d/s	829.91		8484	20.81	32.93	17.07	2.44	47.28	0.70	33.27
		u/s				36.84	47.28	22.10	2.59	63.91	0.75	47.91
		Upper Noor Wah Minor	829.91	1.14								
		Pir Gunio Disty	829.91	7.00								
		Pat Minor	795.43	0.69								
		Pukan Minor	795.43	0.43								
		Werang Channel	784.87	0.36								
		Direct Outlets (D.Os)		3.35								
				12.97								

Q = Discharge to distributary; B = Bed Width; A = Flow Area; Vm = Mean Velocity; Qc = Calculated Discharge; QXR = Discharge at Cross Regulator; FSD = Full Supply Depth; N8 H.G = Hydraulic Gradient

charge of the Dadu Canal at Sukkur Headwork is calculated (**Table 6**) and compared with the headwork discharge as provided in the data. The selection for the design discharge is made from the most relevant data as compared to the headwork discharge. An additional flow of 15% is included in the design discharge to accommodate future increase or any intermittent operational requirements.

The details of these data are presented in **Table 7**. The data at serial number 1 above, sum up to only 76 cumec, showing a short coming of 50% in the total discharge of

152 cumec as given in the **Table 5**. The design discharges from data at serial number 2, show 130 cumec after summation of discharges from each minor/distributary and 152 cumec listed separately as "calculated discharge".

The design discharges are based on the calculations made in the Design Statement of Dadu Main Canal (**Table 3**).

4.2. Section Design

Following the usual approach adopted by various inves

Table 5. Prepared design statement of Dadu canal w.r.to existing data.

S. No.	Name of Regulator	Name of Channel	MRD	Discharge (cumec)	H.G. 1 in	Discharge (cumec)	Bed width B (m)	Depth (m)
1	212th km X.R	d/s	1002.4			3.09	6.0	3.8
		u/s				8.98		
	(Tail Regulator)	Bhambha Distry	1002.4	4.93				
		Daim Branch	1002.4	11.04				
		Bilawal pur Distry	1002.4	1.95				
		Dubi Minor	952.5	0.47				
		Lower Form Minor	937.3	0.80				
		Dircet Outlets (D.Os)		0.82				
				20.10				
2	198th km X.R	d/s	932.4		10000	9.43	7.1	3.8
		u/s				14.78	8.1	4.3
		Phaka Distry	932.4	5.65				
		Upper Form Minor	932.4	1.35				
		Jahir Minor	932.4	2.62				
		Lower Noor Wah Minor	883.9	1.41				
		Direct Outlets (D.Os)		1.14				
				12.17				
3	175th km X.R	d/s	829.9		10000	16.79	8.7	4.3
		u/s				22.03	12.8	6
		Upper Noor Wah Minor	829.9	1.14				
		Pir Gunio Disty	829.9	7.00				
		Pat Minor	795.4	0.69				
		Pukan Minor	795.4	0.43				
		Werang Channel	784.9	0.36				
		Direct Outlets (D.Os)		3.35				
				12.97				

tigators [14-16], in this case the section design for lining option is also based on Manning's formula:

$$V = \frac{1.486}{n} R^{2/3} S^{1/2}$$

where V is velocity of flow (ft/sec); n is Manning's coefficient; S is longitudinal hydraulic slope or canal bed slope (ft per ft); R is hydraulic radius (ft) = A/P; A is area of the flow section (ft^2), and P is wetted perimeter (ft).

4.3. Manning's Coefficient

Different characteristics, to make a lined canal section hydraulically most economical, suggested by Chow [17], French [18] and Guo and Hughes [19], include permeability, Manning's coefficient, durability, cost of construction and maintenance, etc. The permeability of material determines absorption losses and the Manning's coefficient determines the carrying capacity of the channel. Weathering is caused by disruptive action of temperature variation, alternate freezing and thawing, and wetting and drying. The alkali soil causes corrosion of concrete and this can be prevented by the application of sulphate resistant cement (SRC). Cost of construction would vary with the locality and the availability of various materials. The thickness of lined section would depend upon the lining material, the side slope and the existence of hy

Table 6. Comparative statement of dadu canal.

S. No.	Name of Regulator	Name of Channel	MRD	Discharge (cumec)			Bed Width (m)			Depth (m)			Mean Velocity (m)		
				Des.	Exis.	Prop.	Des.	Exis.	Prop.	Des.	Exis.	Prop.	Des.	Exis.	Prop.
1	212th km X.R	d/s	1002.41	3.09	20.27		5.49	12.19	13.72	1.07	1.16	1.98	0.57	0.10	0.65
		u/s		8.98			8.84	14.63		1.52					
	(Tail Regulator)	Bhambha Distry	1002.41	4.93											
		Daim Branch	1002.41	11.04											
		Bilawal pur Distry	1002.41	1.95											
		Dubi Minor	952.50	0.47											
		Lower Form Minor	937.26	0.80											
		Dircet Outlets (D.Os)		0.82											
				20.10											
2	198th km X.R	d/s	932.43	9.43	20.27	9.30	14.63	13.72	1.52	1.16	1.98	0.57	0.28	0.65	
		u/s		14.78	33.27	11.28	19.20	17.07	1.83	1.31	2.44	0.65	0.34	0.70	
		Phaka Distry	932.43	5.65											
		Upper Form Minor	932.43	1.35											
		Jahir Minor	932.43	2.62											
		Lower Noor Wah Minor	883.92	1.41											
		Direct Outlets (D.Os)		1.14											
				12.17											
3	175th km X.R	d/s	829.90	16.79	33.27	12.19	21.64	17.07	1.83	1.31	2.44	0.66	0.36	0.70	
		u/s		22.03	47.91	13.87	24.08	22.10	1.98	1.83	2.59	0.70	0.23	0.75	
		Upper Noor Wah Minor	829.90	1.14											
		Pir Gunio Disty	829.90	7.00											
		Pat Minor	795.42	0.69											
		Pukan Minor	795.42	0.43											
		Werang Channel	784.86	0.36											
		Direct Outlets (D.Os)		3.35											
				12.97											

dro-static pressure. There should be proper drainage system to prevent failure due to back pressure.

Concrete lining is durable and, if laid properly, the absorption losses are reduced by 95%. The Manning's coefficient is low and in view of high permissible velocities, the section is reduced. The construction is carried out in panels and grooves are provided to prevent cracks due to shrinkage and alternate expansion and contraction. Oil paper, crude oil, 1:6 cement plaster or 1:4 cement sand slurry are used at the top of sub grade to avoid its becoming spongy and permeable.

4.4. Canal Section Properties

The remodeling of canal section imposes certain restric-

tion on the selection of various dimensions; the last designed section dictating most of the values like longitudinal slope, full supply depth and full supply levels. These parameters and the values adopted considering the above points are detailed below.

4.5. Depth of Flow

The excessive widening of the existing canal section provides a possibility of selecting a larger bed width with a reduced water depth. In some reaches the silting of canal bed compensates for the reduction in the depth with raised bed levels. The resulting lowering of full supply level, is therefore not so different when compared with full supply levels of last design. A slight reduction in full

Table 7. Proposed Discharges shown at Barrage Headworks, Cross Regulators and Tail Regulator.

RD	Discharge (Cusecs)	At Downstream of Structure
0 + 000	5378	Sukkur Barrage Headworks
19 + 597	5265	39 mile Naudero Cross Regulator
250 + 600	5063	51 mile Pir Sher Cross Regulator
283 + 405	4482	57 mile Pandabad Cross Regulator
331 + 992	4497	67 mile Sonahri Cross Regulator
383 + 460	4127	77 mile Tatri Cross Regulator
420 + 534	3717	95 mile Waha Cross Regulator
473 + 630	3564	95 mile Cross Regulator
546 + 000	1692	109 mile Cross Regulator
611 + 830	1175	123 mile Cross Regulator
657 + 750	716	132 mile Tail Regulator

supply level has been allowed in the design to reduce the extent of the works required for remodeling of the section. The head ponds created upstream of cross regulators provides control of the required full supply levels for feeding into minors and distributaries. The effects of lowering in the full supply depth, at locations away from the head ponds, will be checked to ensure supplies from direct outlets.

The design with reduced flow depth for same flows, results in increased wetted perimeter and a subsequent increase in the cost of expensive lining. The changes in full supply depth are therefore kept to a minimum. The bed widths are selected between the maximum available due to widening of canals and minimum required for maintaining the desired supply depth.

4.6. Side Slopes

Side slope is provided on the consideration of angle of repose of the bank material as it results in a stable bank under dry conditions. The following side slopes are adopted for the lining under dry condition option and the embankment material (sand fill) used in the remodeling:

1) Concrete lined, new parallel canal 2 H:1V

(Utilizing minimum right of way, achieve stability under drawdown conditions and provide ease in construction)

2) Unlined diversion channel In cutting 1 H:1 V
In filling 1.5 H:1 V

4.7. Canal Longitudinal Slope

A lined canal can have steeper longitudinal slopes (L-slopes) than unlined canals, as higher velocities which may cause erosion of unlined canals, are not a problem.

The existing canals have a large number of cross regulators with combination of fall structures, having upstream and downstream controlling levels. The existing bed levels suggest uneven L-slope as well as reverse slopes in small reaches due to uneven silting. The silting is mainly limited to upper reaches and scouring at lower reaches. In reaches having scouring in almost the entire reach, the same L-slopes are adopted as that of last design, and the bed levels are also kept the same.

4.8. Freeboard

Considering the size and discharge of the canal, the value of free board may be roughly taken as one-sixth of flow depth. Recommended values of freeboard are as follows

Discharge (cumec)	Free Board (m)
450 to 255	0.91
255 to 170	0.84
<170	0.76

The freeboard provided in the design includes 80% of the above values for the freeboard plus calculated depths required for passing additional 15% flows. The resulting freeboard is then checked for the depths required during various stages of remodeling, allowing 50% of the freeboard for any such possible increase.

4.9. Velocity of Flow

Since, higher permissible velocities reduce silt deposition in the bed of lined canal and all the silt is passed on to the lower reaches and into the distributaries and minors, it necessitates lining of minors and distributaries to avoid rising of bed levels which may affect the command of the irrigation network.

To limit this phenomenon to a minimum there is no drastic increase in the existing velocities in the remodeled section. The present velocities of around 1.0 m/sec are generally increased by 15% to 25%, to remain below 2.2 m/sec. However in some upper reaches with heavy silting of canal bed, the longitudinal slope has been increased to minimize the required remodeling work. In such reaches, up to 40% increase in the existing flow velocities are allowed with maximum velocity not to exceed 1.4 m/sec.

The velocity of flows during remodeling is also checked. Velocities greater than 1.1 m/sec are not considered suitable for remodeled section with available nearby material or graded soil, before application of lining.

5. Hydraulic Design Worksheets

The design for remodeled sections downstream of cross regulator or fall structures is done through MS Excel worksheet developed for this purpose. There are three main reaches of Dadu Canal first at "123rd Mile X.R", second at "85th Mile X.R", and third at "57th Mile X.R". The worksheet includes design for the plain concrete lining with recommended value of "n" as 0.017.

5.1. Design Procedure

The design worksheet of selected reaches, as shown in **Table 4**, is calculated adopting following procedure.

For the assumed value of bottom width of channel section, discharge (Q) is calculated using continuity equation. The computed discharge is compared with required discharge (Q_{req}); if both discharges are equal it means that the design of selected section is correct and if not then revise the bottom width.

5.2. Checking of Design Parameters

The checking of designed worksheet parameters (*i.e.* Bed width, Flow area, Wetted perimeter, Hydraulic radius, and Flow velocity) has been done by using "Flow Master Software (2009)". One of the trial sections and final remodeled sections computed by the software are shown in **Figures 1** and **2**, respectively. These hydraulic design parameters are compared with those at design stage and existing data as shown in **Table 6**.

6. Quantity and Cost Estimation

Preliminary cost estimates have been prepared for the

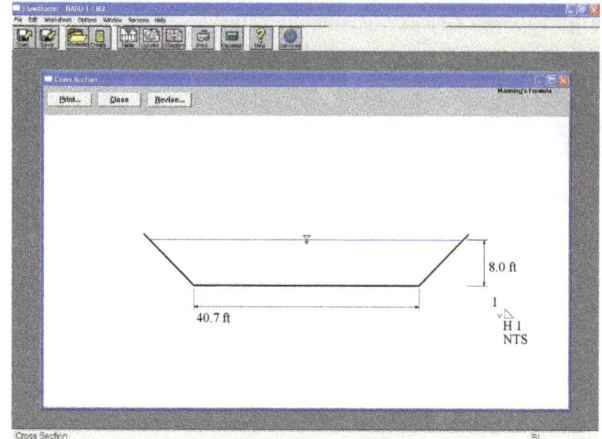

Figure 1. The simplest version of the designed cross section by Flow Master.

Plain cement concrete lining. For this purpose the quantity of cutting (trimming), backfilling at the sides of the canal and quantity of concrete lining perimeter are worked out. The thickness of 13 cm is taken for estimateing the concrete quantity having strength of 3000 psi in selected reaches.

6.1. Quantities

The quantities of the selected reaches up to MRDs 457, 503, 655, 686, 953 and 983 along the length of canal are worked out in MS EXCEL worksheet as shown in **Table 8**.

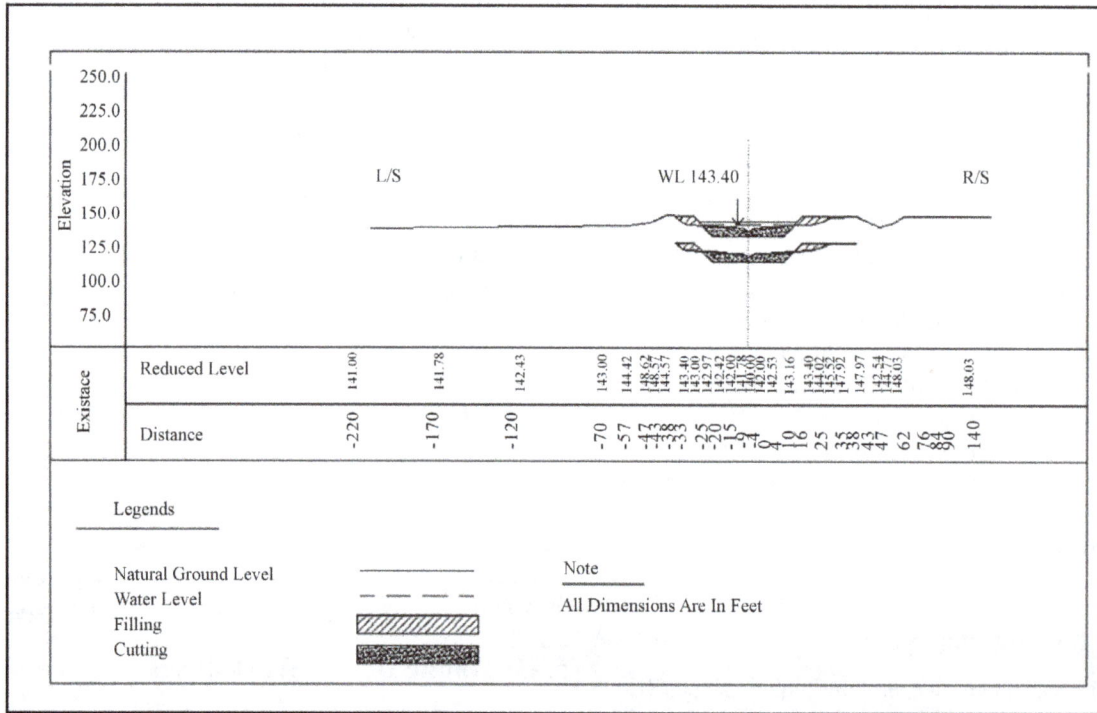

Figure 2. Remodeled cross-section by flow master with complete design.

6.2. Cost Estimation

The cost of the selected reaches as mentioned above are worked out in MS EXCEL worksheet in Millions Rupees as shown in **Table 9**.

7. Results and Discussion

Lining Dadu canal decreases seepage losses from 40% to 50%, consequently water logging becomes negligible. Conveyance efficiency increases from 70% to 90% resulting into significant increase in cropping intensity. Smaller width of lined section increases flow velocity, and that in turn, will reduce silting in canal reach. Lined canal reduces maintenance cost for longer period and improves its performance. Sections designed by MS-EXCEL worksheet are well-matching with those by Flow Master Software, e.g. at 123rd Mile cross regulator the section obtained (shown in **Figure 2**) is same as that obtained through MS-EXCEL worksheet (as shown in **Table 6**). The quantity estimation of the material is done very precisely by using AutoCAD.

8. Conclusion

As a result of proposed lining of the Dadu canal, seepage losses, water logging, silting and maintenance cost of canal can be significantly decreased, consequently, flow velocity, conveyance efficiency and cropping intensity can be increased. The initial investments over canal lining seem to be very high, but canal lining is a sustainable step which proves to be very economical in terms of long term benefits and for a country like Pakistan; it is very necessary to conserve water for its future.

9. Recommendations

All the canals at primary, secondary and tertiary level must be lined step-by-step so as to prevent huge water losses and make the system more and more efficient.

10. Acknowledgements

Authors thank to International Water Management Institute (IWMI) Pakistan and particularly appreciate Dr.

Table 8. Quantity estimation of selected sections of Dadu Canal.

Design Reach	Calculated X-Sections			Quantity for Cut		Quantity for Fill		Quantity for Concrete		
	Selected MRD		Length	At x-section	In total length of section	At x-section	In total length of section	Lined perimeter	At x-section	In total length
	From	To	(m)	(m²)	(m³)	(m²)	(m³)	(m)	(m²)	(m³)
At 92nd km Cross Regulator, length starts from MRD 431.91	431.91	457.21	5058.16	23.28	117,750.80	72.59	367,157.85	52.48	6.67	33,713.75
	457.21	502.93	9144.00	30.00	274,283.60	54.25	496,065.23	52.48	6.67	60,946.81
At 137th km Cross Regulator, length starts from MRD 640.98	640.90	655.33	2885.24	32.89	94,905.73	39.26	113,270.45	45.73	5.81	16,755.07
	655.33	685.81	6096.00	30.09	183,443.93	19.42	118,396.00	45.73	5.81	35,400.53
At 198th km Cross Regulator, length starts from MRD 932.43	932.44	952.51	4014.22	22.27	89,407.34	10.23	41,045.48	23.10	2.93	11,775.71
	952.51	982.99	6096.00	16.02	97,666.29	8.91	54,337.79	23.10	2.93	17,882.63
Summation			33,293.61	154.56	857,457.69	204.66	1,190,272.79		30.81	176,474.50

Table 9. Cost Estimation of Dadu Canal.

Item No:	Item Description	Total Calculated Quantity	Rate	Total Amounts	
		(m³)	(Rs/Unit)	(Rs)	(Millions)
1	Excavation or cutting(triming) of sand or clay (earth material) at the bed of canal to maintain bed slope	857,458	700	600,220,375.6	600,220,376
2	Providing and placing of plain cement concrete of 3000 Psi strenght using Sulphate resistance cement for 5 inches thick Canal lining	176,474	20,000	3,529,489,955	3529,489,96
3	Backfilling of berms of canal sections using the earth material (either excavated or borrowed)	1,190,273	1500	1,785,409,191	1,785,409,19
			Total	5,915,119,523	5,915,119,52

G.V. Skogerboe, the Ex-Director, IWMI-Pak, for providing experimentation facilities.

REFERENCES

[1] M. N. Bhutta and N. Ahmad, "Impact of Watercourse Lining plus Measures on Reducing Drainage Requirements," International Waterlogging and Salinity Research Institute, Lahore, 2006.

[2] WAPDA, "Pakistan Water Sector Strategy," Vol. 5, Government of Pakistan, Ministry of Water and Power, Lahore, 2002.

[3] A. R. Ghumman, M. A. K. Tarrar and A. A. Tahir, "Investigation of Optimal Use of Canal Water in Pakistan," *Proceedings of the International Conference on Water Resources and Arid Environment*, 2004, pp. 1-8.

[4] IDWR, "Augmenting Water Resources: Water in Rajasthan," *Report of the Expert Committee on Integrated Development of Water Resources*, 2005.

[5] M. Arshad, Q. Zaman and A. Madani, "Lining Impact on Water Losses in Watercourses: A Case Study in Indus Basin, Pakistan," *Annual Conference of the Canadian Society for Bioengineering*, North Vancouver, 13-16 July 2008, pp. 1-12.

[6] S. M. S. Shah, Z. M. Maan and M. K. Sarwar, "Impact of the Alternative Lining of Water Course on Cost and Efficiency," *Science, Technology and Development*, Vol. 30, No. 4, 2011, pp. 31-38.

[7] M. A. Kahlown and M. Azam, "Individual and Combined Effect of Waterlogging and Salinity on Crop Yields in the Indus Basin," *Journal of International Commission on Irrigation and Drainage*, Vol. 51, No. 4, 2002, pp. 329-338. doi:10.1002/ird.62

[8] M. A. Kahlown and W. D. Kemper, "Reducing Water Losses from Channels Using Linings: Costs and Benefits in Pakistan," *Agricultural Water Management*, Vol. 74, No. 1, 2005, pp. 57-76. doi:10.1016/j.agwat.2004.09.016

[9] I. Hussain, F. Mariker and W. Jehangir, "Productivity and Performance of Irrigated Wheat Farms across Canal Commands in Lower Indus Basin," International Water Management Institute, Research Report 44, Colombo, 2000.

[10] K. Prabhata, P. K. Swamee, G. C. Mishra and B. R. Chahar, "Minimum Cost Design of Lined Canal Sections," *Water Resources Management*, Vol. 14, No. 1, 2000, pp. 1-12, Netherlands. doi:10.1023/A:1008198602337

[11] "On Farm Water Management Field Manuals," Vol. I, II, and III, *Ministry of Food, Agriculture and Cooperative, Government of Pakistan, Mid-Year Review, July-December* 2008, Planning & Development Department, Government of Sindh, pp. 1-6.

[12] V. L. Streeter, "Economical Canal Cross Sections," *Transactions of ASCE*, Vol. 110, 1945, pp. 421-430.

[13] H. D. Sharma and A. S. Chawla, "Manual of Canal Lining," *Technical Report No.* 14, Central Board of Irrigation and Power, New Delhi, 1975.

[14] M. A. Iqbal, M. Raoof and M. Hanif, "Impact of Waterlogging on Major Crop Yields: A Case Study in Southern Punjab," *Journal of Drainage Water Management*, Vol. 5, No. 2, 2002, pp. 1-7.

[15] P. Monadjemi, "General Formulation of Best Hydraulic Channel Section," *Journal of Irrigation and Drainage Engineering* (*ASCE*), Vol. 120, No. 1, 1994, pp. 27-35.

[16] P. K. Swamee, "Discussion on General Formulation of Best Hydraulic Channel Section," *Journal of Irrigation and Drainage Engineering* (*ASCE*), Vol. 121, No. 2, 1995, p. 222. doi:10.1061/(ASCE)0733-9437(1995)121:2(222)

[17] V. T. Chow, "Open Channel Hydraulics," McGraw Hill Book Co. Inc., New York, 1973.

[18] R. H. French, "Open Channel Hydraulics," McGraw Hill Book Co. Inc., New York, 1994.

[19] C. Y. Guo and W. C. Hughes, "Optimal Channel Cross Section with Free Board," *Journal of Irrigation and Drainage Engineering* (*ASCE*), Vol. 110, No. 3, 1984, pp. 304-313. doi:10.1061/(ASCE)0733-9437(1984)110:3(304)

[20] P. K. Swamee and K. G. Bhatia, "Economic Open Channel Section," *Journal of Irrigation Power* (*CBI&P, New-Delhi*), Vol. 29, No. 2, 1972, pp. 169-176.

23

Prediction of Water Logging Using Analytical Solutions—A Case Study of Kalisindh Chambal River Linking Canal

Dipak N. Kongre, Rohit Goyal
Civil Engineering Department, Malaviya National Institute of Technology, Jaipur, India
Email: dnkongre@gmail.com, rgoyal_jp@yahoo.com

ABSTRACT

The canals are designed to transport water to meet irrigation and other water demands or to divert water from surplus basins to deficient basins to meet the ever increasing water demands. Though the positives of canal network are increase in agricultural output and improvement in quality of life, the negatives of canal introduction and irrigation, along its route, are inherent problems of water logging and salinity due to seepage from canals and the irrigation, when not managed properly. To plan strategies to prevent waterlogging and salinity, it is necessary to predict, in advance, the probable area which would be affected due to seepage. This paper presents a methodology to predict the area prone to water logging due to seepage from canal by using 2D seepage solutions to 3D field problem. The available analytical solutions for seepage from canals founded on pervious medium and asymmetrically placed drains, have been utilized. The area, prone to waterlogging, has been mapped using GIS.

Keywords: Waterlogging; Seepage; Analytical Solutions; GIS; 3D Analyst

1. Introduction

Growing need of food and fibre for increasing population needs more agricultural land to be irrigated, necessitating the transportation of water through canals from the reservoirs, wherever possible. In India there is a large variation in the rainfall in both time and space leading to scenario of droughts and floods simultaneously. To overcome this challenge of droughts and floods and to provide water to the water deficient regions, the Government of India has muted river interlinking project. The Parbati Kalisindh Chambal river interlinking is a part of peninsular river interlinking project. The surplus water of Parbati and Kalisindh basins is to be diverted to meet the demand in upper Chambal basin [1-4]. The construction of this link would surely benefit the recipients of additional water but the problems of water logging and salinity due to seepage from canals and irrigation need consideration for planning preventive measures. In the northwest part of Rajasthan state, canal irrigation was introduced after commissioning Indira Gandhi NaharPariyojna (IGNP) to irrigate nearly 2.2 million ha of arid land. It has increased the food production but it also introduced waterlogging and secondary salinization problems [5,6]. The size of the waterlogged area in the year 1998 was 17,220 ha in stage I of the project and 800 ha in stage II

of the IGNP project [7]. The major reasons for water logging and salinization in this area are indiscriminate use of irrigation water, canal seepage, sandy texture and absence of natural surface drainage [8]. The same factors have led to rise in water table in the north-west region of Haryana, India and water logging problems. About 500,000 ha area is waterlogged and unproductive [9,10]. It is estimated that in India nearly 8.4 million ha is affected by soil salinity and alkalinity, of which about 5.5 million ha is also waterlogged [11]. In the states of Bihar, Gujarat, Madhya Pradesh, Jammu & Kashmir, Karnataka, Kerala, Maharashtra, Odisha and Uttar Pradesh 117.808 thousand hectare waterlogged area has been approved for reclamation under command area development program by Ministry of Water Resources [12]. The total waterlogged area in canal command in India, in the year 1996, was around 2.189 million ha [13]. More than 33% of the world's irrigated land is affected by secondary salinization and/or waterlogging [14]. These statistics clearly show the need to manage canal seepage and irrigation to reduce water logging and secondary salinization.

Canal linings are used to reduce seepage. Yao et al. [15] studied the effect of canal lining and multilayered soil system on canal seepage and found that the combination of canal lining and a low-permeability layer below the canal is effective in reducing seepage. Though perfect canal lining can prevent seepage loss, but cracks can develop in the lining, and the performance of the canal lining deteriorates with time [16]. Wachyan and Ruston [17] studied several canals and concluded that significant seepage losses occur even in a well maintained canal with good lining.

Seepage from canals can be calculated by physical, empirical and mathematical (analytical and numerical) techniques [18-21]. Empirical formulae and graphical solutions are generally used to estimate seepage losses from proposed canals whereas the direct measurements such as inflow-outflow method, ponding method, seepage meter method are used to evaluate seepage from existing canals [22,23]. Various empirical formulae for estimating seepage are discussed by Bakri and Awad [24]. The seepage losses from irrigation canals with different lining materials, subsurface flow and subsurface storage along the canal, channel longitudinal slope, have been estimated by different researchers using direct measurements or electrical resistivity [25-28]. Analytical and electrical analogy solutions of seepage problems related to irrigation canals have been presented by many authors. Zhukovsky was the first to introduce the method for solution of problems involving unconfined seepage using a function which is now well known as Zhukovsky's function. Vedernikov gave an exact mathematical solution to unconfined, steady-state seepage from a triangular and a trapezoidal canal in a homogeneous, isotropic, porous medium of large depth [29]. Vedernikov [30] solved the problem of seepage from a canal to the symmetrically placed collector drainages, neglecting the effect of the canal water depth and side slopes. Many Russian investigators analysed solutions for rectangular, triangular channel of zero depth for different boundary conditions [31,32]. The solution for seepage problem for a rectangular canal was provided by Morel-Seytoux [33]. Sharma and Chawla [34] presented solution of the problem of seepage from a canal to vertical and horizontal drainages, symmetrically located at finite distances from the canal, in a homogenous medium extending up to a finite depth. The water depth in the canal was assumed negligible in comparison to the width. Exact solution of the problem of seepage from a canal in a homogeneous medium to asymmetric drains located at finite distance from the canal was provided by Wolde-Kirkos and Chawla [35]. Goyal [36] provided solutions for seepage from canals founded on pervious soil with asymmetric drainages. Algorithm to solve the nonlinear integral equations to obtain the value of seepage discharges and profile of the free surface was developed. The computer program developed for solutions generates the coordinates of seepage profile and estimates the seepage quantities. Ilyinsky et al. [37] have carried out a comprehensive review of analytical solutions for seepage problems and observed that though numerical techniques have become more significant in solving practical problems of seepage theory but analytical methods are necessary not only to develop and test the numerical algorithms but also to gain a deeper understanding of the underlying physics, as well as for the parametric analysis of complex flow patterns and the optimization and estimation of the properties of seepage fields. Swamee et al. [38] obtained an analytical solution for seepage from a rectangular canal in a soil layer of finite depth overlying a drainage layer using inversion of hodograph and conformal mapping technique. Bardet and Tobita [39] presented finite difference approach for calculating unconfined seepage using spread-sheets. They derived the finite difference equations using flux conservation in the general case of non-uniform and anisotropic permeability and boundary conditions. The flow lines and free surfaces can be obtained using their method but the method cannot be adopted when systems of equations become large.

Sharma and Shakya [40] applied the Bousinessq equation using the Laplace transform and Fourier cosine transform in determining the phreatic surface elevation in horizontal unconfined aquifers along the canal sides. An exact analytical solution for the quantity of seepage from a trapezoidal channel underlain by a drainage layer at a shallow depth obtained by using an inverse hodograph and a Schwarz-Christoffel transformation was presented by Chahar [41]. It provides a set of parametric equations for the location of phreatic line. To reduce the conveyance losses due to seepage, many researchers have provided methods to design the canals for minimum seepage [20,38,42,43]. Ayvaz and Karahan [44] provided spreadsheet application of three-dimensional (3D) seepage modeling with an unknown free surface. They derived governing equation using finite differences method in the general case of anisotropic and non-uniform material properties and variable grid spacing and also modified it by the extended pressure method. Only one finite difference equation was applied to the solution domain instead of derivation of additional finite difference equation to impervious boundary conditions, inclined interfaces, etc. Solution to seepage under the dam has been provided using this method. Ahmed and Bazaraa [45] investigated the problem of seepage under the floor of hydraulic structures considering the compartment of flow that seeps through the surrounding banks of the canal. A computer program, utilizing a finite-element method and capable of handling (3D) saturated-unsaturated flow problems, was used. The results produced from the two-dimensional (2D) analysis were observed to deviate largely from that obtained from 3D analysis of the same problem, despite the fact that the porous medium was isotropic and homogeneous. These solutions may be suitable for a dam with finite length but may not prove suitable when it is applied to canals which are longer in length.

Though, lot of literature about solution to the seepage problem and to estimate seepage is available, no case study about the application of these solutions to map the area, susceptible to waterlogging, has been reported.

The main objectives of this paper are to use available analytical solution to estimate the probable waterlogged area and to develop methodology for adopting 2D solutions to 3D field problem.

2. Analytical Solution

The analytical solution for the seepage problem has been provided by Goyal [36] for the seepage from canals with negligible water depth, founded on finite pervious media and asymmetrically placed drains on the either sides. It has been done using conformal mapping and defining free surface using Zhukovsky's function. Integral equations were obtained by using Zhukovsky's function and Schwarz-Christoffel transformation. It is assumed that the soil medium below the canal is homogeneous and isotropic. The seepage flow is assumed to be steady and furthermore the water depth in the canal and drainages and the level of surface does not fluctuate with time. The soil within the seepage domain is saturated. It is also assumed that there is no infiltration/evaporation in the seepage domain under consideration. Problem of seepage profile has been defined in **Figure 1**.

Where, B is the base width of the canal, h_1 and h_2 are the difference of levels between free surface level in the canal and the drains on the right and left hand sides respectively. L_1 and L_2 are the horizontal distances of the drains from the edge of the canal and T is the depth of impervious layer/bed rock below the canal bed.

To solve the equations for seepage profiles and the seepage losses at a cross section, using this analytical solution, the dimensionless parameters L_1/h_1, L_2/h_1, B/h_1, h_2/h_1 and T/h_1 at that section are required. These are determined from the known values of L_1, L_2, B, T, h_1 and h_2. The user interactive computer program written in C++, on providing the input parameters, calculates profiles of

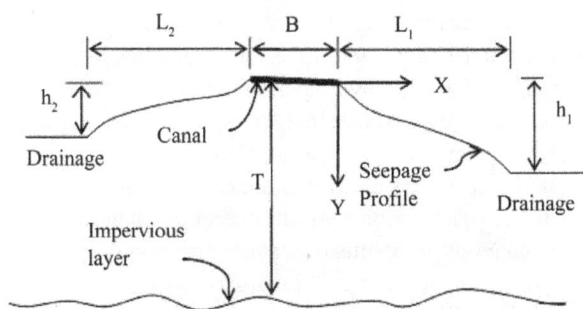

Figure 1. Definition sketch of the seepage problem.

the phreatic surfaces on the left and right sides of the canal and the seepage discharge for that cross section.

The solution obtained is only for a 2D section under consideration. The field problem being a 3-D problem, it is required that the 2D solutions are obtained at different sections along the alignment. Therefore the third dimension of the canal, *i.e.* the length, is discretized by taking sections at regular intervals along its length. The solutions obtained for each section can then be combined to get the solution for a problem in 3D space.

Geographical Information System (GIS) is a very efficient tool for spatial data management and analysis. Mapping the canal, the drains and the sections and then extracting the required data of L_1, L_2, B, T, $h_1$and h_2, for each section, for calculations of input parameters can be efficiently done using GIS. Its versatility in spatial interpolation can be utilized to interpolate the 2-D solutions at each section to get a 3-D solution and then to map the waterlogged area.

3. Study Area

The Parbati-Kalisindh-Chambal (PKC) river interlinking project is one of the projects proposed by National Water Development Agency (NWDA), India [1,2]. The study area, surrounding the Kalisindh-Chambal link canal, lies between 23°29'N and 25°0'N latitudes, and 75°20'E and 76°17'E longitudes. It is bounded by the river Chambal and its tributaries on the west and south west, River Kalisindh and its tributaries forms eastern and south eastern boundary. In its feasibility report, NWDA has prepared feasibility for two of the proposed alternative routes. The alignment of canal between Parbati and Kalisindh rivers is same in both routes. The remaining alignment differs only while joining rivers Kalisindh and Chambal. The two proposed alternative routes are Joining storage dam at Kundaliya on the river Kalisindh with full reservoir level (FRL) of 370 m to:

1) Rana Pratap Sagar dam on river Chambal with FRL of 352.81 m by a 108 Km long gravity canal and 5.25 Km long tunnel, or

2) Gandhi Sagar, with FRL of 399.89 m, on river Chambal involving gravity canal of 76 Km and 20 Km pipelines and intermediate reservoirs with lift of about 50 m. This requires a net power of about 20 MW/annum [1].

Out of the above two, the first alternative requiring no power and maximum length of open channel has been considered for the seepage analysis. The part of the PKC project between Parbati river and Kalisindh river as well as the second route between Kalisindh and Chambal rivers which to a large part coincides with first route and thereafter consists of small reservoirs and tunnels is not being considered for analysis. The alignment has been detailed by NWDA in Technical report in 2004. The alignment passes through the area covered in the Survey of India Toposheets numbered 45P, 46M, 54D and 55A. It mainly passes through Shajapur district of Madhya Pradesh, Jhalawar, Kota and Chittaurgarh districts of Rajasthan. Kalisindh-Chambal link and various districts are shown in **Figure 2**. The hydrogeological fault line on northern side between river Chambal downstream of Rana pratap Sagar and river Kalisindh closes the study area. The study area is drained by rivers Kalisindh and its tributaries, rivers Ahu and Kanthili, Amajar and Takli. These rivers are also shown in **Figure 2**.

The study area is characterized by high hills on south west side and is sloping towards north east. The digital elevation map of the area and the alignment of proposed link are as shown in **Figure 3**. The highest and the lowest elevation in the area are 561 m and 187 m above mean sea level.

4. Methodology

The analytical solution of the seepage problem, as defined in **Figure 1** and presented by Goyal (1994), needs the distances of the drains from the edge of the canal, the thickness of porous media up to the impervious layer and the elevation differences between the canal bed and the drains to define the boundary conditions. To extract this information the canal alignment was mapped using the drawings in the Technical Study under feasibility report of PKC link project. The rivers, viz., Kalisindh, Chambal and their tributaries forming the boundary and the rivers, viz., Ahu and Kanthili, Amajar and Takli draining the area have been mapped using existing maps of the area.

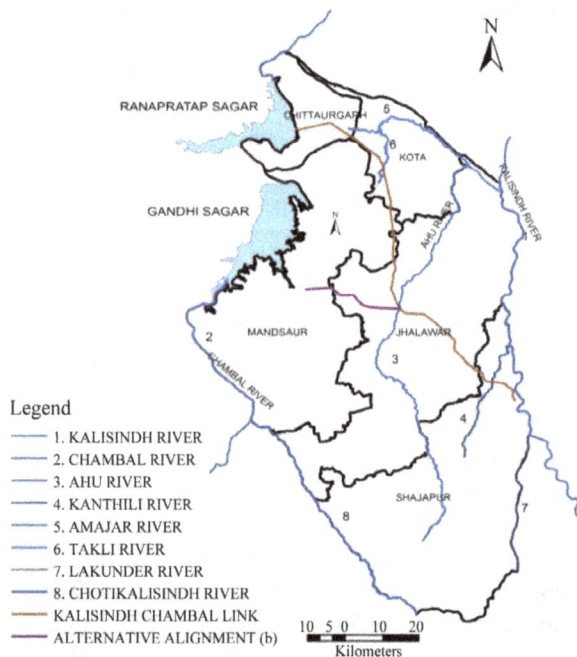

Figure 3. DEM of study area.

The cross sections have been marked at every half kilometer along the alignment. The cross sections start from the drain/reservoir on the left side of alignment cross the alignment and end at the drain on its right side. The Kalisindh-Chambal link alignment can be taken as made of two straight segments neglecting minor deviations. The cross sections have been marked in two sets. In each set, the cross sections have been kept parallel so that solution of one cross section doesn't interfere with the solution of other adjacent section. The sample cross sections are also shown in **Figure 3**. It is assumed that at the change of direction in alignment the sections which intersect with other sections don't have impact on the final solutions.

The area has complex geology and has alluvium, weathered to compact sandstone, limestone, shale and basalt at different depths. The hydrogeological units were idealized by combining similar nature of layers as single layer and conceptual hydrogeologic model was developed. The compact shale has hydraulic conductivity less than $1/10^{th}$ of other formations and therefore shale layer has been taken as impermeable layer. The borehole records of Jhalawar, Kota and Chittaurgarh districts of Rajasthan [46], the details of various hydrogeological formations at piezometer locations in Madhya Pradesh and

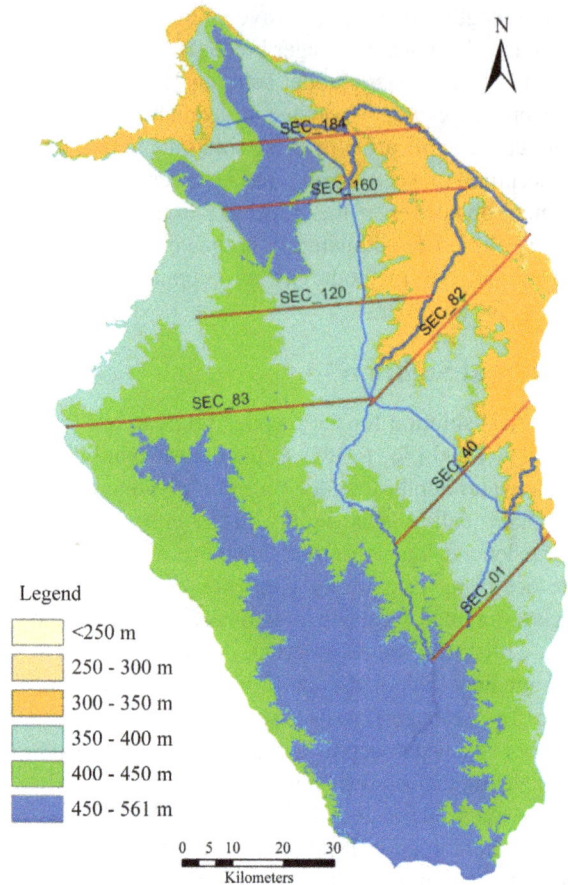

Figure 2. Study area.

the district resource maps of all the districts in the area have been used to create layers of hydro geological formation. The average depth of impervious shale layer from the canal bed for each section was extracted using spatial analyst tool and 3D feature of cross section, in GIS environment. The given analytical solution provides the seepage profile coordinates with origin at the edge of the canal bed. The elevations of various points along the section are required to determine the difference of elevations between the ground level and the seepage profile. These can be extracted from digital elevation models of the area. The ASTER dataset being the latest Global Digital Elevation Model (GDEM) has been used in the study. The GDEM is produced with 30 meter postings and has Z accuracies generally between 10 m and 25 m root mean square error [47]. The ASTER dataset referring to latitudes and longitudes of the study area was downloaded from the ASTER web site. The mosaic of downloaded elevation datasets was made in GIS to get the digital elevation model for the area. The DEM of the area using ASTER data set is shown in **Figure 3**.

The elevations of the ground level and the impervious layer along a section can be extracted in ArcGIS by creating 3D features of the sections which are 2D features. The 3D features were created using 3D Analyst tools in ArcGIS and were then used to get the profiles of ground surface. ArcGIS provides tools to create profile of a particular layer along a line of interest. This profile can be obtained as a graphical representation of distance versus elevation or can be exported as table in excel format. This technique was used to extract the information regarding elevations of ground surface and impervious surface for each section. The sample profile of ground at cross section no. 40 is shown in **Figure 4**.

The distance along the cross section from its starting point is given on X axis and the elevations are shown on Y axis. The vertical scale is exaggerated. The data set exported in excels format gives the elevations of points approximately at 28 to 30 m interval. The distances of

drains, on left and right sides, from the edge of the canal *i.e.* L_1 and L_2 were calculated by using geographical coordinates (in meters) of the points at the intersections of the cross sections with the drains and the canal. The difference of elevations between the canal bed and the drains, h_1 and h_2, have been computed by using the elevations of canal bed and left and right end points of the sections. The elevations of the bed of the canal at each section, along the alignment were calculated using information in the Technical Report of the proposals. The depth of impervious layer from the canal bed, T, at each section was calculated by finding the difference of elevation of canal bed and average elevation of impervious layer at that section. The data of L_1, L_2, h_1, h_2 and T for each section were tabulated for calculation of input parameters. The details of L_1, L_2, h_1 and h_2 for some sample cross sections are presented in **Table 1**.

In the case of Kalisindh Chambal Link, it is evident from the DEM given in **Figure 3** that the elevations of the drains on the left side are higher than the elevation of canal bed at almost all the cross sections. Therefore the value of h_1 works out to be negative. Since the program developed does not work for the negative h_1 and h_2 values, it was not possible to use the program for calculating profile on both sides. It was found that the depth of left hand did not make any significant difference in profile of right hand side and vice-versa and therefore a uniform depth of 1 m for left hand side was artificially chosen to obtain the profile of only right hand side. Profile obtained for left hand side was ignored. **Table 2** shows the details of calculated input parameters for some sample cross sections.

The input text file for all the sections was prepared to run the program. The computer program developed for the solution of seepage problem can be run in two modes, *i.e.* obtaining result on screen for single cross section or for multiple cross sections using input and output text files. The program outputs present the coordinates of

Profile_Section 40

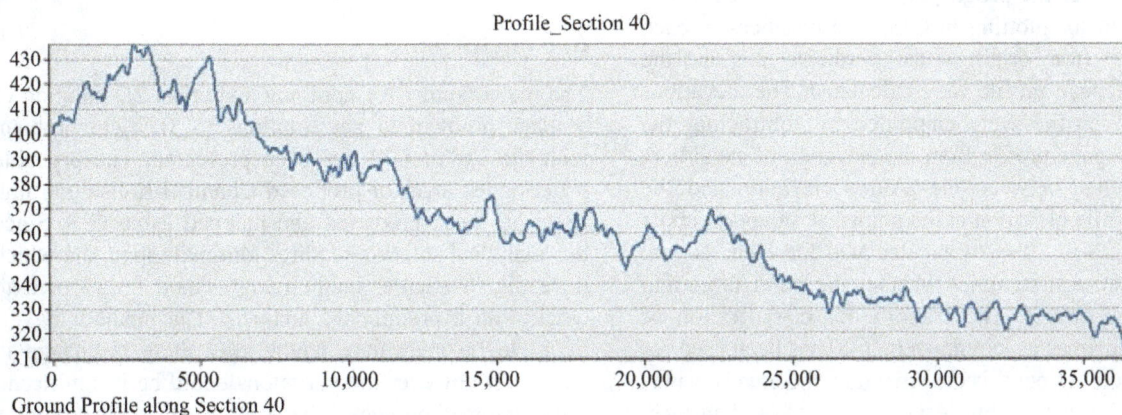

Ground Profile along Section 40

Figure 4. Sample profile at cross-section no. 40.

Table 1. Selected cross-section details.

Section No.	L_1, L_2 in m	Elevation of canal bed in m	Elevation of left & right drain in m	h_1, h_2 in m
1	30,455 1893	364.7	451.0 330.0	−86.3 34.7
50	19,727 24,185	362.0	396.0 301.0	−34.0 61.0
150	21,408 17,066	358.1	398.5 315.0	−40.5 43.1
184	17,135 8171	356.9	352.8 332.0	4.0 24.9

Table 2. Calculated input parameters.

Section No.	L_1/h_1	L_2/h_1	B/h_1	h_2/h_1	T/h_1
1	54.6	877.7	0.231	0.029	1.03
10	129.9	736.7	0.198	0.025	1.21
30	344.6	535.8	0.192	0.024	1.75
50	396.6	323.5	0.131	0.016	1.26
70	490.0	258.3	0.103	0.013	0.86
90	110.0	2841.2	0.410	0.051	1.90
110	150.5	422.7	0.116	0.015	0.68
130	441.2	677.6	0.228	0.029	1.12
150	396.0	496.7	0.186	0.023	0.68
170	366.3	555.4	0.144	0.018	0.53
184	328.8	689.4	0.322	0.163	1.28

L_1/h_1	150.47	Left free surface	
L_2/h_1	422.68	x/L_2	y/h_2
B/h_1	0.12	−1.00028	1
h_2/h_1	0.01	−0.990284	0.992401
T/h_1	0.68	−0.980284	0.985305
q/kh_1	0.004507	.	.
q_1/kh_1	0.004496	.	.
		−0.0202839	0.304121
		−0.0102839	0.297026
Right free surface		−0.000283903	0
x/L_1	y/h_1		
0	0	Impervious	Layer
0.01	0.013201	x/L_1	y/h_1
0.02	0.02315	−5.61893	0.690019
0.03	0.033098	−5.05711	0.689958
.	.	−4.4953	0.690033
.	.		
0.98	0.978121	.	.
0.989999	0.988068	0.79992	1.47901
1	1	0.89996	1.57853
		1.2	1.68

Figure 5. Sample output.

m and safe if it is below 3 m.

Layer of points, at approximately 28 to 30 m spacing along every section, with difference of elevation of ground and elevation of seepage profile as attribute, was created in GIS. The difference between elevation of ground and elevation of seepage profile was spatially interpolated. The raster output was then reclassified as per the MOWR classification of water logged area.

5. Result

The results show that the total area which is prone to water logging is around 256 Sq. Km out of the area of 1846 Sq. Km which lies between the right side of the canal and the boundary. The probable area susceptible to water logging is shown in **Figure 6** and the area in square kilometers lying under each category is tabulated in **Table 3**.

6. Conclusion

The methodology for applying analytical 2D solution of seepage problem to get solutions to 3D field problem with the use of GIS has been presented. The probable waterlogged area for proposed Chambal-Kalishindh link canal has been predicted and mapped using GIS. Since the analytical solution requires minimal input data which is readily available, methodology could be adopted to quickly do a preliminary analysis. The methodology is likely to be inaccurate where ever there is change in alignment however the variations could be insignificant. Using the methodology it has been predicted that an area of 214.36 Sq. Km is likely to be waterlogged and 42.07

seepage profile, with origin at the edge of canal, in terms of non dimensional ratios x/L and y/h. Where x and y are the horizontal and vertical distance of the seepage profile form the origin, L is length of cross section on one side of the canal and h is the difference of elevation of canal bed and the drain on that side. The program also calculates the rate of seepage. The output of seepage profile coordinates, required for plotting water logged area, was used for further processing. Sample output for one set of input parameters is shown in **Figure 5**.

The output of the program was post processed to obtain the data for plotting in GIS. The distances at each output point from origin of its coordinate axis and the depth of seepage profile were calculated. The elevations of seepage profile were computed by subtracting the depth of seepage profile from the elevation of canal bed. The difference between the ground elevation and the seepage profile elevation at the points, at an approximate spacing of 28 to 30 m, was calculated for each section. This difference determines whether the point is getting waterlogged or not. The criterion for classification of waterlogged areas as laid down by [48] has been used for classification. As per this classification, the area is waterlogged, if the water table is within 2 m of land surface, potential area for waterlogging, when it is between 2 to 3

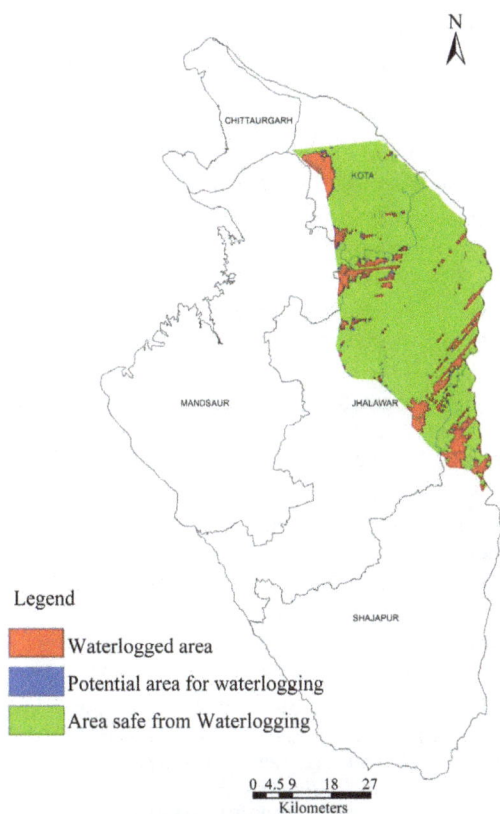

Figure 6. Waterlogged area.

Table 3. Waterlogged area (Sq. Km).

District, state	Area falling in study area, area analyzed for waterlogging	MOWR (Govt. of India) classification		
		Water logged area	Potential area for waterlogging	Safe area
Jhalawar, Rajasthan	2838.58, 1178.49	110.25	24.18	1044.06
Kota, Rajasthan	740.77, 478.98	45.38	9.54	424.06
Chittaurgarh, Rajasthan	485.63, 0.00	0.00	0.00	0.00
Mandsaur, M.P.	2439.01, 20.56	7.00	1.62	11.94
Shajapur, M.P.	2500.72, 168.17	51.72	6.73	109.72
Total	9004.71, 1846.20	214.36	42.07	1589.77

Sq. Km is likely to become potentially waterlogged if the Chambal-Kalisindh link canal project is adopted. However it is important to compare the result with another solution, such as mathematical groundwater model.

REFERENCES

[1] National Water Development Agency (NWDA), "Feasibility Report of ParbatiKalisindh Chambal Link Project," 2004. http://www. nwda. gov.in

[2] C. D. Thatte, "Inter-Basin Water Transfer for Augmentation of Water Resources in India—A Review of Needs, Plans, Status and Prospects," 2006.

http://hdr.undp.org/en/reports/global/hdr2006/papers/cdth atte_interbasin_watertransfer_india. pdf

[3] J. Bandopadhyay and S. Parveen, "The Interlinking of Indian Rivers: Some Questions on Scientific, Economic and Environmental Dimensions of the Proposal," 2008. http://www.soas.ac.uk/water/publications/papers/file3840 3. pdf

[4] N. Pasi and R. Smardon, "Interlinking of Rivers: A Solution for Water Crisis in India or a Decision in Doubt?" *The Journal of Science Policy and Governance*, Vol. 2, No. 1, 2013, pp. 1-42. http://www.Sciencepolicyjournal.org/current-edition.html

[5] A. N. Arora and R. Goyal, "Groundwater Model of Waterlogged Area of Indira Gandhi NaharPariyojna, Stage I," *ISH Journal of Hydraulic Engineering*, Vol. 18, No. 1, 2012, pp. 64-76.

[6] A. K. Mandal and R. C. Sharma, "Delineation and Characterization of Waterlogging and Salt Affected Areas in A Canal Irrigated Semiarid Region of North West India," *Geocarto International*, Vol. 23, No. 3, 2008, pp. 181-195.

[7] K. D. Sharma, "Indira Gandhi Nahar Pariyojana—Lessons Learnt from Past Management Practices in the Indian Arid Zone Regional Management of Water Resources," *Proceedings of a Symposium Held during the 6th IAHS Scientific Assembly at Maastricht*, IAHS Publications, No. 268, July 2001, pp. 49-55.

[8] H. S. Shankarnarayana and V. K. Gupta, "Soils of the Region," In: J. Venkateswarulu and I. P. Abrol, Eds., *Prospect of Indira Gandhi Canal Project*, ICAR, New Delhi, 1991, pp. 19-35.

[9] K. K. Dutta and C. de Jong, "Adverse Effects of Waterlogging and Soil Salinity on Crop and Land Productivity in North West Region of Haryana," *Agricultural Water Management*, Vol. 57, No. 3, 2002, pp. 223-238 doi:10.1016/S0378-3774(02)00058-6

[10] A. Singh, S. N. Panda, W. A. Flugel and P. Krause, "Waterlogging and Farmland Salinisation: Causes and Remedial Measures in an Irrigated Semi-Arid Region of India," 2012.

[11] H. P. Ritzemaa, T. V. Satyanarayanab, S. Ramanc and J. Boonstraa, "Subsurface Drainage to Combat Waterlogging and Salinity in Irrigated Lands in India: Lessons Learned in Farmers' Fields," *Agricultural Water Management*, Vol. 95, No. 3, 2008, pp. 179-189. doi:10.1016/j.agwat.2007.09.012

[12] Anonymous, "Annual Report 2011-12," Ministry of Water Resources, Government of India, New Delhi, 2012, p. 4.

[13] N. K. Tyagi, "Salinity Management: The CSSRI Experience and Future Research Agenda," *Proceedings of the Jubilee Symposium* (25-26 *November* 1996) *at the Occasion of the Fortieth Anniversary of ILRl and Thirty-Fifth Anniversary of the ICLD*, W. B. Snellen Ed. ILRl Wageningen, April 1997, pp. 17-26. http://edepot.wur. nl/149440

[14] A. F. Heuperman, A. S. Kapoor and H. W. Denecke, "Biodrainage—Principles, Experiences and Applications," Knowledge Synthesis Report No. 6, International Pro-

gramme for Technology and Research in Irrigation and Drainage, IPTRID Secretariat, Food and Agriculture Organization of the United Nations, Rome, 2002, p. 79.

[15] L. Yao, S. Feng, X. Mao, Z. Huo, S. Kang, S. and D. A. Barry, "Coupled Effects of Canal Lining and Multi-Layered Soil Structure on Canal Seepage and Soil Water Dynamics," 2012.

[16] P. K. Swamee, G. C. Mishra and B. R. Chahar, "Design of Minimum Seepage Loss Canal Sections," *Journal of Irrigation and Drainage Engineering*, Vol. 126, No. 1, 2000, pp. 28-32.
doi:10.1061/(ASCE)0733-9437(2000)126:1(28)

[17] E. Wachyan and K. R. Rushton, "Water Losses from Irrigation Canals," *Journal of Hydrology*, Vol. 92, No. 3-4, 1987, pp. 275-288. doi:10.1016/0022-1694(87)90018-7

[18] H. Bouwer, "Theory of Seepage from Open Channel," *Advances in Hydroscience*, Vol. 5, Academic Press, New York, 1969.

[19] A. Mirnateghi and J. C. Bruch Jr., "Seepage from Canals Having Variable Shape and Partial Lining," *Journal of Hydrology*, Vol. 64, No. 1-4, 1983, pp. 239-265.
doi:10.1016/0022-1694(83)90071-9

[20] A. R. Kachimov, "Seepage Optimization for Trapezoidal Channel," *Journal of Irrigation and Drainage Engineering*, Vol. 118, No. 4, 1992, pp. 520-525.
doi:10.1061/(ASCE)0733-9437(1992)118:4(520)

[21] S. Yussuff, H. Chauhan, M. Kumar and V. Srivastava, "Transient Canal Seepage to Sloping Aquifer," *Journal of Irrigation and Drainage Engineering*, Vol. 120, No. 1, 1997, pp. 97-109.

[22] R. V. Worstell, "Estimating Seepage Losses from Canal Systems," *Journal of Irrigation and Drainage Engineering*, Vol. 102, No. IR1, 1976, pp. 949-956.

[23] A. Sarki, S. Q. Memon and M. Leghari, "Comparison of Different Methods for Computing Seepage Losses in an Earthen Watercourse," *Agricultura Tropica et Subtropica*, Vol. 41, No. 4, 2008, pp. 197-205.

[24] M. W. Bakry and A. B. E. Awad, "Practical Estimation of Seepage Losses along Earthen Canals in Egypt," *Water Resources Management*, Vol. 11, No. 3, 1997, pp. 197-206. doi:10.1023/A:1007921403857

[25] C. Santhi, R. S. Muttiah, J. G. Arnold and R. Srinivasan, "A GIS-Based Regional Planning Tool for Irrigation Demand Assessment And Savings Using Swat," *Transactions of the ASAE*, Vol. 48, No. 1, 2005, pp. 137-147.

[26] S. K. Singh, C. Lal, N. C. Shahi and C. Khan, "Estimation of Canal Seepage under Shallow Water Table Conditions," *Journal of Acdemic & Industrial Research*, Vol. 1, No. 9, 2013, pp. 571-576.

[27] H. E. M. Moghazi and E. S. Ismail, "A Study of Losses from Field Channels under Arid Region Conditions," *Irrigation Science*, Vol. 17, No. 3, 1997, pp. 105-110.
doi:10.1007/s002710050028

[28] R. H. Hotchkiss, C. B. Wingert and W. E. Kelly, "Determining Irrigation Canal Seepage with Electrical Resistivity," *Journal of Irrigation and Drainage Engineering*, Vol. 127, No. 1, 1979, 2001, pp. 20-26.

[29] M. E. Harr, "Groundwater and Seepage," McGraw Hill,

New York, 1962.

[30] V. V. Vedernikov, "Seepage Theory and Its Application in the Fields of Irrigation and Drainage," State Press, Gosstroiizdat, 1939.

[31] P. Y. Polubarinova-Kochina, "Theory of Ground Water Movement," Princeton University Press, Princeton, 1962.

[32] V. I. Aravinand S. N. Numerov, "Theory of Fluid Flow in Undeformable Porous Media," Israel Program for Scientific Translations, Jerusalem, 1965.

[33] H. J. Morel-Seytoux, "Domain Variations in Channel Seepage Flow," *Journal of the Hydraulics Division, ASCE*, Vol. 90, No. HY2, 1964, pp. 55-79.

[34] H. D. Sharma and A. S. Chawla, "Canal Seepage with Boundary at Finite Depth)," *Journal of the Hydraulics Division, ASCE*, Vol. 105, No. 7, 1979.

[35] A. T. Wolde-Kirkos and A. S. Chawla, "Seepage from Canal to Asymmetric Drainages," *Journal of Irrigation and Drainage Engineering*, Vol. 120, No. 5, 1994, pp. 949-956. doi:10.1061/(ASCE)0733-9437(1994)120:5(949)

[36] R. Goyal, "Seepage from Canals Founded on Pervious Soil with Asymmetric Drainages," Ph.D. Thesis, University of Roorkee, Roorkee, 1994.

[37] N. B. Ilyinsky, A. R. Kacimov and N. D. Yakimov, "Analytical Solutions of Seepage Theory Problems. Inverse Method, Variational Theorems, Optimization and Estimates (A Review)," *Fluid Dynamics*, Vol. 33, No. 2, 1998, pp. 157-168. doi:10.1007/BF02698697

[38] P. K. Swamee and D. Kashyap, "Design of Minimum Seepage Loss in Non-Polygonal Canal Sections," *Journal of Irrigation and Drainage Engineering*, Vol. 127, No. 2, 2001, pp. 113-117.
doi:10.1061/(ASCE)0733-9437(2001)127:2(113)

[39] J. P. Bardet and T. Tobita, "A Practical Method for Solving Free-Surface Seepage Problems," *Computers and Geotechnics*, Vol. 29, No. 6, 2002, pp. 451-475.
doi:10.1016/S0266-352X(02)00003-4

[40] S. Sharma and S. K. Shakya, "Rise of Water Table Induced by Seepage from Canal," In: V. Kumar, M. Kothari and R. C. Purohit, Eds., *37th ISAE Annual Convention, Udaipur, India, 29-30 January* 2003, *and 38th ISAE Annual Convention*, Dapoli, 16-18 January 2004, 2005, pp. 280-288.

[41] B. R. Chahar, "Analysis of Seepage from Polygon Channels," *Journal of Hydraulic Engineering*, Vol. 133, No. 4, 2007, pp. 451-460.
doi:10.1061/(ASCE)0733-9429(2007)133:4(451)

[42] P. K. Swamee, G. C. Mishra and B. R. Chahar, "Design of Minimum Water-Loss Canal Sections," *Journal of Hydraulic Research*, Vol. 40, No. 2, 2002, pp. 215-220.
doi:10.1080/00221680209499864

[43] P. K. Swamee, G. C. Mishra and B. R. Chahar, "Optimal Design of a Transmission Canal," *Journal of Irrigation and Drainage Engineering*, Vol. 128, No. 4, 2002, pp. 234-243.
doi:10.1061/(ASCE)0733-9437(2002)128:4(234)

[44] M. T. Ayvaz and H. Karahan, "Modeling Three-Dimensional Free-Surface Flows Using Multiple Spreadsheets," *Computers and Geotechnics*, Vol. 34, No. 2, 2007, pp.

112-123.

[45] A. Ahmed and A. Bazaraa, "Three-Dimensional Analysis of Seepage below and around Hydraulic Structures," *Journal of Hydraulic Engineering*, Vol. 14, No. 3, 2009, pp. 243-247. doi:10.1061/(ASCE)1084-0699(2009)14:3(243)

[46] Unpublished Basic Data Reports of Bore Logs of Jhalawar, Kota and Chittaurgarh Districts. State Ground Water Department, Government of Rajasthan.

[47] Earth Remote Sensing Data Analysis Center (ERSDAC), "ASTER GDEM," 2010. http://gdex.cr.usgs.gov/gdex/

[48] Working Group, Ministry of Water Resources, "Report on Identification in Irrigated Areas with Suggested Remedial Measures," Government of India, New Delhi, 1991.

24

The Influence of Land-Use on Water Quality in a Tropical Coastal Area: Case Study of the Keta Lagoon Complex, Ghana, West Africa

Angela M. Lamptey[1*], Patrick K. Ofori-Danson[1], Stephen Abbenney-Mickson[2],
Henrik Breuning-Madsen[3], Mark K. Abekoe[4]

[1]Department of Marine and Fisheries Sciences, University of Ghana, Legon, Ghana; [2]Department of Agriculture Engineering, Faculty of Engineering, University of Ghana, Legon, Ghana; [3]Department of Geography and Geology, University of Copenhagen, Copenhagen, Denmark; [4]Department of Soil Science, Faculty of Agriculture, University of Ghana, Legon, Ghana.
Email: [*]amlamptey@ug.edu.gh, [*]angela.lamptey@yahoo.com

ABSTRACT

The Keta Lagoon and its catchment areas in Ghana are influenced by intensive agriculture and the use of agro-chemicals. It has therefore, become necessary to assess the quality of water in the lagoon and the surrounding fresh water aquifers. In this study, a water quality index (*WQI*), indicating the water quality has been adopted. The *WQI* was determined on a basis of various physico-chemical parameters like pH, conductivity, turbidity, dissolved oxygen, calcium, magnesium, chloride, nitrates, ammonium and sodium. The index was used both for tracking changes at one site over time, and for comparisons among sites. The *WQI* was also employed to wells used for irrigation on farms along the Keta Sand Spit as well as that of the Keta Lagoon Complex and its surrounding floodplains, in order to ascertain the quality of water for public and livestock consumption, irrigation, recreation and other purposes. The *WQI* of the wells, Keta lagoon and its floodplains showed various degrees of poor water quality and therefore considered unsuitable for drinking and recreation. By WHO standards, this calls for intensive physical and chemical treatment of the water for human consumption.

Keywords: Tropical Coastal Area; Keta Lagoon Complex; Floodplains; Water Quality Index; WHO Standards; Physico-Chemical Parameters; Ghana

1. Introduction

The rapid development of sprinkler irrigation through the shallot (Allium spp.) horticulture system on the Keta sand spit (**Figure 1**) of Ghana has introduced a heavy nutrient load on the soils [1] due to the application of manure and fertilizers. This is partly because the soil is naturally very poor in plant nutrients and partly due to the truncation of nutrient when crops are harvested and sold. In the past, bat droppings and rotten fish were used as manure, but today imported cow dung and poultry droppings are the main sources of organic manure used in the farmlands [1]. Cow dung is normally applied before planting, and in every growing season about 1.3 kg/m^2 of cow dung is applied. Since the 1970s chemical fertilizers such as NPK have been applied by some farmers [1]. This has increased the transport of nutrients from run-offs to coastal waters including the Keta Lagoon Complex, and therefore the increased risk of eutrophication, thereby reducing the water quality of the lagoon and subsequently a decline in the fishery [2].

Formerly the people of this area were mainly fishermen and grew a few vegetables for domestic consumption and coconuts for sale. In the 1930s the coconut production collapsed because of the Cape St. Paul disease, and the rapidly increasing population compelled the people to give up the old agricultural systems and develop intensive horticulture based on vegetable production [1,3]. Typical vegetable production systems of the Keta area are shallots, pepper, okra, tomatoes, carrots, which are grown all year round based on irrigation with groundwater from small wells using the rope and bucket method [1]. This system is highly dependent on the application of organic manure and irrigation, the later due to the semiarid climate of the region [2,4].

---

[*]Corresponding author.

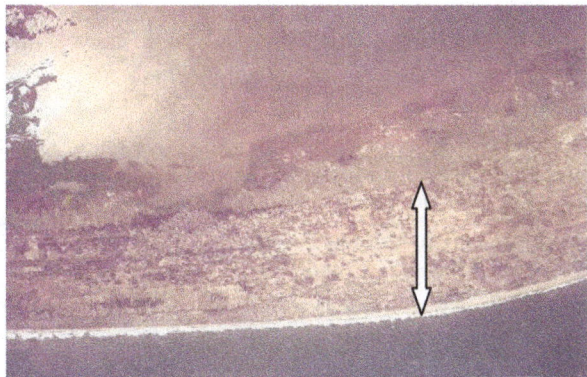

Figure 1. Satellite image of Keta Sand Spit (arrowed) (Google Maps, 2003).

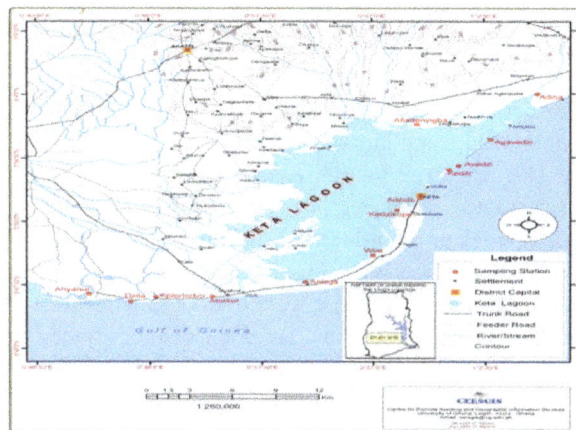

Figure 2. Map of Ghana showing the Keta Lagoon Complex and its surrounding floodplains (after CERSGIS, 2010).

The most important nutrients of concern in coastal waters are nitrates and phosphates. In excessive quantities these can cause the rapid growth of marine plants, and result in algal blooms. Sewage discharges, household and commercial waste that is carried to the sea by storm runoff, add excess nutrients to coastal waters. Detergents and fertilizers supply high quantities of nutrients to streams and rivers and ultimately the marine environment [5]. The main objectives of this study were to assess the quality of water and to develop a water quality index (*WQI*) of the Keta Lagoon and its surrounding floodplains as well as the surrounding fresh water aquifers (well water).

2. Materials and Methods

2.1. Site Description

Keta Lagoon (**Figure 2**) with co-ordinates 5°55'N 0°59'E lies in the far south-east of the country, near the international frontier with Togo. The lagoon is about 140 km east-northeast of the capital city of Accra, Ghana [6,7]. The lagoon is an extensive, brackish water-body situated to the east of the Volta river estuary. The site comprises the open water of the lagoon and the surrounding floodplains and mangrove swamps. Although considered to be an open lagoon, it is effectively closed for most of the year. The area of open water varies with the season, but is estimated to be over 164,000 ha, stretching for 40 km along the coast and separated from the sea by a narrow ridge. Inflow into the lagoon includes the Volta River at Anyanui [7] with an unknown discharge rates. The Keta Lagoon is part of the Volta system; 27 km long with a variable width of up to 16 km and a surface area of 28,400 ha and an average depth of 0.8 m (maximum 2 m) and an average salinity of 1.87‰ (18.7 PSU) [6,8]. The lagoon is bordered by numerous settlements and the surrounding flood-plain consists of marsh, scrub, farmland and substantial mangrove stands, which are heavily ex-

ploited for fuelwood. The commonest economic activities are vegetable farming and fishing [2].

From a geological point of view there are two main formations in the region around the Keta Lagoon. The northern half of the region is dominated by a tertiary formation of limonitic residuals, both consolidated and unconsolidated. The general appearance of the formation is reddish with widely spread ferrunginised sheets [6,9]. The rest of the region consists mainly of Pleistocene to recent formations of mud, clays and gravels. The shallow aquifers occur in the Recent Sediments consisting mainly of beach sand, gravels, silt and clay [9]. The Recent Sediments vary in thickness from 30 m to over 100 m [9].

The climate in the region is semi-arid tropical with average daily temperatures of 27°C - 28°C and no pronounced variation during the year. The winds on this part of coastal West Africa are generally weak and regular with predominant south-easterly winds between March and November [10]. During the Harmattan season (December-February), winds occasionally blow from the northwest. These winds follow the Inter-Tropical Convergence Zone (ITCZ) and create a seasonal pattern of rainfall with the main rainy season from April to July [6]. This part of Ghana is one of the driest parts of the country with a mean annual rainfall of 783 mm and a mean annual evaporation of 1964 mm. The relative humidity in the area is generally more than 90% during the night and early morning. During the day the humidity decreases to as low as 65% with a seasonal variation of 15% [6,9] making it different from many tropical regions.

2.2. Methodology

2.2.1. Collection of Water Samples and Analysis

The study was intended to calculate the Water Quality Index (*WQI*) of wells used for irrigation on farms along the Keta Sand Spit as well as that of the Keta Lagoon Complex and its surrounding floodplains, in order to ascertain the quality of water for public and livestock consumption, irrigation, recreation and other purposes.

Monthly field sampling was undertaken from August 2010 to August 2012 for the Keta Lagoon; from November 2010 to November 2011 for the wells. Water samples were taken from three fish-landing sites (namely Anyanui, Anloga, and Woe, **Figure 2**, in Keta) and the surrounding floodplains of the Keta Lagoon in Atorkor, Kplortorkor, Dzita, Kedzikope, Adzido, Kedzi, Afiadenyigba, Avedzi, Agavedzi and Adina were sampled every three months from February 2011 to February 2012. Water quality data sources included ten (10) physicochemical parameters such as pH, Electrical Conductivity, salinity, turbidity and dissolved oxygen which were monitored *in-situ* at the various sampling sites using VWR EC 300 and Hanna Multi-Parameter Probes; and Ionic composition such as (Na^+, NH_4^+, Ca^{2+}, K^+, Mg^{2+}, PO_4^{3-}, NH_3^-, and Cl^-). Water samples from selected wells, the lagoon, floodplains and harvested rainwater were collected and analysed for the various physico-chemical parameters by following various established procedures. All the samples were frozen and transported to the Ecological Laboratory of the University of Ghana, Legon for analyses. Before running laboratory tests, they were warmed to room temperature by allowing them to thaw, and neutralized to an approximate pH of 7.0 with 5.0 N Sodium Hydroxide standard solution. The parameters such as nitrate, chloride, ammonium, and phosphates were analysed in the laboratory following procedures described by the American Public Health Association (APHA) [11]. Other parameters such as sodium and potassium were analysed according to standard procedures established by [12,13]. In addition, calcium and magnesium were analysed according to standard procedures described by [14-17]. Statistically, the interactions between the environmental factors in explaining the distribution of the seasons and rain, wells, the lagoon and floodplains were explored using Principal Component Analysis (PCA).

2.2.2. The Water Quality Index

Water Quality Index (*WQI*) is defined as a rating reflecting the composite influence of different water quality parameters [18]. In the formulation of *WQI*, the importance of various parameters depends on the intended use of water; here, water quality parameters are studied from the point of view of suitability for public consumption. The "standards" (permissible values of various parameters) for the drinking water used in this study are those recommended by the World Health Organization [19]. When the WHO standards were not available, a

combination of drinking water standards for developed countries, [20] was applied.

Determination of the Water Quality Index (*WQI*)

The calculation and formulation of the *WQI* involved the following steps:

1) In the first step, each of the ten parameters were assigned a weight (AW_i) ranging from 1 to 4 depending on the collective expert opinions taken from different previous studies. However, a relative weight of 1 was considered as the least significant and 4 as the most significant.

2) In the second step, the relative weight (RW) was calculated by using the following equation:

$$RW = \frac{AW_i}{\sum_{i=1}^{n}(AW_i)} \quad (1) \, [18]$$

where, RW = the relative weight, AW = the assigned weight of each parameter, n = the number of parameters. The calculated relative weight (*RW*) values of each parameter are given in **Table 1**.

3) In the third step, a quality rating scale (Q_i) for all the parameter except pH and DO was assigned by dividing its concentration in each water sample by its respective standard according to the drinking water guideline recommended by the World Health Organization [19]. The result was then multiplied by 100.

$$Q_i = \frac{C_i}{S_i} * 100 \quad (2) \, [18,20]$$

While, the quality rating for pH or DO (QpH, DO) was calculated on the basis of

Table 1. Relative weight and WHO 2004 standards of the water quality parameters.

Parameter	[19]	Assigned Weight	Relative Weight
pH	6.5 - 8.5	1	0.0345
Electrical Conductivity (µS/cm)	250	3	0.1034
Turbidity (FTU)	5	3	0.1034
Dissolved Oxygen (mg/l)	5	4	0.1379
NO_4^+ (mg/l)	0.5	4	0.1379
NO_3^-	50	4	0.1379
Na^+ (mg/l)	200	4	0.1379
Cl^- (mg/l)	250	4	0.1379
Ca^{2+} (mg/l)	75	1	0.0345
Mg^{2+} (mg/l)	30	1	0.0345
		$\Sigma AW_i = 29$	$\Sigma RW_i = 0.9998$

[4,20].

$$QpH, DO = \frac{(C_i - V_i)}{(S_i - V_i)} * 100 \qquad (3) \; [18,20]$$

where, Q_i = the quality rating, C_i = value of the water quality parameter obtained from the laboratory analysis, S_i = value of the water quality parameter obtained from recommended WHO, V_i = the ideal value which is considered as 7.0 for pH and 14.6 for DO.

Equations (2) and (3) ensures that Q_i = 0 when a pollutant is totally absent in the water sample and Q_i = 100 when the value of this parameter is just equal to its permissible value. Thus the higher the value of Q_i, the more polluted the water is [21]

4) Finally, for computing the WQI, the sub-indices (SIi) were first calculated for each parameter, and then used to compute the WQI as in the following equations:

$$SI_i = RW \times Q_i \qquad (4) \; [18,20]$$

$$WQI = \sum_{i=1}^{n} SI_i \qquad (5) \; [18,20]$$

The computed WQI values was classified as <50 = Excellent; 50 - 100 = Good; 100 - 200 = Poor; 200 - 300 = Very poor; >300 = Unsuitable [18].

3. Results

Table 2 shows the mean physico-chemical parameters of various wells, the Keta lagoon and its surrounding floodplains in comparison with rain water, FAO drinking water standards and FAO irrigation water standards. pH ranged between 7.3 - 8.3, temperature was from 25.8 to 29.2°C, and phosphates ranged between 0.1 to 3.5 mg/l which all fell within WHO standards and FAO irrigation water standards, whilst all the other parameters such as electrical conductivity, nitrates, ammonium, sodium, chloride exceeded WHO and FAO irrigation water standards, except for calcium which although exceeded WHO standards, fell within FAO irrigation standards. Also, nitrate pollution poses a serious threat to the domestic use as measurements of nitrate in the experimental well showed an average value (72.8 mg/l) above the WHO Standard for drinking water of 50 mg/l [19]. Eutrophication of aquifers and water bodies are normally determined by the amount of nitrates and phosphates they receive from their surroundings. Nitrate is easily leached from the farmland while phosphate is normally retained in the soils fixed with aluminium, iron and calcium, and for that matter not easily leached. Thus phosphates are usually the limiting factors for algal blooms. The sandy soils on the Keta sand spit (farmlands) have a low content of iron and aluminium and therefore a low phosphorus-retention capacity and it is not clear whether the phosphorus-retention in some of the soils is exhausted and hereby increases the potential eutrophication of the lagoon, in the next few years [23,24].

In comparing the physico-chemical parameters to data from [22], sodium and chloride concentrations have reduced over the years. The most imminent risk is the intrusion of salt water during floods and the salinization of the aquifer that make the ground water unsuitable for irrigation. It was therefore, not uncommon that most farmers had raised the walls of most of their wells on the farmlands to prevent them from being submerged under lagoon water, during floods. In so doing, the inflow of sodium and chloride from the lagoon is curbed. The pH of the wells has become more alkaline over the years with the increase in calcium and magnesium within a period of ten years. All the water quality parameters fall within FAO Irrigation Standards, except for ammonium, nitrates and phosphates which are the main determining factors for possible eutrophication [5]. This confirms the fact that most of the water in the wells are of poor quality based on the water quality index (WQI) values. In comparison to WHO drinking water standards, all the parameters except for ammonium, nitrate and calcium of the wells, Keta Lagoon and floodplains fall within. This also supports the view that, the water quality of the wells, Keta Lagoon and its surrounding floodplains are poor and unsuitable for drinking by both livestock and humans. The water quality parameters of the rainwater all fell within the WHO and FAO water standards, therefore, it is of an excellent quality and suitable for drinking.

The results of the PCA (**Figure 3**) showed that rainwater was the least influenced by the interactions of the

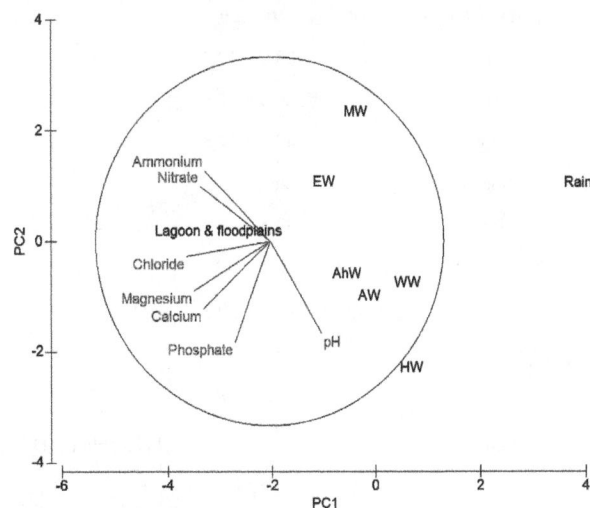

Figure 3. PCA ordination diagram for the relationship between the physico-chemical parameters of rainwater, wells, Keta Lagoon and its surrounding floodplains. Legend: PC1 = the axis which maximizes the variance of points projected perpendicularly onto it; PC2 = perpendicular to PC1 and direction in which the variance of points projected onto it is maximized.

Table 2. Comparison of water quality parameters of rainwater, the Keta Lagoon complex and floodplains, and surrounding wells obtained from August 2010 to March 2012 with WHO Standards and FAO irrigation standards.

	Keta Lagoon		
	Rain	**& floodplains**	**[22]**
pH	8.0 ± 0.6	7.3 ± 0.4	5.1 - 7.2
Temp. (°C)	25.8 ± 0.7	28.4 ± 1.0	28.5 - 29.8
EC (µS/cm)	125 ± 88	29484 ± 6276	200 - 8300
Salinity (‰)	0.1 ± 0.1	17.3 ± 3.7	No Data
NH_4^+ (mg/l)	0.78 ± 0.1	18.2 ± 8.8	No Data
NO_3^- (mg/l)	1.9 ± 1.9	41.6 ± 17.2	0 - 173
Na^+ (mg/l)	0.1 ± 0.0	7961.4 ± 1852.3	17.1 - 196.2
K^+ (mg/l)	0	179.6 ± 38.2	0.5 - 27.1
Cl^- (mg/l)	0.1 ± 0.0	9594.5 ± 2041.6	57.9 - 1,239
PO_4^{3-} (mg/l)	0.2 ± 0.2	1.5 ± 1.1	0.1 - 0.4
Ca^{2+} (mg/l)	0	181.0 ± 38.5	1.3 - 202.9
Mg^{2+} (mg/l)	0	927.9 ± 789.9	No Data
WQI	6.98	3639.08	
Status	Excellent	Unsuitable	
	MW	**EW**	**AW**
pH	7.4 ± 0.4	7.6 ± 0.4	8.0 ± 0.2
Temp. (°C)	27.8 ± 0.4	29.2 ± 0.4	28.7 ± 0.3
EC (µS/cm)	1432 ± 756	759 ± 290	1282 ± 436
Salinity (‰)	0.3 ± 0	0.4 ± 0.1	0.7 ± 0.1
NH_4^+ (mg/l)	38.9 ± 19.2	38.9 ± 19.2	8.5 ± 6.4
NO_3^- (mg/l)	36.9 ± 25.3	72.8 ± 57.6	38.7 ± 10.1
Na^+ (mg/l)	48.2 ± 14.5	30.9 ± 9.4	43.7 ± 12.1
K^+ (mg/l)	26.2 ± 8.3	19.7 ± 7.8	22.5 ± 6.9
Cl^- (mg/l)	33.5 ± 15.4	62.6 ± 34.1	67.6 ± 26.1
PO_4^{3-} (mg/l)	0.1 ± 0.1	1.6 ± 1.6	2.3 ± 2.0
Ca^{2+} (mg/l)	44.8 ± 39.7	39.3 ± 26.3	108 ± 53.5
Mg^{2+} (mg/l)	8.3 ± 7.2	17.6 ± 11.7	36.2 ± 15.3
WQI	887.02	405.93	268.55
Status	Unsuitable for Drinking		Very Poor
	HW	**AW**	**WW**
pH	8.3 ± 0.1	8.2 ± 0.2	8.3 ± 0.2
Temp. (°C)	29.2 ± 0.5	28.7 ± 0.5	28.6 ± 0.5
EC (µS/cm)	718 ± 436	784 ± 342	867 ± 449
Salinity (‰)	0.3 ± 0	0.4 ± 0.1	0.4 ± 0.1
NH_4^+ (mg/l)	1.0 ± 1.2	12.7 ± 15.9	3.6 ± 1.7

Continued

NO_3^- (mg/l)	1.59 ± 0.6	17.5 ± 10.3	14.4 ± 19.6
Na^+ (mg/l)	24.4 ± 9.9	33.3 ± 14.5	25.8 ± 10.6
K^+ (mg/l)	15.6 ± 2.1	16.9 ± 9.3	8.5 ± 4.9
Cl^- (mg/l)	26.6 ± 9.9	67.0 ± 35.0	54.7 ± 41.4
PO_4^{3-} (mg/l)	2.3 ± 1.3	3.5 ± 7.0	1.4 ± 1.5
Ca^{2+} (mg/l)	337.2 ± 125.6	58.0 ± 36.0	61.5 ± 44.2
Mg^{2+} (mg/l)	78.8 ± 57.5	18.0 ± 9.5	15.6 ± 8.3
WQI	108.52	344.41	135.33
Status	Poor	Unsuitable	Poor

Legend: NG = No Guidelines; FAO I.S = FAO Irrigation Standards; WHO = World Health Organization; MW = Meteo. Well, which is located on a farmland close to the Weather station in Anloga, has been possibly cultivated for over 100 years; EW = Experiment well, which is located on a farmland that has been cultivated for over 15 years. HW = Household well, which is located in households where there land is uncultivated, used for cleaning and drinking purposes. AW = Ahiabor well, which is located on a farmland very close to the sea (Atlantic Ocean) and cultivated for over 5 years. AW and WW = Anloga and Woe wells, which are located on farmlands that have been cultivated for over 100 years in Anloga and Woe, respectively.

various physico-chemical parameters. For PC1, WW (Woe wells) and HW (Household wells) were influenced most by the physico-chemical parameters. For PC2, the physico-chemical parameters that showed a strong positive correlation or association were ammonium and nitrate. The Keta Lagoon and its surrounding floodplains were influenced most, followed by MW and EW. In terms of the water quality of the wells, Keta Lagoon and its surrounding floodplains, the most important parameters were pH, magnesium, calcium, phosphate, nitrate, ammonium and chloride. The percentage of contributions of the various physico-chemical parameters to the water quality of wells, Keta Lagoon and its surrounding floodplains were estimated using a dominance plot (**Figure 4**). The results of the Dominance Plot (**Figure 4**) of physico-chemical parameters of rainwater, wells, the Keta Lagoon and its surrounding floodplains showed that percentage contribution of pH and temperature to the water quality of the rainwater, wells, Keta Lagoon and its floodplains was 12%, followed by phosphates with 28%, nitrate and ammonium contributed 30%, followed by calcium with 40%. The parameters that contributed most were potassium with 62%, followed by electrical conductivity and magnesium with 84%, followed by salinity with 86%, sodium and chloride contributed highest to the water quality parameters with 96%. This supports the view that, for the Keta area, the main physico-chemical parameters that affect the water quality of the fresh water aquifers (wells), the Lagoon Complex and its surrounding floodplains are salinity, electrical conductivity, magnesium, potassium, sodium and chloride.

Physico-chemical parameters are ranked in order of importance along the x-axis, and their percentage contributions to the total are plotted along the y-axis.

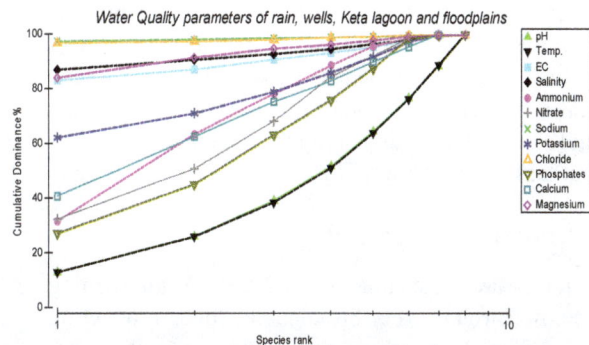

Figure 4. Dominance Plot diagram for the relationship between the physico-chemical parameters of rainwater, wells, the Keta Lagoon and its surrounding floodplains.

4. Conclusions

The major environmental impacts associated with crop farming in Anloga, Ghana result from the use of fertilizers (N: P: K) mainly nitrogen compounds, manure, to a small extent pesticides, soil erosion and storm run-offs [1]. The causes of fertilizers and pesticides loss into the environment need to be addressed, and measures taken to help reduce this particular area of environmental impact.

In this study, the *WQI* has been demonstrated as a possible tool to communicate information on the quality of water to the concerned citizens and policy makers [18]. From this study, the *WQI* of the wells in Anloga and Woe, the Keta Lagoon Complex and its surrounding floodplains provide an in-depth knowledge into the state of their water quality. The *WQI* of the wells, lagoon and floodplains currently, suggest that they are either of poor water quality or unsuitable for drinking (**Table 2**). In order to improve their quality to a drinkable status, treatment interventions include simple and intensive physical and chemical treatment and disinfection, such as rapid filtration and disinfection, pre-chlorination, decantation, filtration, adsorption (activated carbon), ozonation and final chlorination [25] are recommended. Finally, the PCA analysis (**Figure 3**) and dominance plots (**Figure 4**) suggest that in the Keta Area, the most important water quality parameters that need to be constantly monitored include pH, phosphate, calcium, magnesium, chloride, nitrate and ammonium. According to WHO organization, about 80% of all the diseases in human beings are caused by water. Once the groundwater is contaminated, its quality cannot be restored by stopping the pollutants from the source. It therefore becomes imperative to regularly monitor the quality of groundwater and to device ways and means to protect it.

In conclusion, management approaches such as water quality auditing, [23], are recommended to assess their potential use for drinking and irrigation. In addition, the increased use of computers in aquatic sciences provide opportunities for developing environmental management and decision support systems to assist farmers, aquatic biologists and environmentalists address environmental demands on water quality through sustainable land-use without loss of profitability [5].

5. Acknowledgements

The authors are most grateful to DANIDA for funding this study through the Sustainable food production through irrigated intensive farming systems in West Africa (SIFA) Project which forms part of a research project carried out in collaboration with University of Ghana and University of Copenhagen, Denmark. Many thanks go to the then Director of the Ecological Laboratory, Professor Patrick K. Ofori-Danson for making available the laboratory space and instruments for the laboratory analysis of samples as well as the technicians of the Department of Marine and Fisheries Sciences and Ecological laboratory of the University of Ghana for diverse assistance during the field and laboratory investigations.

REFERENCES

[1] T. W. Awadzi, E. Ahiabor and H. Breuning-Madsen, "The Soil-Land Use System in a Sandspit Area in the Semi-Arid Coastal Savanna Region of Ghana. Sustainability and Threats," Technical Paper, 2007.

[2] P. K. Ofori-Danson, M. Entsua-Mensah and C. A. Biney, "Monitoring of Fisheries in Five Lagoon Ramsar Sites in Ghana," Final Report to the Department of Wildlife under the Ghana Coastal Wetlands Management Project, Forestry Commission, Accra, 1999.

[3] R. D. Asiamah, "Soils of the Ho-Keta Plains, Volta Region, Ghana," Soil Research Institute, Kumasi, 1995.

[4] T. Piersma and Y. Ntiamoa-Baidu, "Waterbird Ecology and the Management of Coastal Wetlands in Ghana," NIOZ Report 1995-1996, Royal |Netherlands Institute for Sea Research (NIOZ), Texel, 1995.

[5] J. A. Skinner, K. A. Lewis, K. S. Bardon, P. Tucker, J. A. Catt and B. J. Chambers, "An Overview of the Environmental Impact of Agriculture in the UK," *Journal of Environmental Management*, Vol. 50, No. 2, 1997, pp. 111-128. http://dx.doi.org/10.1006/jema.1996.0103

[6] T. H. Sorensen, G. Volund, A. K. Armah, C. Christensen, L. B. Jensen and J. T. Pedersen, "Temporal and Spatial Variations in Concentrations of Sediment Nutrients and Carbon in the Keta Lagoon, Ghana," *West African Journal of Applied Ecology*, Vol. 4, 2003, pp. 91-105.

[7] A. K. Armah, M. Awumbila, S. Clark, A. Szietror, R. Foster-Smith, R. Porter and E. M. Young, "Coping Responses and Strategies in the Coastal Zone of South-Eastern Ghana. A Case Study in the Anloga Area," In: S. M. Evans, C. J. Vanderpuye and A. K. Armah, Eds., *The Coastal Zone of West Africa—Problems and Management*, Proceedings of an International Seminar, Accra, 23-28 March 1996, Penshaw Press, Sunderland, 1997, pp. 17-27.

[8] Anonymous, "Ramsar Library: The Directory of Wetlands of International Importance," 4th Edition, The Ramsar Convention on Wetlands, Ramsar Secretariat, Gland, 1993. www.ghanaweb.com

[9] G. O. Kesse, "The Mineral and Rock Resources of Ghana," A. A. Balkema, Rottendam/Boston, 1985.

[10] R. Arfi, D. Gurial and M. Bouvy, "Wind-Induced Resuspension in a Shallow Tropical Lagoon," *Estuarine, Coastal and Shelf Science*, Vol. 36, No. 6, 1993, pp. 587-604. http://dx.doi.org/10.1006/ecss.1993.1036

[11] APHA, "Water Analysis Handbook," 2nd Edition, HACH Company World, Washington DC, 1995.

[12] D. C. Harris, "Quantitative Chemical Analysis," 4th Edition, W. H. Freeman and Company, New York, 1995.

[13] H. B. Sawyer, "Chemistry Experiments for Instrumental Methods," Wiley, New York, 1984.

[14] International Standard Organization, "Sodium Chloride for Industrial Use—Determination of Matter Insoluble in Water or in Acid, and Preparation of Principal Solutions for Other Determinations," *ISO*, Vol. 2479, No. 1, 2011, pp. 1972-1974.

[15] US Environmental Protection Agency, "Calcium: Method 215.1 (Atomic Absorption, Direct Aspiration)," In: *Methods for Chemical Analysis of Water and Wastes*, U.S.E.P.A., Cincinnati, 1983, pp. 215.1-215.2.

[16] US Environmental Protection Agency, "Metals (Atomic Absorption Methods)—General Procedure for Analysis by Atomic Absorption," In: *Methods for Chemical Analysis of Water and Wastes*, U.S.E.P.A., Cincinnati, 1983, pp. 67-70.

[17] European Union Salt Producers Association (EuSalt), "Determination of Calcium and Magnesium. Flame Atomic Spectrophotometric Method," EuSalt, Brussels, 2005, p. 5.

[18] C. R. Ramakrishnaiah, C. Sadashivaiah and C. Ranganna, "Assessment of Water Quality Index for the Groundwater in Tumkur Taluk, Karnataka State, India," *E-Journal of Chemistry*, Vol. 6, No. 2, 2009, pp. 523-530. http://dx.doi.org/10.1155/2009/757424

[19] WHO (World Health Organization), "Guidelines for Drinking Water Quality," 3rd Edition, WHO, Geneva, 2004.

[20] A. H. M. J. Alobaidy, H. S. Abid and B. K. Maulood, "Application of Water Quality Index for Assessment of Dokan Lake Ecosystem, Kurdistan Region, Iraq," *Journal of Water Resource and Protection*, Vol. 2, No. 9, 2010, pp. 792-798. http://dx.doi.org/10.4236/jwarp.2010.29093

[21] S. K. Mohanty, "Water Quality Index of Four Religious Ponds and Its Seasonal Variation in the Temple City, Bhuvaneshwar," In: A. Kumar, Ed., *Water Pollution*, APH Publishing Corporation, New Delhi, 2004, pp. 211-218.

[22] B. K. Kortatsi, E. Young and A. Mensah-Bonsu, "Potential Impact of Large Scale Abstraction on the Quality of Shallow Groundwater for Irrigation in the Keta Strip, Ghana," *West African Journal of Applied Ecology*, Vol. 8, 2005, pp. 1-12.

[23] O. K. Borggaard, C. Szilas, A. L. Gimsing and L. H. Rasmussen, "Estimation of Soil Phosphate Adsorption Capacity by Means of a Pedotransfer Function," *Geoderma*, Vol. 118, No. 1-2, 2004, pp. 55-61. http://dx.doi.org/10.1016/S0016-7061(03)00183-6

[24] S. E. A. T. M. Van der Zee, M. M. Nederlof, W. H. van Riemsdijk and F. A. M. de Hann, "Spatial Variability of Phosphate Adsorption Parameters," *Journal of Enviromental Quality*, Vol. 17, No. 4, 1988, pp. 682-688. http://dx.doi.org/10.2134/jeq1988.00472425001700040027x

[25] EC (EUROPEAN COUNCIL), Consolidated Text Produced by the CONSLEG System of the Office for Official Publications of the European Communities. Council Directive of 16 June 1975, "Concerning the Quality Required of Surface Water Intended for the Abstraction of Drinking Water in the Member States (75/440/EEC)," Office for Official Publications of the European Communities, Luxembourg City, 1991.

25

Effect on Treatment of the Landfill Leachate with the Furrow Irrigation in Onland Planting Reed (*Phragmites*)

Kun Shi[1], Ming Zou[1], Hongxiang Cai[2]

[1]School of Environmental and Chemical Engineering, Dalian Jiaotong University, Dalian, China
[2]Extensive Center of Agricultural Technology, Dalian, China
Email: skshikun@sohu.com

ABSTRACT

The furrow irrigation tests were done to estimated the efficiency of the HRT (Hydraulic Retention Time) and landfill leachat collected from Dalian Maoyimgzi Municipal Solid Waste Landfill, which contained high level of COD (Chemical Oxygen Demand, 3.8×10^4 mg·L^{-1}), TOC (total carbon, 4.8×10^3 mg·L^{-1}), TN (total nitrogen, 2.9×10^3 mg·L^{-1}) and SS (Suspended Solids, 6.5×10^2 mg·L^{-1}), using the reed (*phragmites*) cultivated onland located in south area of Dalian Jiaotong University. The results showed that: 1) The TN concentration was decreased from 9.8×10^2 mg·L^{-1} in the landfill leachate to 7.6×10^2 mg·L^{-1} in the soil water, and the 22.4% of the removal rate; 2) The TOC concentration was decreased from 4.8×10^3 mg·L^{-1} in the landfill leachate to 1.0×10^3 mg·L^{-1} in the soil water, and the 79.2% of removed rate; 3) The water concentration in the soil was no significant difference of irrigation between the water and the landfill leachate; 4) ΔHRT was 2.1 hours in irrigation 39 L of the water and landfill leachate and 1.3 hours in the 9 L.

Keywords: Cultivated Onland; Landfill Leachate; Total Nitrogen (TN); Total Carbon (TOC)

1. Introduction

There are many project cases and researches on treating the landfill leachate with constructed wetland around the world. For example, Ithaca Landfill in New York in the USA started to treat the landfill leachate with constructed wetland in 1989 [1]. Chunchula Landfill in the city named Mobile in Alabama in the USA mixed the wastewater and landfill leachate together and treated with constructed wetland, and expelled at the standard [2]. Des Moines area in Aihua Florida in the USA treated the landfill leachate with constructed wetland, and the removal rate was significant [3]. Perdido Landfill in Escambia in Finland is running and treating the landfill leachate with constructed wetland for 13 years in cold temperature. Canada, United Kingdom, Australia and Poland and some other countries applied to the constructed wetland system process for landfill leachate [4]. Laogang Land fill in Shanghai in China has a certain scale and effective application on treating the landfill leachate with constructed wetland by the way of "Facultative anaerobic pond + Facultative anaerobic and aerated ponds + Aeration pond + Constructed Wetland" [5], and the average removal rate of COD was 66% [6]. The city named Monroe in New York in the USA treated the landfill leachate from Northeast-quadrant Solid Waste Facility

after sealing of landfill with constructed wetland, and the average removal rate of COD was 68%. Dragonja Landfill in Slovenia treated the landfill leachate with constructed wetland, and the average removal rate of COD was 68% [7]. Esval Landfill in Norway treated the landfill leachate by the way of "Oxidation pond + constructed wetland", the HRT was 40d, and the average removal rate of COD was 88% [8,9]. But there is little reach on treating the landfill leachate with the method of land fill leachate preferential flow and reflowing tests on undisturbed soil cuboids in cultivated onland.

Cultivated onland treatment of landfill leachate is a relatively new wastewater treatment technology, and was attention and gradually applied in recent decades [10-16]. Landfill leachate has a large concentration of pollutants, black, dark green and chocolate color, and complex components, so it's hard to treat by tradition ways. As investigation that the main treatment of landfill leachate in China is reverse osmosis, and it has good treatment effect, but there are running unstable, expensive, consume large and some other shortcomings [17-22]. In addition, there is no effective disposal traditional way for the production of concentrated solution, and any mistake will cause secondary pollution. Therefore, research on cultivated onland treatment of landfill leachate is of great significance. the applicable criteria that follow.

2. Materials and Methods

2.1. Materials

The landfill leachate was collected from Dalian Mao-yingzi Municipal Solid Waste Landfill, which contained high level TN (9.8×10^2 mg·L^{-1}) and TOC (4.8×10^3 mg·L^{-1}). Its color was dark brown and smelled bad. The cultivated onland located in the South Area of Dalian Jiaotong University. The main onland plant was reed (*Phragmites*) with a spacing of 0.5 m, and the roots extended about 30 cm depth under the onland surface. Preparing instruments and medicines used for analyzing the TN and TOC by international standard methods, and Marriott Bottles, polyethylene plastic films, collecting basins, tape measure, shovel, pickaxe, scraper knife, adhesive tapes, sieve, bezel and some other equipment and tools.

2.2. Methods

The schematic diagram of experiment ground can be seen from Figures. Digging the soil in between the reed of 2 ridges with long (1 m), wide (0.34 m) and depth (0.25 m), and became the furrow, and dripping the landfill leachate (**Figure 1**) and the deionized water (**Figure 2**) into the furrow with two times, irrigating the volume

Figure 1. Landfill leachate treatment of furrow irrigation.

Figure 2. Deionized water treatment of the furrow irrigation.

of the water and the landfill leachate was 39 L and 9 L, respectively. The experiment was 4 times of repeats. While the water and the landfill leachate had permeated the ground, having been taken 1 kg of the soil nearby 5 cm of the furrow were 12 samples, and N, C and water concentrations in the samples were measured by semi-micro-Kjeldahl azotometer, external heating potassium dichromate and drying and weighing method.

3. Results and Discussion

3.1. Total Nitrogen Contents in the Landfill Leachate and the Soil

Total nitrogen in the 48 L of the landfill leachate is 47.14 g, and the soil (volume $1.00 \times 0.25 \times 0.34$ m, density 2.65 g/cm^3 and background value 0.58 g/kg) is 130.65 g. It is 177.79 g of both in the landfill leachate and the soil. The total nitrogen content in the soil might be 0.79 g/kg as the theory. In 1 day after treatment, the 0.76 g/kg its content was measured (**Figure 3**), The content was very significantly different ($P < 0.01$) from the deionized water (0.53 g/kg) and background value (0.58 g/kg). The landfill leachate surely increases the nitrogen concentrations in the soil. No significant difference ($P > 0.05$) of the contents was measured between the deionized water and the background value. The TN concentration was decreased from 9.8×10^2 mg·L^{-1} in the landfill leachate to 7.6×10^2 mg·L^{-1} in the soil water, and the 22.4% of the removal rate.

3.2. Total Carbon Contents in the Landfill Leachate and the Soil

The total carbon of 48 L of the landfill leachate is 228.77 g, and the soil (volume $1.00 \times 0.25 \times 0.34$ m, density 2.65 g/cm^3 and background value 4.67 g/kg) is 1051.92 g. It is 1280.69 g of both in the landfill leathate and the soil. The total carbon content might be 5.69 g/kg as the theory. In 1 day after treatment, the 10.01 g/kg of its content was measured (**Figure 4**). The content was very significantly different ($P < 0.01$) from it in the treatment of the deionized water (7.81 g/kg) and background value (4.67 g/kg); the contents in the treatment of the deionized water was very significantly different ($P < 0.01$) from the background value. Water and landfill leachate surely increase the total carbon contents. The TOC concentration was decreased from 4.8×10^3 mg·L^{-1} in the landfill leachate to 1.0×10^3 mg·L^{-1} in the soil water, and the 79.2% of removed rate.

3.3. Water Contents and Hydraulic Retention Time

In the 1 day after the treatments, the average water contents of the background value, the deionized water and

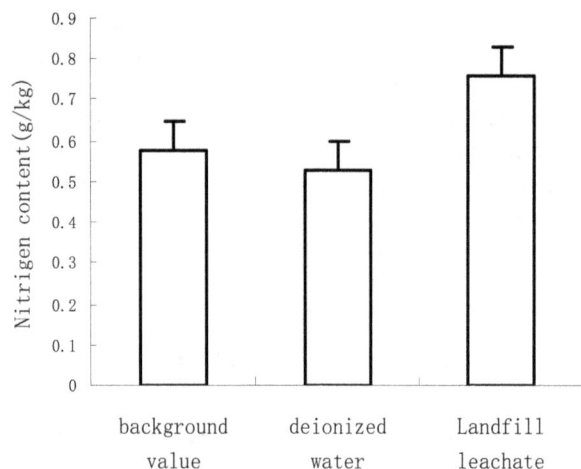

Figure 3. Total nitrogen contents.

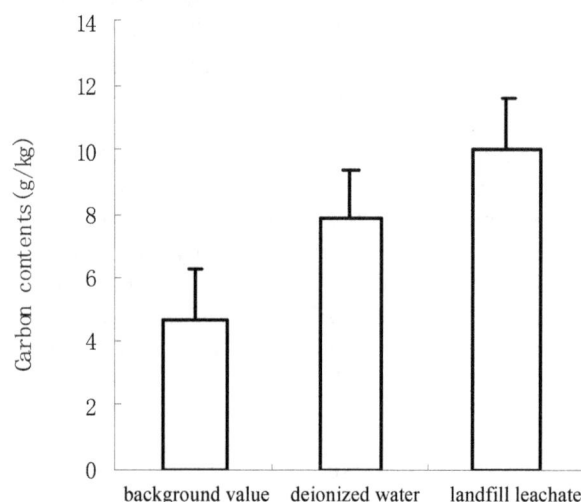

Figure 4. Total carbon contents.

the landfill leathate in the soil were 9.51%, 17.23% and 17.18%, respectively. It ranged from 7.41% to 10.71% in the soil of the background value, from 12.50% to 20.00% in soil of the deionized water and from 15.79% to 18.52% in the soil of the landfill leathate. It was no significant difference ($P > 0.05$) of irrigation between the water and the landfill leachate, and background value. First time, HRT of irrigating 39 L the deionized water was 15 hours and it of the landfill leathate was 12.9 hours; then it of the 9 L of the deionized water was 8.3 hours and the landfill leathate was 7.0 hours. ΔHRT was 2.1 hours in irrigation 39 L of the water and landfill leachate and 1.3 hours in the 9 L. To some extent, the landfill leathate increased the leakage of water and solute loss, and reduced the HRT of pollutants in soil. It was in accordance with Chen *et al*. [10]. Hydraulic load of the landfill leathate was 1.70×10^3 m^3/(ha·d) and it of the deionized water was 1.45×10^3 m^3/(ha·d), it was 16.9% higher than hydraulic load of the deionized water.

4. Conclusion

In 1 day after the treatment of furrow irrigation the landfill leathate, total nitrogen and total carbon remove 22.4 and 79.2% of efficiency rate from 9.8×10^2 and 4.8×10^3 mg·L^{-1} in the landfill leachate to 7.6×10^2 and 1.0×10^3 mg·L^{-1} in the soil water, respectively. Hydraulic load in the landfill leathate of 1.70×10^3 m^3/(ha·d) was 16.9% higher than it in the water of 1.45×10^3 m^3/(ha·d). As time goes by, the removal efficiency will be very higher than the 1 day. Furrow irrigating the landfill leathate in growing Reed onland may be a better method of treating the landfill leathate, which more and more will have being permeated by city solid waste.

REFERENCES

[1] K. D. Johnson and C. D. Martin, "Constructed Wet Lands for the Treatment of Landfill Leachate," Lewis Publishers, Chelsea, Vol. 1, 1998, pp. 63-69.

[2] D. M. Craig, D. J. Keith and A. M. Gerald, "Performance of a Constructed Wetland Landfill Leachate Treatment System at the Chunchula Landfill. Mobile County. Alabama," *Water Science and Technology*, Vol. 40, No. 3, 1999, pp. 67-74. doi:10.1016/S0273-1223(99)00453-9

[3] S. Rasmussen, "Locals Build Wetland to Ttreat Landfill Leachate," *American City and County*, Vol. 1, No. 10, 2000, p. 54.

[4] Y. Shi and X. N. Wan, "Constructed Wetland System in the Landfill Leachate Treatment," *Soil and Water Conservation*, Vol. 12, No. 1, 2005, pp. 56-58.

[5] US EPA, "Subsurface Flow Constructed Wetland for Waster Water Treatment a Technology Assessment," US EPA, Washington DC, 1993.

[6] C. Wu, "Municipal Landfill Leachate Treatment," *Water and Wastewater*, Vol. 22, No. 5, 1996, pp. 11-12.

[7] D. A. V. Eckhardt, "Constructed Wetland for the Treatment of Landfill Leachates," Lewis Publishers, Chelsea, Vol. 1, 1998, pp. 205-217.

[8] T. Maehlum, "Treatment of Landfill Leachate in On-Site Lagoons and Constructed Wetland," *Water Science and Technology*, Vol. 32, No. 3, 1995, pp. 129-135. doi:10.1016/0273-1223(95)00613-3

[9] G. W. Thomas and R. E. Philips, "Consequences of Water Movement in Macropores," *Journal of Environmental Quality*, Vol. 8, No. 2, 1979, pp. 149-152. doi:10.2134/jeq1979.00472425000800020002x

[10] C. C. Chen, R. J. Roseberg and J. S. Selker, "Using Microsprinkler Irrigation to Reduce Leaching in a Shrink/Swell Clay Soil," *Agriculture Water Management*, Vol. 54, No. 2, 2002, pp. 159-171. doi:10.1016/S0378-3774(01)00150-0

[11] L. W. Dekker and C. J. Ritsema, "Uneven Moisture Patterns in Water Repellent Soil," *Geoderma*, Vol. 70, No. 2-4, 1996, pp. 87-99. doi:10.1016/0016-7061(95)00075-5

[12] F. Hagedorn and M. Bundt, "The Age of Preferential Flow Paths," *Geoderma*, Vol. 108, No. 1-2, 2002, pp.

119-132. doi:10.1016/S0016-7061(02)00129-5

[13] R. J. Glass, T. S. Steenhuis and J. Y. Parlange, "Wetting Front Instability as a Rapid and Far Reaching Hydrologic Process in the Vadose Zone," *Journal of Contaminant Hydrology*, Vol. 3, No. 2-4, 1998, pp. 207-226. doi:10.1016/0169-7722(88)90032-0

[14] B. Gjettermann, K. L. Nielsen, C. T. Petersen, *et al.*, "Preferential Flow in Sandy Loam Soils as Affected by Irrigation Intensity," *Soil Technology*, Vol. 11, No. 2, 1997, pp. 139-152. doi:10.1016/S0933-3630(97)00001-9

[15] S. Reichenberger, W. Amelung, V. Laabs, *et al.*, "Pesticide Displacement along Preferential Flow Pathways in a Brazilian Oxisol," *Geoderma*, Vol. 110, No. 1-2, 2002, pp. 63-86. doi:10.1016/S0016-7061(02)00182-9

[16] K. Lipsius and S. J. Mooney, "Using Image Analysis of Tracer Staining to Examine the Infiltration Patterns in a Water Repellent Contaminated Sandy Soil," *Geoderma*, Vol. 136, No. 3-4, 2006, pp. 865-875. doi:10.1016/j.geoderma.2006.06.005

[17] K. Täumer, H. Stoffregen and G. Wessolek, "Seasonal Dynamics of Preferential Flow in a Water Repellent Soil," *Vadose Zone Journal*, Vol. 5, No. 1, 2006, pp.

405-411. doi:10.2136/vzj2005.0031

[18] K. J. Winter and D. Goetz, "The Impact of Sewage Composition on the Soil Clogging Phenomena of Vertical Flow Constructed Wetland," *Water Science and Technology*, Vol. 45, No. 5, 2003, pp. 9-14.

[19] C. Platzer and K. Mauch, "Soil Clogging in Vertical Flow Reed Beds-Mechanisms, Parameters, Consequences and Solutions," *Water Science and Technology*, Vol. 35, No. 5, 1997, pp. 175-181. doi:10.1016/S0273-1223(97)00066-8

[20] J. Zhang, C. F. Shao, X. Huang, *et al.*, "Land Treatment of Soil Clogging Process," *China Water & Wastewater*, Vol. 19, No. 3, 2003, pp. 17-20.

[21] X. D. Cao, "Strengthen the Pond Wetland Complex Ecological Pond System to Remove Nitrogen and Phosphorus in the Law," *Environmental Sciences*, Vol. 13, No. 2, 2000, pp. 15-19.

[22] L.-H. Cui, X.-Z. Zhu and S.-M. Luo, "Infiltration Wetland Wastewater Treatment System Technology Research," *Journal of Applied Ecology*, Vol. 14, No. 4, 2003, pp. 623-626.

26

Assessing Sprinkler Irrigation Performance Using Field Evaluations at the Medjerda Lower Valley of Tunisia

Samir Yacoubi[1], Khemaies Zayani[2], Adel Slatni[1], Enrique Playán[3]

[1]Institut National de Recherches en Génie Rural, Eaux et Forêts, University of Carthage, Tunis, Tunisia
[2]High Institute of Environmental Science and Technology, University of Carthage, Tunis, Tunisia
[3]Department Soil and Water, Estación Experimental de Aula Dei, Consejo Superior de Investigaciones Cientificas, Zaragoza, Spain
Email: yacoubi.samir@iresa.agrinet.tn, slatni.adel@iresa.agrinet.tn, khemaies.zayani@isste.rnu.tn, enrique.playan@eead.csic.es

ABSTRACT

Irrigation uniformity and wind drift and evaporation losses (WDEL) are major concerns for the design and management of sprinkler irrigation systems under arid or semi-arid conditions. Field trials were carried out to assess irrigation uniformity and WDEL under various wind velocities, sprinkler spacings and operating pressure heads. Based on experimental data, a frequency analysis was performed to infer the occurrence probability of a given uniformity coefficient (UC). In addition, statistical regressions were used to model WDEL as a function of different climatic variables. Increasing the operating pressure head improved uniformity at low wind speeds. It was shown that UC has been severely impaired at wind speeds above 4 m/s. In the prevailing wind conditions, the frequency analysis showed that a sprinkler spacing of 12 m × 12 m provided the best uniformity. In the local conditions, it is recommended to stop irrigation when wind velocity exceeds 4 m/s. Moreover, it was shown that wind speed and relative humidity were the main significant variables influencing WDEL.

Keywords: Sprinkler Irrigation; Uniformity; Wind Drift and Evaporation Losses; Wind Speed; Pressure Head

1. Introduction

Arid and semi-arid regions are characterized by an acute imbalance between rainfall and evapotranspiration. The tough competition between the water demands of agricultural, industrial and urban water sectors is becoming a serious concern [1]. It is against this backdrop that the improvement of on-farm irrigation efficiency is becoming a must: it permits to conserve water and better valorizes it. This objective may be achieved by using pressurized irrigation systems. It is worth emphasizing that sprinkler irrigation is practiced over 32% of the irrigated acreage in Tunisia. In the lower valley of Medjerda (located at northern Tunisia), near 27,000 ha of irrigated land is supplied by canals. Surface irrigation is by far the most common irrigation method in the region [2]. The high cost of switching from surface to pressurized irrigation systems currently limits the adoption of modern irrigation systems in the study area [3]. Slatni *et al.* [4], Zairi *et al.* [5] and AHT-Group/SCET [6] showed that irrigation systems in the region are characterized by low irrigation efficiency due to the lack of farmers' skills and inappropriate irrigation scheduling. A strategy for improving irrigation efficiency at the Medjerda lower valley has been approved by Tunisian public authorities [7]. Likewise, about 4500 ha are currently being converted from surface to pressurized irrigation systems.

When properly designed and managed, sprinkler irrigation produces high uniformity, reduces water supply and protects farmer's income. Pereira [8] stressed that on-farm sprinkler irrigation performance depends on wind speed, pressure head variation, nozzle diameter and shape, sprinkler layout and spacing, soil infiltration and farmer skills. Unfortunately, a limited number of studies have been devoted to assess the impact of hydraulic and climatic factors on sprinkler irrigation performance under Tunisian conditions. Field evaluations are required to calibrate sprinkler irrigation simulation models recently developed for predicting irrigation performance under various operating and environmental conditions. Therefore, field trials are essential in upgrading on-farm sprinkler irrigation design and management.

According to Keller and Bliesner [9], distribution uniformity, wind drift and evaporation losses determine sprinkler irrigation performance. Burt *et al.* [10] reported that pressure head variation at the hydrant, sprinkler design, spacing and layout, as well as climate conditions, are critical sources of heterogeneity in water application under sprinkler irrigation. Out of these, wind speed is the most uncontrollable factor, and has the greatest impact on distribution uniformity. Wind disrupts the trajectory of raindrops and induces droplet evaporation, therefore enhancing wind drift and evaporation losses (WDEL). According to Vories et al. [11], ignoring wind speed and direction in sprinkler design leads to the underestimation of the peak flows and the required capacity of the irrigation system. Highly variable wind speed disfavors the reliable design and management of sprinkler irrigation systems [12]. Zapata et al. [13] presented a method to estimate the fraction of suitable time for irrigation operation. This method was adapted to the design of collective sprinkler irrigation networks in windy areas, guaranteeing maximum on-farm performance through the use of a historical wind speed database and a ballistic solid-set irrigation simulation model. The operating pressure head determines sprinkler discharge, wetted diameter, drop diameter and hence irrigation uniformity [14]. Kincaid et al. [15] reported that nozzle diameter and operating pressure head determined drop diameter. Among the series of drops emitted by a sprinkler, the smallest drop diameter influences WDEL, while the largest one determines soil surface compaction. In the literature, values of WDEL have been reported ranging between 2% and 50% [16]. This variability is due to climatic and hydraulic conditions prevailing at the experimental sites. Indeed, it has been reported that WDEL is dependent on relative humidity, air temperature, wind speed, pressure head at the nozzle, drop diameter and riser height [17]. In arid and semi-arid areas, Yazar [18] showed that WDEL may represent an important fraction of the supplied water. Yacoubi et al. [19] found that WDEL amounted to 24% of annual water depth applied to tomato crops in northern Tunisia. Using simulation models, these authors emphasized the interest of nocturnal irrigation to reduce water losses, improve irrigation uniformity and sustain crop yield. The identification of factors influencing WDEL is crucial inasmuch as they determine the water stewardship strategy. Playán et al. [16] demonstrated how wise management of sprinkler irrigation systems can lead to the reduction of WDEL and to significant water conservation.

This study intends to elucidate the combined effect of the abovementioned factors on irrigation uniformity and WDEL. Experiments were carried out under combinations of operating pressure, sprinkler spacing and climatic conditions to assess the local influence of each factor on irrigation performance. Experimental data were used to infer the occurrence probability of a given distribution uniformity and to identify practices leading to adequate on-farm irrigation performance standards for the Medjerda lower valley region.

2. Material and Methods

2.1. Characterization of the Field Experiments

Trials were carried out at the experimental station of INRGREF in the lower valley of the Medjerda in northeast Tunisia. The region is characterized by a Mediterranean semi-arid climate with a mild winter and hot and dry summer. Annual evapotranspiration and average rainfall amount to 1112 and 440 mm, respectively. Field trials were designed in accordance with local sprinkler irrigation practices, namely the range of operating pressure heads and the sprinkler spacing. The overwhelming majority of irrigated parcels in the area has a small size, and use hand-move laterals under moderate pressure head. This technique doesn't require high technical skills and properly valorizes water resources and local manpower.

The impact sprinklers used in this research were IR30, manufactured by Irriline (Vancouver, Canada). Sprinklers were equipped with circular nozzles (4.4 mm in diameter) set up at 1 m height and 27° as trajectory angle. Three sprinklers' spacings were analyzed in this research: 12 m × 12 m, 12 m × 18 m and 18 m × 18 m. Nozzle pressures of 200, 300 and 400 kPa were experimented. **Table 1** summarizes the main hydraulic characteristics of used sprinklers.

2.2. Irrigation Uniformity

The evaluation of irrigation uniformity was performed at the plot scale on bare soil. For the sake of convenience, three plots were set up with the abovementioned spacings. Plots were separated 24 m to avoid interferences. Inside each plot, a battery of rain gauges was set up on a square grid of 3 m × 3 m, in accordance with ISO standard 7749/2 [20] and Merriam and Keller [21]. Rain gauges were made out of plastic, and had an open diameter of 79 mm and a height of 240 mm. Consequently 16, 24 and 36 rain gauges were used in plots with spacings 12 m × 12 m, 12 m × 18 m and 18 m × 18 m, respectively. Gauges were mounted on plastic supports, so that the upper part of the gauge was placed at 0.50 m above the soil surface.

Table 1. Hydraulic characteristics of used sprinkler as determined in laboratory conditions.

Pressure (kPa)	Range (m)	Flow rate (m³/h)	Rainfall intensity (mm/h)		
			12 m × 12 m	12 m × 18 m	18 m × 18 m
200	13.5	1.050	7.31	4.87	3.25
300	14.5	1.285	8.96	5.95	3.96
400	14.5	1.480	10.28	6.85	4.57

All experiments lasted for 2 hours. A total of 264 field trials were performed.

The water depths collected in the rain gauges were used to infer the uniformity coefficient, UC (%) as defined by Christiansen [22]:

$$UC = 100\left(1 - \frac{\sum_{i=1}^{n}|h_i - h_m|}{nh_m}\right) \quad (1)$$

where h_m, h_i and n refer to the average water depth within the plot, the collected water depths in the rain gauges and the number of rain gauges, respectively.

A meteorological station, installed at the edge of the field provided the climatic parameters, viz. wind speed, wind direction (both at 2 m height), air temperature and relative humidity, at a 15 min interval.

2.3. Frequency Analysis of Distribution Uniformity

A frequency analysis was performed on the water depths collected in the rain gauges at the pressure heads of 200 and 300 kPa. The pressure of 400 kPa was not used for this purpose, since the number of experiments at high wind speed was limited. The recorded average wind speed was within 0.05 and 6.5 m/s. The frequency analysis provides the probability of occurrence of certain uniformity according to wind speed for the different sprinkler spacing and pressures. Three probability distribution laws were assessed: normal, log-normal and Weibull. The density functions of these laws are:

Normal law:

$$f(UC) = \frac{1}{\sigma\sqrt{2\pi}} e^{-\frac{(UC-\mu)^2}{2\sigma^2}} \quad (2)$$

Log-normal law:

$$f(UC) = \frac{1}{UC\sigma\sqrt{2\pi}} e^{-\frac{(\ln(UC)-\mu)^2}{2\sigma^2}} \quad (3)$$

Weibull law:

$$f(UC) = (\beta/\lambda)(UC/\lambda) e^{-(UC/\lambda)^\beta} \quad (4)$$

where μ and σ are the average and the standard deviation of UC_i values (for the normal law) and $\log(UC_i)$ values (for the log-normal law), respectively. Likewise, β and λ are shape and scale factors: $\beta > 0$ and $\lambda > 0$.

The distribution function F of UC values was used to calculate the probability of occurrence of a given uniformity as follows:

$$F_{UC}(x) = \int_0^x f_{UC}(t)\, dt \quad (5)$$

$$F_{UC}(x) = P(UC \le x) \quad (6)$$

The adequacy of these laws can be assessed using the χ^2 and Kolmogorov-Smirnov tests with 5% as level of significance. Dagnelie [23] stated that agreement is not unanimous regarding the application of χ^2 test if the number of observations is insufficient. For this reason, the following parameters are often used to assess the adequacy of a given distribution law:

Bias:

$$Bias = \sum_{i=1}^{N}(n_i - n_i')/N \quad (7)$$

Mean Absolute Error:

$$MAE = \sum_{i=1}^{N}|n_i - n_i'|/N \quad (8)$$

Mean Square Error:

$$MSE = \sum_{i=1}^{N}(n_i - n_i')^2/N \quad (9)$$

where N, n_i and n_i' are the number of observations, the observed and the theoretical number of UC data, respectively. The law which best fits the recorded data is the one minimizing the abovementioned parameters.

2.4. WDEL Evaluation

WDEL measurements were performed at the 12 m × 12 m spacing during April-August, 2009.

WDEL was estimated as:

$$WDEL = 100\,\frac{h_b - h_m}{h_b} \quad (10)$$

where h_b and h_m are the supplied and the average measured water depths, respectively. The supplied water depth was determined by:

$$h_b = \frac{qt}{L_s S_s} \quad (11)$$

where q is the sprinkler discharge, t the duration of the test, L_s the spacing between adjacent laterals and S_s the spacing between two consecutive sprinklers. Thus, estimated WDEL represents the amount of water discharged by the sprinklers which does not reach the crop canopy. The effect of individual climatic parameters on WDEL at pressure heads of 200 and 300 kPa was analyzed through simple linear regression. Multiple linear and stepwise regressions were then performed for proper prediction of WDEL using the regression models presented in **Table 2**. In this table, W, RH, $(e_s - e_d)$ and T refer to the wind speed, relative humidity, vapor pressure deficit and air temperature, respectively.

The analysis was carried out using the XLSTAT software, version 17. Following Tarjuelo *et al.* [24], three

Table 2. Multiple linear and stepwise regression models used for modeling WEDL.

Equation	
$WEDL = a \cdot W + b \cdot RH + c \cdot (e_s - e_a) + d \cdot T + e$	(2a)
$WEDL = a \cdot W + b \cdot RH + c$	(2b)
$WEDL = a \cdot W + b \cdot (e_s - e_a) + c$	(2c)
$WEDL = a \cdot W + b \cdot T + c$	(2d)

Table 3. Distribution of the relative frequency of wind speed.

Wind speed (m/s)	0 - 2	2 - 4	4 - 6	6 - 8	8 - 10
Relative frequency (%)	23.1	56.6	18.0	2.2	0.1

wind speed classes were considered: 0 - 2 m/s (low wind speed), 2 - 4 m/s (moderate wind speed) and high wind speed (W ≥ 4 m/s). A frequency analysis was performed using hourly data of a time series covering the period 1998-2008. Results are summarized in **Table 3**, showing that moderate wind speeds are the most frequent in the area, while high speeds have a probability of occurrence of about 20%.

3. Results and Discussion

3.1. Uniformity Analysis

Table 4 summarizes the core statistics of UC data obtained with the three sprinkler spacings, the three operating pressure heads and the three prevailing wind speed classes. In the table, N, UC_m and CV refer to the number of field trials, the mean uniformity coefficient and the coefficient of variation.

3.1.1. Effect of Pressure on Uniformity

For the same wind class, **Table 4** shows that UC increases with the pressure head. However, this improvement should be balanced against the increase in energy consumption. Conversely, for the same pressure head, UC decreases as wind speed increases. A threshold value of UC less than 75% has often been used to indicate unsatisfactory uniformity, as is often the case of sprinkler irrigation at high wind speeds [9]. Likewise, **Table 4** shows that the increase in pressure head from 200 to 400 kPa improves UC at low and moderate wind speeds. This effect was clear for 12 m × 18 m and 18 m × 18 m spacing, the effect on the spacing 12 m × 12 m being less evident. Increasing the pressure head from 200 to 300 kPa showed a less discernible impact at moderate wind speeds. These results bolster those of Tarjuelo et al. [25] who stated that increasing pressure improves UC only in the presence of low wind speeds.

3.1.2. Effect of Spacing on Uniformity

The frequency analysis showed that UC exceeds 80% in 60%, 46% and 25% of cases for the spacings of 12 m × 12 m, 12 m × 18 m and 18 m × 18 m, respectively.

Thus, 12 m × 12 m would be the only spacing producing UC greater than the threshold of 75% in most cases. Examination of **Table 4** shows that small sprinkler spac-

ings improved UC for all pressure heads and wind speed classes. Similar findings were reported by Moazed et al. [26] and Tarjuelo et al. [25]. At wind speeds lower than 4 m/s, the spacing 12 m × 12 m provided the highest values of UC (76.6% ≤ UC ≤ 86.1%) for the entire pressure head range. This result corroborates the findings of Tarjuelo et al. [24], who showed that the 12 m × 12 m spacing produced the maximum uniformity for a wide range of wind speeds.

By cons, at low wind speeds the 12 m × 18 m spacing produced UC values greater than 75% for the entire range of pressure heads. This result is due to the alignment of the sprinklers with respect to the dominant wind direction. Similar results were reported by Vories and Von Bernuth [27]. The combination of 18 m × 18 m spacing and 200 kPa pressure head should be avoided for all wind speeds because of the poor sprinkler overlap. Using rotating sprinklers operated at low pressure heads (100 - 250 kPa), Sahoo et al. [28] obtained results discouraging the use of large sprinkler spacings. Likewise, the combination of 12 m × 18 m spacing and 300 kPa pressure is only prescribed at low wind speeds.

3.1.3. Effect of Wind Speed on Uniformity

Experimental results show a consistent decrease of uniformity with increasing wind speed (**Table 4**) in all spacings and pressures. Uniformity was significantly affected by wind speeds higher than 4 m/s, regardless of the spacing. For wind speeds below 4.4 m/s, a pressure head of 400 kPa produced the highest UC values in all spacings. Wind effect on UC can be appreciated for speeds higher than 2 m/s. Similar results were reported by Kincaid [29] who stated that UC dwindles when the wind speed exceeds 2.2 m/s for the spacings 12 m × 12 m, 12 m × 15 m and 12 m × 18 m. Under pressures of 200 and 300 kPa, a wind speed of 4 m/s stands as a threshold beyond which irrigation should be cut-off, since UC values would fall below 75%.

3.2. Modeling the Distribution of UC

Experiments performed at 400 kPa were not considered in this analysis given the small variability in wind speed. In order to obtain a sufficiently large data set, data obtained at 200 and 300 kPa for a given sprinkler spacing were pooled. The analysis was also performed for fixed operating pressure head and variable sprinkler spacing. Based on χ^2 and Kolmogorov-Smirnov tests at a probability value of 5% and minimum values of Bias, MAE and MSE, the Weibull law is the best suited to fit ex-

Table 4. Effect of wind speed and sprinkler spacing on UC at different pressure heads.

Pressure head kPa		Sprinkler spacing								
		12 m × 12 m			12 m × 18 m			18 m × 18 m		
	N	21	8	4	21	8	4	21	8	4
200	W (m/s)	0 - 2	2 - 4	4 - 6.5	0 - 2	2 - 4	4 - 6.5	0 - 2	2 - 4	4 - 6.5
	UC_m (%)	83.8	76.6	68.5	78.9	72.8	62.5	68.4	66.8	59.4
	CV (%)	4.6	4.1	7.0	4.5	5	9.9	9.9	6.8	8.0
	N	13	10	7	8	9	4	9	7	5
300	W (m/s)	0 - 2	2 - 4	4 - 6.4	0 - 2	2 - 4	4 - 6.4	0 - 2	2 - 4	4 - 6.4
	UC_m (%)	86.1	76.8	74.1	79.8	71.9	63.7	76.3	71.3	64.7
	CV (%)	3.5	7.2	3.3	5.5	9.1	13.1	5.4	7.5	6.7
	N	19	12		19	12		19	12	
400	W (m/s)	0 - 2	2 - 4.4		0 - 2	2 - 4.4		0 - 2	2 - 4.4	
	UC_m (%)	84.2	79.6		85.0	80.0		82.7	79.8	
	CV (%)	5.5	5.9		3.5	5.8		4.5	5.5	

perimental data for each spacing. When results were separated according to the operating pressure head, Weibull law provided the best fit for 200 kPa, while normal law provided the best fit for 300 kPa. The optimum parameters are reported in **Table 5** for the three wind speed ranges.

Table 6 summarizes the occurrence probabilities corresponding to the acceptable distribution uniformity (Keller and Bliesner, 1990).

Indeed, these authors considered that UC equal to 75% and 80% are deemed to be relatively low and satisfactory, respectively. Spacings 12 m × 18 m and 18 m × 18 m should be avoided in the local conditions since the probability to obtain UC higher than 75% is relatively small (less than 68% for wind speed higher than 4 m/s). In the prevalent wind conditions, the 12 m × 12 m spacing provides the best water distribution inasmuch as the occurrence probability is higher than 90% and 69% for UC values of 75% and 80 %, respectively. Under low wind speeds, results show that fair uniformity can be achieved by the 12 m × 18 m spacing (probability of 88% for UC higher than 75%) while the 18 m × 18 m spacing produces the lowest irrigation uniformity (probability of 30% for UC higher than 75%).

Figure 1 shows that the cumulative distribution functions are virtually coalescing at 200 kPa and are well distinguishable at 300 kPa. For a pressure head of 200 kPa, **Figure 1** shows that the probability to obtain UC in excess of 75% drops from 66% to 58% in passing from the interval 0 - 2 to 0 - 4 m/s. Conversely, the pressure head of 300 kPa produces a clearer drop in probability (22.6%) for the same ranges of wind speed.

Figures 2(a) exhibits a clear effect of pressure head on UC at low wind speeds. However, when the wind speed exceeds 2 m/s, the effect of pressure head on UC fades

(**Figure 2(b)**). Under calm weather, the probability that UC lies within 80% and 90% is 37% for a pressure head of 200 kPa (**Figure 2(a)**). The same probability increases to 52% when the pressure head is 300 kPa. For wind speeds below 4 m/s, **Figure 2(b)** shows that the increase of probability when the pressure head increases from 200 to 300 kPa is very moderate (31% and 32%).

3.3. WDEL Analysis

Tables 7 and **8** present basic statistics for the climatic factors controlling WDEL at 200 and 300 kPa (respectively) on the basis of thirty experimental trials. Field trials performed at a pressure head of 200 kPa correspond to more stern climatic conditions in terms of air temperature and vapor pressure deficit than those carried out at 300 kPa. Notwithstanding similar average losses for 200 and 300 kPa, the corresponding CV values were quite different

Figures 3(a)-(d) illustrate the effect of individual climatic parameters on WDEL at 300 kPa. Wind speed was the most explicative variable, followed by relative humidity. Vapor pressure deficit and air temperature were the less relevant variables.

The parameters of the regression models are summarized in **Table 9**. Model (2b) produced the best results with wind speed and relative humidity as input variables. Based on field evaluations under semiarid conditions, Seginer et al. [30] showed that WDEL generated by a single sprinkler operating at a pressure head of 300 kPa were related to wind speed and the wet-bulb depression.

Figure 4 displays measured versus predicted WDEL. The regression line is virtually identical to the first bisector, which upholds the accuracy of the model.

Figures 5(a)-(d) show the effect of isolated climatic

Table 5. Fitting parameters of the Weibull and normal laws.

Wind speed (m/s)	Sprinkler spacing								
	12 m × 12 m			12 m × 18 m			18 m × 18 m		
	N	β	λ	N	β	λ	N	β	λ
≤6.5 m/s	63	15.2	83.1	54	13.6	77.7	54	11.4	72.0
≤4 m/s	52	18.7	84.3	46	18.1	79.0	45	12.5	73.1
≤2 m/s	34	28.9	86.3	29	27.2	80.8	30	12.2	73.9

	Pressure head					
	200 kPa			300 kPa		
	N	β	λ	N	μ	σ^2
≤6.5 m/s	99	10.4	77.9	72	75.8	63.6
≤4 m/s	87	11.8	79.0	56	77.8	49.8
≤2 m/s	63	12.4	80.4	30	81.5	31.9

Table 6. Occurrence probability UC > 80% and UC > 75%.

Wind speed	≤6.5 m/s	≤4 m/s	≤2 m/s
	12 m × 12 m		
P (UC > 80 %)	0.57	0.69	0.90
P (UC > 75 %)	0.81	0.90	0.98
	12 m × 18 m		
P (UC > 80 %)	0.23	0.29	0.47
P (UC > 75 %)	0.54	0.68	0.88
	18 m × 18 m		
P (UC > 80 %)	0.04	0.04	0.07
P (UC > 75 %)	0.20	0.25	0.30

Table 7. Basic statistics of climatic factors controlling WDEL at 200 kPa.

	T (°C)	W (m/s)	RH (%)	$(e_s - e_a)$ (kPa)	WDEL (%)
Number of observations	30	30	30	30	30
Mean	30.7	1.89	49.4	2.3	23.4
Minimum	27.2	0.29	28.4	1.4	10.3
Maximum	35.4	6.53	66.3	4.1	45.5
Range	8.2	6.24	37.9	2.7	35.2
Standard deviation	2.1	1.47	9.9	0.7	8.4
Coefficient of variation (%)	6.7	77.8	20.1	29.7	35.8

Table 8. Basic statistics of climatic factors controlling WDEL at 300 kPa.

	T (°C)	W (m/s)	RH (%)	$(e_s - e_a)$ (kPa)	WDEL (%)
Number of observations	30	30	30	30	30
Mean	23.8	2.53	57.1	1.5	21.8
Minimum	15.4	0.11	26.3	0.2	1.0
Maximum	33.4	6.43	91.6	3.5	36.5
Range	18.0	6.31	65.4	3.4	35.1
Standard deviation	6.1	1.94	18.3	1.1	10.5
Coefficient of variation (%)	25.7	76.7	32.0	69.6	48.1

Table 9. Parameters of the multiple linear and the stepwise regression models at 300 kPa.

Model	a	b	c	d	e	SE (%)	R^2	F
(2a)	3.309***	−0.369*	0.889 ns	−0.533 ns	45.841**	4.29	0.856	37.1***
(2b)	3.434***	−0.262***	28.089***			4.42	0.835	68.2***
(2c)	3.877***	3.765**	6.288**			4.97	0.791	51.1***
(2d)	4.032***	0.482*	0.098 ns			5.63	0.695	

***Fisher test significant at 1‰; **Fisher test significant at 1 %; *Fisher test significant at 5 %; nsFisher test insignificant; SE = standard error.

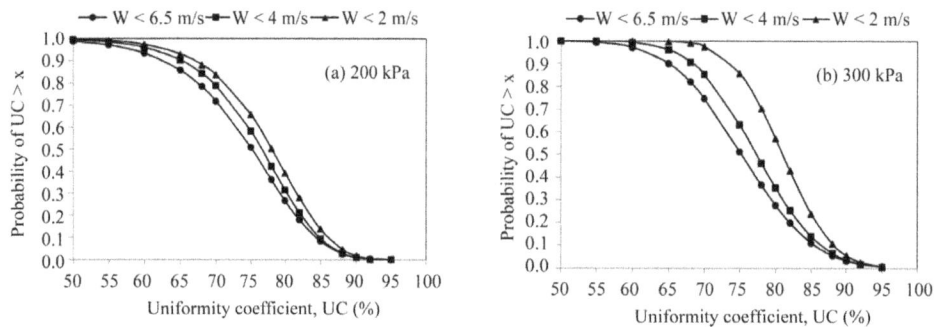

Figure 1. Effect of wind speed on cumulative distribution function at 200 kPa and 300 kPa.

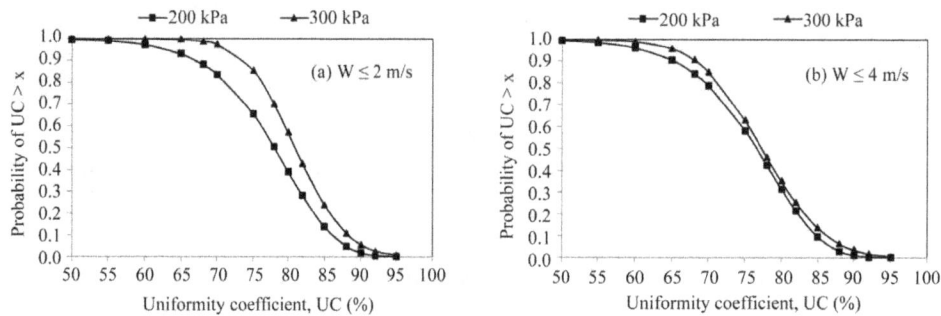

Figure 2. Effect of pressure head on cumulative distribution function for fixed wind speed: (a) W < 2 m/s and (b) W < 4 m/s.

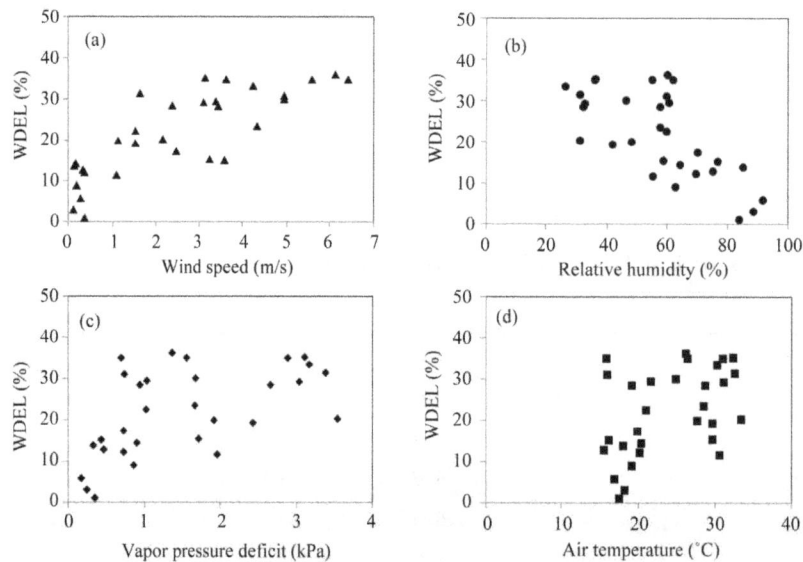

Figure 3. WDEL as function of wind speed, relative humidity, vapor pressure deficit and air temperature at 300 kPa.

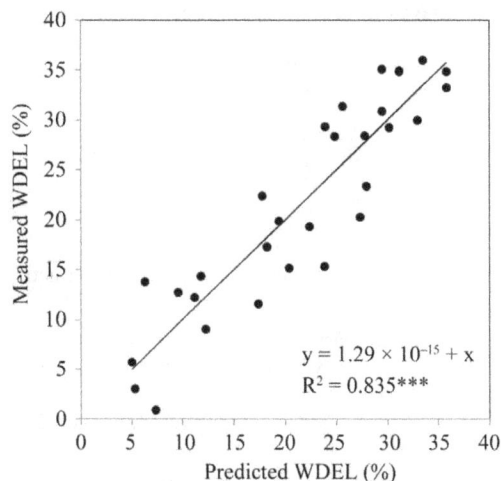

Figure 4. Measured vs predicted WDEL at 300 kPa.

parameters on WDEL at 200 kPa. Apart from wind speed, other climatic parameters showed no significant effect on WDEL (5% level). Regression parameters are summarized in **Table 10**. WDEL could only be explained by wind speed. Edling [31] showed that small droplets evaporate faster than large ones. According to Keller and Bliesner [9], decreasing the pressure head increases the drop diameter and therefore reduces the water-air contact area per unit of water mass. This result upholds the finding of Montero *et al.* [32], who showed that drop diameter, measured by optical spectropluviometer, increased as pressure head decreased. Consequently, large drops contribute less than small ones in alleviating the atmospheric evaporation demand [33].

Figure 6 exhibits a fair agreement between measured and predicted WDEL. The regression line is virtually identical to the first bisector, which upholds the accuracy of the model. The non-zero intercept in the regression line of the model WDEL = $a \cdot W + b$ suggests prominent water losses even under calm weather (14.95%). This result is at odds of the statement that wind speed is the only significant factor determining WDEL. Consequently, this model should be used with watchfulness at very low wind speed.

4. Conclusion

Field trials were carried out to assess irrigation uniformity under three sprinkler spacings and three operating pressure heads. Increasing pressure from 200 to 400 kPa improved UC at low wind speeds (W < 2 m/s). This improvement is less distinguishable for wind speeds in the interval of 2 - 4 m/s. For pressures in the range of 200 - 400 kPa and wind speeds below 4 m/s, the 12 m × 12 m spacing produced the highest UC values. The ill-effect of wind on UC was clear for speeds higher than 2 m/s. Water distribution was largely distorted by wind speeds higher than 4 m/s. For a given sprinkler spacing, the Weibull law was the most suitable to fit lumped experimental UC data obtained at 200 and 300 kPa. Similarly, the normal and Weibull laws were the best suited to model UC data obtained at 300 and 200 kPa separately. In the prevalent wind conditions, 12 m × 12 m spacing provided the best water distribution since the probability of UC higher than 80% exceeded 69%. Likewise, for the 12 m × 18 m and 18 m × 18 m spacings, the probability

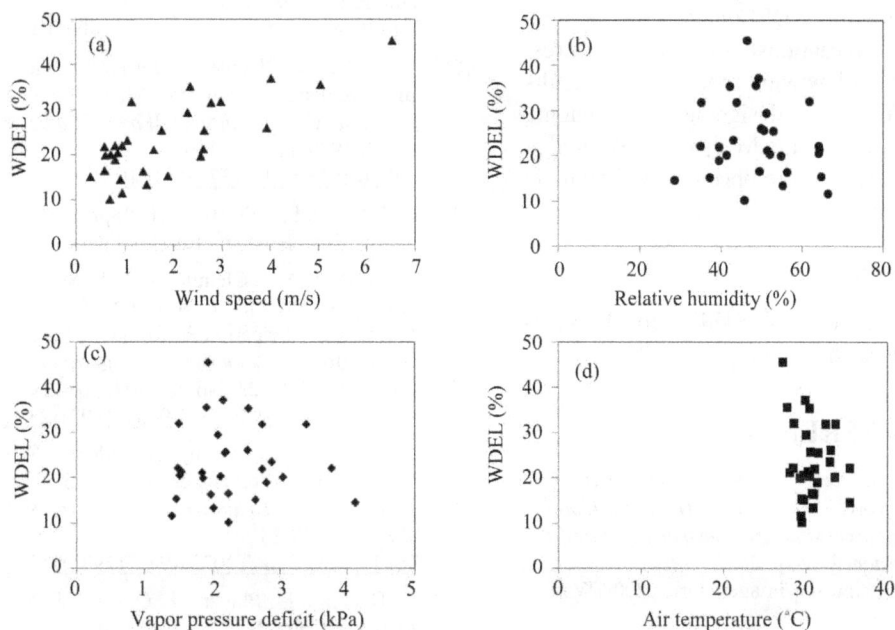

Figure 5. WDEL as function of wind speed, relative humidity, vapor pressure deficit and air temperature at 200 kPa.

Table 10. Parameters of the multiple linear and the stepwise regression models at 200 kPa.

Model	a	b	c	SE (%)	R^2	F
(2b)	4.467^{***}	-0.128^{ns}	21.229^{**}	5.26	0.631	23.1^{***}
(2c)	4.548^{***}	1.434^{ns}	11.471^{**}	5.34	0.622	22.2^{***}
(2d)	4.479^{***}	0.091^{ns}	12.077^{ns}	5.43	0.609	21^{***}
$a \cdot W + b$	4.443^{***}	14.954^{***}	-	5.33	0.608	43.5^{***}

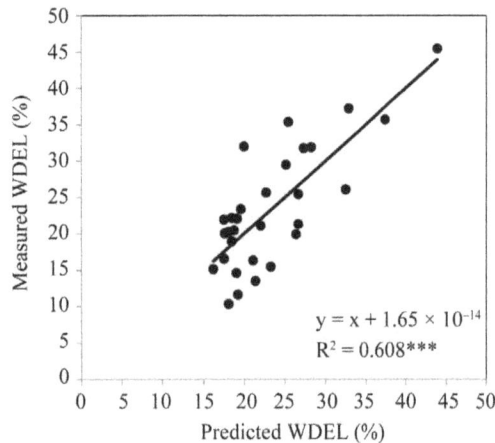

Figure 6. Measured versus predicted WDEL at 200 kPa.

of UC higher than 75% was less than 68% for wind speeds higher than 4 m/s. Under these circumstances, it is recommended to cut-off sprinkler irrigation. It was shown that wind drift and evaporation losses are dependent on the prevalent hydraulic and climatic conditions. Operating at 300 kPa, WDEL was conditioned by wind speed and relative humidity; operating at 200 kPa WDEL was only conditioned by wind speed. Using wind speed as the only variable for calculating WDEL under 200 kPa pressure head should be taken cautiously since it may overestimate WDEL under very low wind speeds. These results may be used as guidelines for designing and managing sprinkler irrigation systems in the Medjerda lower valley and in other irrigated areas where operating and climatic conditions could be similar.

5. Acknowledgements

The authors are thankful to the INRGREF for the valuable support of this research.

REFERENCES

[1] M. Abu-Zeid and A. Hamdy, "Coping with Water Scarcity in the Arab World," *The 3rd International Conference on Water Resources and Arid Environments and the 1st Arab Water Forum*, 2008, p. 26. http://www.icwrae-psipw.org/images/stories/2008/Water/1.pdf

[2] HYDROPLAN/SCET-Tunisie, "Etude du Projet de Modernisation des Périmetres Publics Irrigués de la Basse Vallée de la Medjerda. Phase 1: Analyse de la Situation Actuelle," Tunis, 2002, p. 196.

[3] M. Rebai and A. Zairi, "Les Périmètres Irrigués de la Basse Vallée de la MEDJERDA: Problématiques et Perspectives," 2006. http://www.eau-sirma.net/les_rencontres/marrakech-29-31-mai-2006-maroc/les-actes

[4] A. Slatni, J. C. Mailhol, A. Zairi, G. Château and T. Ajmi, "Analyse et Diagnostic de la Pratique de l'Irrigation Localisée dans les Périmètres Publics Irrigués de la Basse Vallée de la Medjerda en Tunisie," *Actes du séminaire Euro-Méditerranéen "La Modernisation de l'Agriculture Irriguée"*, Tome1, IAV Hassan II, Rabat-Institus, Rabat, 2004, pp. 112-122.

[5] A. Zairi, A. Slatni, J. C. Mailhol, R. Boubaker, H. El Amami, M. Ben Ayed and M. Rebai, "Analyse-Diagnostic de l'Irrigation de Surface dans les Périmètres Publics Irrigués de la Basse Vallée de la Medjerda," *Numéro Spécial des Annales de l'INRGREF, Actes du Séminaire "Economie de l'eau en irrigation"*, Hammamet, 2000, pp. 10-26.

[6] AHT-Group/SCET Tunisie, "Projet de Modernisation des Périmètres Publics Irrigués de la Basse Vallée de la Medjerda. Analyse Diagnostic des Equipements d'Irrigation," Tunis, 2009, p. 52.

[7] R. Al Atiri, "Les Efforts de Modernisation de l'Agriculture Irriguée en Tunisie," 2004. http://www.wademed.net/ Articles/005DGGRO.pdf

[8] L. S. Pereira, "Higher Performances through Combined Improvements in Irrigation Methods and Scheduling: A Discussion," *Agricultural Water Management*, Vol. 40, No. 2, 1999, pp. 153-169. doi:10.1016/S0378-3774(98)00118-8

[9] J. Keller and R. D. Bliesner, "Sprinkler and Trickle Irrigation," Van Nostrand Reinhold, New York, 1990.

[10] C. M. Burt, A. J. Clemmens, T. S. Strelkoff, K. H. Solomon, R. D. Bliesner, L. A. Hardy, T. A. Howell and D. E. Eisenhauer, "Irrigation Performance Measures: Efficiency and Uniformity," *Journal of Irrigation and Drainage Engineering*, Vol. 123, No. 6, 1997, pp. 423-442. doi:10.1061/(ASCE)0733-9437(1997)123:6(423)

[11] E. D. Vories, R. D. Von Bernuth and R. H. Mickelson, "Simulating Sprinkler Performance in Wind," *Journal of Irrigation and Drainage Engineering*, Vol. 113, No. 1, 1987, pp. 119-130. doi:10.1061/(ASCE)0733-9437(1987)113:1(119)

[12] F. Dechmi, E. Playán, J. Cavero, J. M. Faci and A. Martínez-Cob, "Wind Effects on Solid Set Sprinkler Irrigation Depth Yield of Maize," *Irrigation Science*, Vol. 22, No. 2, 2003, pp. 67-77.

doi:10.1007/s00271-003-0071-9

[13] N. Zapata, E. Playan, A. Martinez-Cob, I. Sanchez, J. M. Faci and S. Lecina, "From On-Farm Solid-Set Sprinkler Irrigation Design to Collective Irrigation Network Design in Windy Areas," *Agricultural Water Management*, Vol. 87, No. 2, 2007, pp. 187-199. doi:10.1016/j.agwat.2006.06.018

[14] N. Lamaddalena, U. Fratino and A. Daccache, "On-Farm Sprinkler Irrigation Performance as Affected by the Distribution System," *Biosystems Engineering*, Vol. 96, No. 1, 2007, pp. 99-109. doi:10.1016/j.biosystemseng.2006.09.002

[15] D. C. Kincaid, K. H. Solomon and J. C. Oliphant, "Drop Size Distributions for Irrigation Sprinklers," *Transactions of the ASAE*, Vol. 39, No. 3, 1996, pp. 839-845.

[16] E. Playán, R. Salvador, J. M. Faci, N. Zapata, A. Martínez-Cob and I. Sánchez, "Day and Night Wind Drift and Evaporation Losses in Sprinkler Solid-Sets and Moving Laterals," *Agricultural Water Management*, Vol. 76, No. 3, 2005, pp. 139-159. doi:10.1016/j.agwat.2005.01.015

[17] J. M. Tarjuelo, J. F. Ortega, J. Montero and J. A. de Juan, "Modelling Evaporation and Drift Losses in Irrigation with Medium Size Impact Sprinklers under Semi-Arid Conditions," *Agricultural Water Management*, Vol. 43, No. 3, 2000, pp. 263-284. doi:10.1016/S0378-3774(99)00066-9

[18] A. Yazar, "Evaporation and Drift Losses from Sprinkler Irrigation Systems under Various Operating Conditions," *Agricultural Water Management*, Vol. 8, No. 4, 1984, pp. 439-449. doi:10.1016/0378-3774(84)90070-2

[19] S. Yacoubi, K. Zayani, N. Zapata, A. Zairi, A. Slatni, R. Salvador and E. Playán, "Day and Night Time Sprinkler Irrigated Tomato: Irrigation Performance and Crop Yield," *Biosystems Engineering*, Vol. 107, No. 1, 2010, pp. 25-35. doi:10.1016/j.biosystemseng.2010.06.009

[20] ISO Standard 7749/2, "Agricultural Irrigation Equipment. Rotating Sprinklers. Part 2. Uniformity of Distribution and Test Methods," Geneva, 1990.

[21] J. L. Merriam and J. Keller, "Farm Irrigation System Evaluation: A Guide for Management," Utah State University, Logan, 1978.

[22] J. E. Christiansen, "Irrigation by Sprinkling," California Agricultural Experiment Station Bulletin 670, University of California, Berkeley, 1942.

[23] P. Dagnelie, "Théorie et Méthodes Statistiques. Applications Agronomiques. Vol. 2," Les Presses Agronomiques de Gembloux, Gembloux, 1978.

[24] J. M. Tarjuelo, J. Montero, P. A. Carion, F. T. Honroubia and M. A. Calvo, "Irrigation Uniformity with Medium Size Sprinklers. Part II: Influence of Wind and Other Factors on Water Distribution," *Transactions of the ASAE*, Vol. 42, No. 3, 1999, pp. 677-689.

[25] J. M. Tarjuelo, J. Montero, F. T. Honroubia, J. Ortiz and J. F. Ortega, "Analysis of Uniformity of Sprinkler Irrigation in Semi-Arid Area," *Agricultural Water Management*, Vol. 40, No. 2, 1999, pp. 315-331. doi:10.1016/S0378-3774(99)00006-2

[26] H. Moazed, A. Bavi, S. Boroomand-Nasab, A. Naseri and M. Albaji, "Effects of Climatric and Hydraulic Parameters on Water Uniformity Coefficient in Solid Set Systems," *Journal of Applied Sciences*, Vol. 10, No. 16, 2010, pp. 1792-1796. doi:10.3923/jas.2010.1792.1796

[27] E. Vories and R. D. von Bernuth, "Single Nozzle Sprinkler Performance in Wind," *Transactions of the ASAE*, Vol. 29, No. 5, 1986, pp. 1325-1330.

[28] N. Sahoo, P. L. Pradhan, N. K. Anumala and M. K. Ghosal, "Uniform Water Distribution from Low Pressure Rotating Sprinklers," *The CIGR Ejournal*, Manuscript LW 08014, Vol. X, 2008, p. 10. http://www.cigrjournal.org/index.php/Ejounral/article/viewFile/1231/1088

[29] D. C. Kincaid, "Minimizing Energy Requirements for Sprinkler Laterals," ASAE Paper No. 84-2585, St. Joseph, 1984.

[30] I. Seginer, D. Kantz and D. Nir, "The Distortion by Wind of the Distribution Patterns of Single Sprinklers," *Agricultural Water Management*, Vol. 19, No. 4, 1991, pp. 341-359. doi:10.1016/0378-3774(91)90026-F

[31] R. Elding, "Kinetic Energy, Evaporation and Wind Drift of Droplets from Low Pressure Irrigation Nozzles," *Transactions of the ASAE*, Vol. 28, No. 5, 1985, pp. 1543-1550.

[32] J. Montero, J. M. Tarjuelo and P. Carrion, "Sprinkler Droplet Size Distribution Measured with an Optical Spectropluviometer," *Irrigation Science*, Vol. 22, No. 2, 2003, pp. 47-56. doi:10.1007/s00271-003-0069-3

[33] D. L. Martin, D. C. Kincaid and W. M. Lyle, "Design and Operation of Sprinkler Systems," In: Glenn J. Hoffman, Ed., *Design and Operation of Farm Irrigation Systems*, American Society of Agricultural and Biological Engineers*, St. Joseph, 2007, pp. 557-631.

Permissions

All chapters in this book were first published by Scientific Research Publishing; hereby published with permission under the Creative Commons Attribution License or equivalent. Every chapter published in this book has been scrutinized by our experts. Their significance has been extensively debated. The topics covered herein carry significant findings which will fuel the growth of the discipline. They may even be implemented as practical applications or may be referred to as a beginning point for another development.

The contributors of this book come from diverse backgrounds, making this book a truly international effort. This book will bring forth new frontiers with its revolutionizing research information and detailed analysis of the nascent developments around the world.

We would like to thank all the contributing authors for lending their expertise to make the book truly unique. They have played a crucial role in the development of this book. Without their invaluable contributions this book wouldn't have been possible. They have made vital efforts to compile up to date information on the varied aspects of this subject to make this book a valuable addition to the collection of many professionals and students.

This book was conceptualized with the vision of imparting up-to-date information and advanced data in this field. To ensure the same, a matchless editorial board was set up. Every individual on the board went through rigorous rounds of assessment to prove their worth. After which they invested a large part of their time researching and compiling the most relevant data for our readers.

The editorial board has been involved in producing this book since its inception. They have spent rigorous hours researching and exploring the diverse topics which have resulted in the successful publishing of this book. They have passed on their knowledge of decades through this book. To expedite this challenging task, the publisher supported the team at every step. A small team of assistant editors was also appointed to further simplify the editing procedure and attain best results for the readers.

Apart from the editorial board, the designing team has also invested a significant amount of their time in understanding the subject and creating the most relevant covers. They scrutinized every image to scout for the most suitable representation of the subject and create an appropriate cover for the book.

The publishing team has been an ardent support to the editorial, designing and production team. Their endless efforts to recruit the best for this project, has resulted in the accomplishment of this book. They are a veteran in the field of academics and their pool of knowledge is as vast as their experience in printing. Their expertise and guidance has proved useful at every step. Their uncompromising quality standards have made this book an exceptional effort. Their encouragement from time to time has been an inspiration for everyone.

The publisher and the editorial board hope that this book will prove to be a valuable piece of knowledge for researchers, students, practitioners and scholars across the globe.

List of Contributors

Kozue Yuge and Shinogi Yoshiyuki
Faculty of Agriculture, Kyushu University, Fukuoka, Japan

Keiki Shigematsu
Graduate School of Bioresource and Bioenvironment Science, Kyushu University, Fukuoka, Japan

Mitsumasa Anan
Takasaki Sogo Consultant Co., Ltd., Kurume, Japan

Rachid Razouk
Department of Agronomy, National Institute of Agronomic Research, Meknès, Morocco
Department of Biology, Faculty of Sci- ences, University of Moulay Ismail, Meknès, Morocco

Jamal Ibijbijen
Department of Biology, Faculty of Sci- ences, University of Moulay Ismail, Meknès, Morocco

Abdellah Kajji
Department of Agronomy, National Institute of Agronomic Research, Meknès, Morocco

Mohammed Karrou
Integrated Water and Land Management Program, International Center for Agricultural Research in the Dry Areas, Allepo, Syria

Baburao Kamble, Ayse Irmak and Kenneth Hubbard
University of Nebraska-Lincoln, Lincoln, USA

Prasanna Gowda
USDA-ARS, Bushland, USA

Anitta Fanish Sundara Raj, Purushothaman Muthukrishnan and Pachamuthu Ayyadurai
Department of Agronomy, Tamil Nadu Agricultural University, Tamil Nadu, India

Basel Natsheh, Sameer Amereih and Mazen Salman
Palestine Technical University-Kadoorei, Tullkarm, Palestine

Zaher Barghouthi
Department of Natural Resources Research, National Agricultural Research Center (NARC), Jenin, Palestine

D. G. Regulwar and V. S. Pradhan
Department of Civil Engineering, Government College of Engineering, Aurangabad, India

Jiale Li, Caixiang Zhang, Yihui Dong, Xiaoping Liao and Linlin Yao
State Key Lab of Biogeology and Environmental Geology, China University of Geosciences, Wuhan, China

Bin Du
China National Administration of Coal Geology General Prospecting Institute, Beijing, China

Nadhir A. Al-Ansari
Department of Civil, Environmental and Natural Resources and Engineering, Luleå University of Technology, Luleå, Sweden

Azam Arabi and Amin Alizadeh
Ferdowsi University of Mashhad, Mashhad, Iran

Yaser Vahab Rajaee
Water Engineering Department, Azad University of Ferdows, Iran

Kazem Jam
Regional Water Authority Company of Razavi Khorasan, Mashhad, Iran

Naser Niknia
Water Engineering Department, Shiraz University, Shiraz, Iran

Noor Ul Hassan Zardari
Institute of Environmental and Water Resources Management (IPASA), University of Technology Malaysia (UTM), Skudai, Malaysia

Ian Cordery
School of Civil and Environmental Engineering, University of New South Wales, Sydney, Australia

Hamed K. Abbas
United States Department of Agriculture-Agricultural Research Service, Biological Control of Pests Research Unit, Stoneville, USA;

Henry J. Mascagni Jr.
Northeast Research Center, Louisiana State University AgCenter, St. Joseph, USA

H. Arnold Bruns
United States Department of Agricul-ture-Agricultural Research Service, Crop Production Systems Research Unit, Stoneville, USA

W. Thomas Shier
Department of Medicinal Chemistry, College of Pharmacy, University of Minnesota, Minneapolis, USA;

Kenneth E. Damann
Department of Plant Pathology & Crop Physiology, Louisiana State University AgCenter, Baton Rouge, USA

Sandeep Kumar
Carbon Management & Sequestration Center, School of Environment & Natural Resources, Ohio State University, Columbus, USA

Pradeep K. Sharma
Department of Soil Science, CSK HPKV, Palampur, India

Stephen H. Anderson
Department of Soil, Environmental and Atmospheric Sciences, Univer-sity of Missouri, Columbia, USA

Kapil Saroch
Department of Agronomy, Forage and Grassland Management, CSK HPKV, Palampur, India

Preeti Parashar and Fazal Masih Prasad
School of Chemical Sciences, Department of Chemistry, St. John's College, Agra, India

Hani A. Mansour and Hany M. Mehanna
Water Relations Field Irrigation Department, Agricultural and Biological Division, National Research Center, Cairo, Egypt

Mohamed E. El-Hagarey
Soil and Water Resources Conservation Department, Desert Research Center, Cairo, Egypt

Ahmehd S. Hassan
Irrigation Department, Agriculture Engineering Research Institute, Agricultural Research Center, Cairo, Egypt

Mahmoud Rahil and Hajaj Hajjeh
Faculty of Agricultural Science and Technology, Palestine Technical University-Kadoorie, Tulkarm, Palestine

Alia Qanadillo
Palestinian National Agriculture Research Center, Jenin, Palestine

John M. Halloran, Robert P. Larkin, O. Modesto Olanya and Zhongqi He
USDA Agricultural Research Service, New England Plant, Soil, and Water Laboratory, Orono, ME, USA

Sherri L. DeFauw
Department of Agricul-tural Economics and Rural Sociology, The Pennsylvania State University, University Park, PA, USA

Muhammad Mubeen, Ashfaq Ahmad, Aftab Wajid, Tasneem Khaliq, Syeda Refat Sultana and Shahid Hussain
Agro-Climatology Laboratory, Department of Agronomy, University of Agriculture, Faisalabad, Pakistan;

Amjed Ali
University College of Agriculture, University of Sargodha, Sargodha, Pakistan;

Hakoomat Ali
Department of Agronomy, Faculty of Agricultural Sciences and Technol-ogy, Bahauddin Zakariya University, Multan, Pakistan

Wajid Nasim
Department of Environmental Sciences, COMSATS Institute of Information Technology, Vehari, Pakistan

Raúl Leonel Grijalva-Contreras, Rubén Macías-Duarte, Gerardo Martínez-Díaz, Fabián Robles-Contreras and Manuel de Jesús Valenzuela-Ruiz
National Research Institute for Forestry, Agricultural and Livestock (INIFAP), Caborca, Sonora, México

Fidel Nuñez-Ramírez
Agricultural Science Institute, University Autónoma of Baja California (ICA-UABC), Mexicali, B.C. México

Samuel Jerry Cobbina
Department of Ecotourism & Environmental Management, Faculty of Renewable Natural Resources, University for Development Studies, Nyankpala, Ghana
School of the Environment, Jiangsu University, Zhenjiang, China

Mohammed Clement Kotochi
Department of Ecotourism & Environmental Management, Faculty of Renewable Natural Resources, University for Development Studies, Nyankpala, Ghana

Joseph Kudadam Korese
Department of Agricultural Mecha- nization & Irrigation Technology, Faculty of Agriculture, University for Development Studies, Nyankpala, Ghana

Mark Osa Akrong
Environmental Biology & Health Division, CSIR Water Research Institute, Accra, Ghana

Yasser El-Nahhal
Environmental Protection and Research Institute, Gaza, Palestine
The Islamic University of Gaza, Gaza, Palestine

Khalil Tubail and Mohamad Safi
Environmental Protection and Research Institute, Gaza, Palestine

Jamal Safi
Environmental Protection and Research Institute, Gaza, Palestine
Al-Azhar University of Gaza, Gaza, Palestine

Ahmed Melegy
Department of Geological Sciences, National Research Centre, Dokki, Egypt

Ahmed El-Kammar and Ghadir Miro
Department of Geology, Faculty of Science, Cairo University, Giza, Egypt

Mohamed Mokhtar Yehia
Central Laboratory for Environmental Quality Monitoring, National Water Research Center, Kanater El-Khairia, Egypt

Noha Samir Donia
Environmental Studies and Researches Institute, Ain Shams University, Cairo, Egypt

Ashfaque A. Memon, Khalifa Q. Leghari, Agha F. H. Pathan and Kameran Ansari
Department of Civil Engineering, Mehran UET, Jamshoro, Sindh, Pakistan

Kanya L. Khatri, Sadiq A. Shah and Rabia Soomro
Department of Civil Engineering, Mehran UET Khairpur Campus, Sindh, Pakistan

Kanwal K. Pinjani
Water Resources Division, National Engineering Services, Lahore, Pakistan

Dipak N. Kongre and Rohit Goyal
Civil Engineering Department, Malaviya National Institute of Technology, Jaipur, India

Angela M. Lamptey and Patrick K. Ofori-Danson
Department of Marine and Fisheries Sciences, University of Ghana, Legon, Ghana

Stephen Abbenney-Mickson
Department of Agriculture Engineering, Faculty of Engineering, University of Ghana, Legon, Ghana

Henrik Breuning-Madsen
Department of Geography and Geology, University of Copenhagen, Copenhagen, Denmark

Mark K. Abekoe
Department of Soil Science, Faculty of Agriculture, University of Ghana, Legon, Ghana

Kun Shi and Ming Zou
School of Environmental and Chemical Engineering, Dalian Jiaotong University, Dalian, China

Hongxiang Cai
Extensive Center of Agricultural Technology, Dalian, China

Samir Yacoubi and Adel Slatni
Institut National de Recherches en Génie Rural, Eaux et Forêts, University of Carthage, Tunis, Tunisia

Khemaies Zayani
High Institute of Environmental Science and Technology, University of Carthage, Tunis, Tunisia

Enrique Playán
Department Soil and Water, Estación Experimental de Aula Dei, Consejo Superior de Investigaciones Cientificas, Zaragoza, Spain